ASTER FAMILY FLOWER HEAD

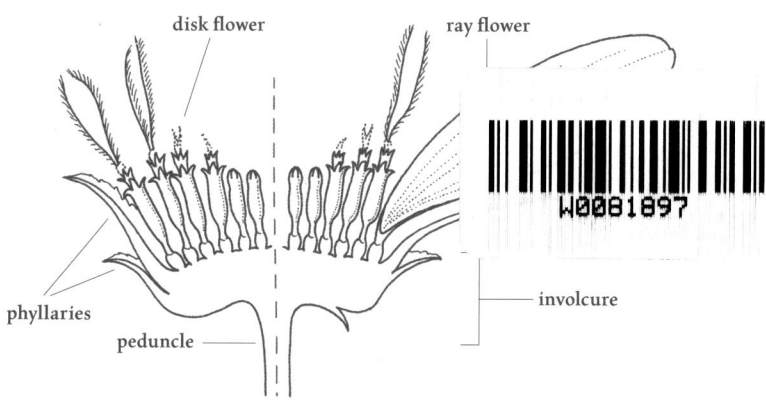

disk flower

ray flower

phyllaries

peduncle

involcure

INFLORESCENCES

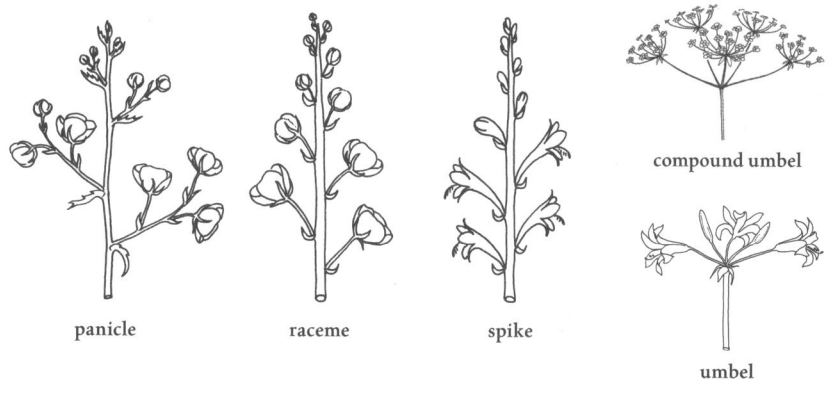

panicle

raceme

spike

compound umbel

umbel

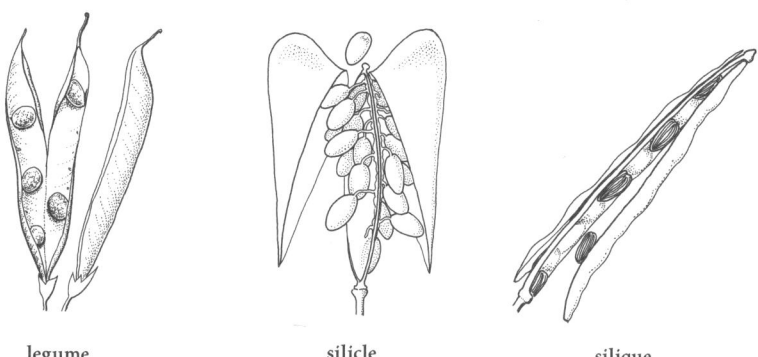

legume

silicle

silique

WILDFLOWERS
OF THE MIDWEST

A TIMBER PRESS

**FIELD
GUIDE**

To my sons, Aaron and Wes —MH

To my wife, Lindsay —SN

Copyright 2022 by Michael Homoya and Scott Namestnik.
All rights reserved.

Endpapers by Alan Bryan
Map by Michele Angel
Title page: *Chamaecrista fasciculata*

Published in 2022 by Timber Press, Inc.
The Haseltine Building
133 S.W. Second Avenue, Suite 450
Portland, Oregon 97204-3527
timberpress.com

Printed in China on paper from responsible sources
Text and cover design by Adrianna Sutton
Text is set in Maiola, a typeface designed by Veronika Burian in 2005

ISBN 978-1-60469-907-4
A catalog record for this book is available from the Library of Congress.

Michael Homoya
and Scott Namestnik

WILDFLOWERS
OF THE MIDWEST

CONTENTS

PREFACE

*What I want now is a chance to get out in the woods and see
the birds and flowers. I like to study them, and the rocks and
ferns and other things of the woods; it rests me.*
—Thomas Edison, 1905

MIDWESTERN WILDFLOWERS are an amazingly diverse group, and no wonder, as the Midwest is truly a botanical crossroads. Here is where floras of eastern North America come together on a grand scale, resulting in a wonderful mingling of plants from the northern woods and wetlands with those of the grassy plains of the west and the profusely rich, verdant forests of the south and east. Despite all the agricultural and industrial pursuits for which the Midwest is famous, over 3300 native species of vascular plants have been recorded here, plus about 1400 non-native and sometimes invasive ones that have been introduced and naturalized to a degree. There are species of every size, shape, and color, from showstoppers like the resplendent royal catchfly to the minuscule terrestrial starwort that requires a refined taste to appreciate. Wild orchids—over 75 species!—occur here too, including the fiery orange fringed orchid and the magnificent showy lady's-slipper.

Wildflowers are most commonly found in natural areas, but even if you have never set foot in one there are many species out there you might already know, such as our native species of iris, lily, phlox, and rose. Conversely, some might be strangers, like the rare Midwest endemic glade mallow, or the bizarre Indian pipe, a pale, mysterious-looking plant that "robs" nutrition from other plants via subterranean fungi, or purple pitcherplant, whose specialized leaves operate as carnivorous vessels of death for unsuspecting insects.

Whether it be the colorfully ornamental or the bizarre and deadly, all these plants and more are presented in this easy-to-use field guide, which features text and photographs of over 1000 Midwestern species. While the emphasis is on plant identification, we have included information on the region's "lay of the land" along with a map and information about the major ecoregions and natural communities. There are also sections on plant families and a helpful guide to understanding information in the main body of the book— the species accounts.

As we worked on this book, our lives changed with the appearance of COVID-19. Facial masks and social distancing became the norm, and many activities, events, and businesses came to a halt. As horrible as the pandemic was, one positive was that it gave many people an opportunity to slow down and reconnect with nature. We hope you are one of those—and that this book will whet your appetite to see and learn even more about the Midwest's cornucopia of botanical wealth.

ACKNOWLEDGMENTS

CREATING A BOOK such as this requires the help of a great number of people. We greatly appreciate all that everyone did on our behalf, and we wish to offer our gratitude and thanks to each of you. Please accept our sincerest apologies if we failed to include your name. Any omissions are completely accidental.

Many photographers graciously donated their images for use in this book. Their names and initials are listed on page 579, and their initials conclude each photo caption and/or species account throughout this field guide. Thank you all so very much. The quality and scale of this book could not have happened without you.

In addition to procuring photos we consulted many people throughout the process for various kinds of help and information, including, but not limited to, topics dealing with species distribution, life history and identification, climate, and natural communities. We are most fortunate for your help. Our thanks go out to Adam Balzer, Rebecca Dolan, Tony Fleming, Henry Gray, Barbara Homoya, Paul Nelson, Corey Raimond, Tony Reznicek, Paul Rothrock, and Brad Slaughter for their most helpful reviews of our draft manuscripts; Brandon Board for providing images taken by the late Keith Board; the Missouri Botanical Garden, current host of missouriplants.com, and Karen Hoksbergen for providing images taken by the late Dan Tenaglia; John Kartesz, Misako Nishino, and the Biota of North America Program for state distribution information; illustrators Michele Angel for the map of Midwest ecoregions and Alan Bryan for the endpapers; Brent Baker, Dan Boone, Kevin Doyle, Rick Gardner, Andrew Gibson, Paul Marcum, Paul Nelson, John Pearson, Welby Smith, and Theo Witsell for species distribution information; Howard Diamond, National Oceanic and Atmospheric Administration, and Bryan Peake, Midwestern Regional Climate Center, for climate data; Harvey Ballard, Carol Baskin, Rich Blatz, Michael Eason, Ted Elliman, Mary Ann Feist, Sarada Krishnan, Doug Ladd, Mark Leoschke, Paul Marcum, Chuck Miller, Robert Mohlenbrock, Kelly Randall, Steve Turner, Justin Thomas, Allison Vaughn, and George Yatskievych for counsel, assistance, and assorted useful information; and Franni Farrell, Alex Fus, Will McKay, and Sarah Milhollin of Timber Press for their considerable patience and help as we navigated the various aspects of the publication process.

Above all, we are greatly indebted to our wives, Barbara Homoya and Lindsay Namestnik, for their unending support, assistance, and patience during the book's preparation. Their love and understanding are beyond measure and did not go unnoticed by these two inveterate botanists. Thank you, dear wives!

INTRODUCTION
GETTING TO KNOW THE MIDWEST

WHAT EXACTLY is the "Midwest"? The precise boundary is arguable, as some have a broad view that includes more than a dozen states, including the Dakotas, Nebraska, Kansas, and even Kentucky. For our purposes it consists of eight states, namely Illinois, Indiana, Iowa, Michigan, Minnesota, Missouri, Ohio, and Wisconsin.

Our eight-state Midwest is quite a large area, nearly the size of Alaska, covering slightly more than half a million square miles. It extends from its southernmost point in the "Bootheel" of southeast Missouri to the "Northwest Angle" at the northernmost tip of Minnesota (and of the contiguous 48 states!), a distance of nearly 1000 miles. The distance from east to west, from Ohio's eastern border to Iowa's western edge, is almost as long at around 800 miles.

LANDFORM
The Midwest has a reputation of being rather unremarkable and flat, or at best rolling, and such perceptions are not entirely unfounded, as elevational change in most of the region is not significant. The elevation of the Missouri Bootheel, at around 230 feet above sea level, is the lowest in our region, and much of the terrain elsewhere gradually increases by only a few hundred feet. However, the Midwest does have its rugged "mountainous" areas, such as the Allegheny Plateaus in Ohio, the Ozark Highlands of Missouri, and areas in the western Great Lakes region. At just over 2300 feet

elevation, Eagle Mountain in northeastern Minnesota is the highest point in the Midwest. That area and the Porcupine Mountains of Michigan provide the most topographic relief, in places as much as 1400 feet. Much farther south the greatest elevational change occurs in the St. Francois Mountains of Missouri, where approximately 1000 feet separate the lowest valley floor and the highest peak.

Continental glaciation is responsible for most of the Midwest's topography. Massive glaciers are reported to have occurred during several periods of advancement and retreat during the Pleistocene Epoch, which began approximately 2 million years ago and ended about 10,000 years ago. The ice sheets were thought to have been as much as a mile or more thick in places, leveling hills and filling valleys as they deposited glacial "drift" a few feet thick near their termini to hundreds of feet thick around the Great Lakes. Aside from the leveling effect, the glaciers left us lake plains, outwash plains, till plains, various types of moraines, and kame and kettle topography. The glaciers reached their terminus in the southern Midwest, leaving sizable areas unglaciated. In addition, the Driftless Area is a mostly unglaciated "island" located primarily in southwestern Wisconsin.

Another major influence that has shaped the Midwestern landscape comes from bedrock that has been folded, shifted, uplifted, and eroded over geologic time. The upper levels consist primarily of sedimentary limestone and sandstone along with lesser amounts of dolomite and shale. For the most part it is buried. Where outcrops do occur, they are mostly in unglaciated areas. Outcrops of igneous and metamorphic rock are overall quite localized, found mostly in areas near Lake Superior and Missouri's St. Francois Mountains.

The Midwest is a large watershed, with most of the runoff going into the Mississippi River and eventually the Gulf of Mexico. The Mississippi is augmented by the flow of other large waterways, especially the Missouri, Ohio, Illinois, and Wabash rivers, with watersheds of the latter two entirely within our Midwestern region. Water from Michigan and portions of surrounding states flows into the North Atlantic via the Great Lakes and St. Lawrence River. Lastly, a small area of northern Minnesota ultimately drains into Canada's Lake Winnipeg before continuing to Hudson Bay and the Arctic Ocean. These waterways and their corridors provide pathways for both animals and plants.

With some exceptions, Midwestern soils are quite fertile and productive, making the region famous for its agricultural production, especially for growing corn and soybeans. Much of this fertility can be attributed to glaciation, particularly in areas where mineral-rich till was deposited, or in outwash plains and sites occupied by former glacial lakes. The fertility is particularly good in the deep organic soils once occupied by prairie vegetation. Glaciation has also given us extensive areas of fertile loess (wind-deposited silt). Exceptional deposits of loess, as much as 200 feet thick, border the east side of the Missouri River in Iowa and Missouri in an area known as the Loess Hills.

Soil nutrients, pH, and especially moisture play a significant role in determining plant distribution. The soils in much of our glaciated region are circumneutral, but several areas with lower pH (acidic) or higher pH (alkaline) exist. Acid soils are commonly developed on sandstone, shale, and igneous (or related rock) parent materials, or on peat developed from partially decayed sphagnum moss and other organic materials. Areas of alkaline soils are most prominently associated with limestone and dolomite outcrops, certain glacially deposited gravels, and in springy areas with marl and tufa.

CLIMATE AND FIRE

Given its broad expanse the Midwest climate is understandably quite varied, with a temperature gradient from north (cooler) to south (hotter) and a pronounced moisture gradient from west (drier) to east (wetter). In general the region is considered temperate with hot, humid summers and cold winters.

During the past 30 years the average minimum temperatures for the main winter months (December, January, February) range from -5–15°F in the north to 20–30°F in the south. For the summer months (June, July, August), the average maximum temperatures for most of the north is 70–75°F; this increases to 85–90°F in the south.

Annual accumulated precipitation is greatest in our far southern areas at 40–50 inches. Precipitation generally decreases north and westward, with the far northwest receiving about half that of the south. Not surprisingly, snowfall amounts are greatest in the north, averaging 25–50 inches per year, but isolated areas near the Great Lakes, such as the Keweenaw Peninsula in Lake Superior, average over 150 inches per year. In the southern half of our region, average annual snowfall amounts of 25 inches or less are the norm, the least occurring in the Missouri Bootheel with 6 inches or less per year.

Fire is a natural phenomenon that has shaped our Midwest landscape for millennia. It is now largely missing, occurring mostly as prescribed burns on managed lands. Fire maintains certain natural communities and species that are adapted to and/or require fire for various aspects of their life cycle. This is especially true for those species found in prairie, savanna, and woodland ecosystems. Without fire, species richness declines as sites become more shaded and mesic due to the encroachment of fire-intolerant trees and shrubs.

DIVERSITY AND ENDEMISM

According to Biota of North America Program, the Midwest is home to an amazing 4757 species of vascular plants in 1314 genera and 192 families. Almost half the species fall into only ten plant families. In order by greatest number of species, they are the aster family, grass family, sedge family, rose family, pea family, mustard family, mint family, pink family, plantain family, and buttercup family. All these top ten families (with the exception of the grass and sedge families) are represented in our species accounts.

While the Midwest is richly diverse in plant species, almost all are distributed well beyond its borders. That is expected, as most plants have extensive ranges. A few, however, are endemic, that is, known to occur naturally only in the Midwest. Those given their own species account in this book include kittentails (*Besseya bullii*), glade mallow (*Napaea dioica*), and green trillium (*Trillium viride*). Also noteworthy are near-endemics, that is, ones that occur mostly in our region but sparingly elsewhere—Hill's thistle (*Cirsium hillii*), for example.

ECOREGIONS AND NATURAL COMMUNITIES

AN ECOREGION is a landscape unit that possesses a distinctive combination of natural features. While it reflects natural conditions, interpreting them as a unit is somewhat subjective. Probably no two maps created by individual ecologists would be the same, in part because the features emphasized by them often differ. Nonetheless, general patterns observable in the landscape will help give some sense of place. Ecological features taken into consideration for defining an ecoregion include climate, diversity of plant and animal species, hydrology, glacial history, topography, and dominant vegetation.

For the purposes of this book we have identified 4 basic ecoregions. They are intended to give the reader a broad view of the Midwest landscape, one that emphasizes major characteristics of the vegetation and affinities of their flora and fauna. The ecoregions are Eastern Forests, Northern Lakes, Ozark Highlands, and Tallgrass Prairie. The reader should keep in mind that ecological make-up in the ecoregions is more complex than depicted by the mapped units, and that species and vegetation types are rarely confined to just one of them. This is especially true of areas close to the boundaries. For example, while prairie vegetation occurs extensively in the Tallgrass Prairie it can be found in other ecoregions as well. And while an ecoregion may be best known for certain natural communities, such as natural lakes and wetlands in the Northern Lakes ecoregion, other communities exist there as well.

Familiarity with ecoregions can help in the pursuit of finding certain wildflowers. For example, the Northern Lakes clearly has the greatest occurrence of sphagnum bogs and their associated species, such as bog rosemary (*Andromeda glaucophylla*), dragon's mouth (*Arethusa bulbosa*), and purple pitcherplant (*Sarracenia purpurea*). Similarly, only in the Ozark Highlands will you find yellow coneflower (*Echinacea paradoxa*), false aloe (*Manfreda virginica*), and Missouri evening primrose (*Oenothera macrocarpa*) growing together.

The best way to find a plant is to seek out its habitat, that is, the community that provides the growing conditions it requires. Natural communities are where native organisms live, interact, and obtain resources in order to survive. In a sense they are smaller versions of an ecoregion, utilizing some of the same ecological factors characterizing them. They are often classified by dominant vegetation, substrate type, moisture condition, and topographic position. Ruderal sites (those heavily disturbed by human activity—agricultural fields, industrial sites, roadsides, waste areas) often include invasive non-natives.

The major types of natural communities referenced in this book are as follows.

Barrens. A usually dry, nutrient-poor site with sparse vegetation. Trees, if present, are often stunted and twisted.

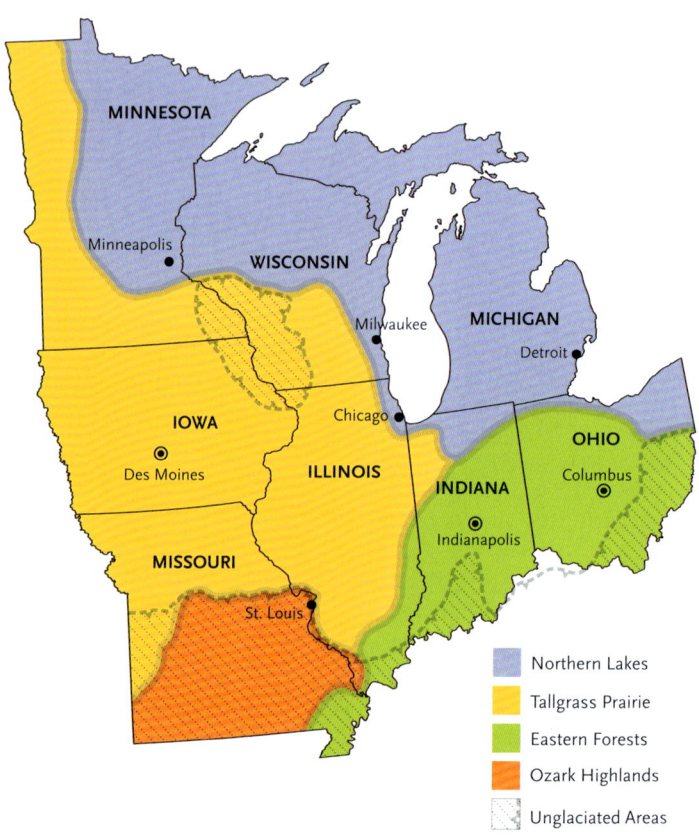

Map of Midwest ecoregions.

Bog. A typically open and acidic, nutrient-poor wetland with no outlet, commonly with a floating mat of living and decaying sphagnum moss, dominated by heath family shrubs.

Cliff. A vertical outcrop of rock. Includes the rock face and edges.

Fen. A site with highly organic soil saturated with groundwater flowing in a slow, diffuse manner. In the Midwest, fens are usually alkaline but can range to acidic, open, and graminoid or forested.

Forest. A site where trees dominate, usually with tall, straight trunks and commonly forming a closed upper canopy. Layers of smaller trees and shrubs form mid- and understories.

Glade. A natural opening in a woodland or forest, typically dry, with extensive bedrock at the surface and little or no soil. Sites may possess scattered, stunted trees.

Lake. A relatively large body of water, often so deep that only aquatic plants can survive. Ponds have similar characteristics but are smaller.

Marsh. A wetland dominated by sedges, grasses, cattails, bulrushes, and arrowheads.

Prairie. A mostly treeless area dominated by native grasses and forbs (herbaceous wildflowers).

Savanna. An area dominated by grasses and forbs but interspersed with open-grown and widely branched trees. Canopy cover is generally less than 30%.

Spring. A natural, continuous or intermittent flow of water from the ground.

Stream/River. Flowing waters confined within channels and banks occurring in valleys and floodplains. Associated communities include gravel bar, sandbar, and mudflat.

Swamp. A forested wetland, usually with standing water, the source of which comes from direct precipitation, flooding, or groundwater seepage. Some swamps are dominated by shrubs.

Woodland. Similar to forests but trees typically less dense and typically not forming a closed canopy. Mid- and understories are usually sparse.

EASTERN FORESTS

The Eastern Forests ecoregion makes up the southeastern portion of the Midwest and includes areas that are unglaciated (in eastern Ohio, southcentral Indiana, southern Illinois, and the Missouri Bootheel), areas manipulated by the Illinois glaciation 220,000 to 110,000 years ago (in southwestern Ohio, southeastern and southwestern Indiana, and southeastern Illinois), and areas scoured by the Wisconsin glaciation 75,000 to 10,000 years ago (in Ohio and Indiana). This varied glacial history has resulted in a diversity of soil depths, topography, substrates, and geological features, including flat lands covered mostly by forests, rugged and rolling hills covered mostly with a mix of forests, woodlands, glades, and cliffs, and low swamps similar to those of the Mississippi Embayment. Karst features such as sinkholes and caves are common in portions of the unglaciated regions. The lowest elevation in the Midwest is found within this ecoregion in the Missouri Bootheel. The easternmost portion of the ecoregion is an extension of the Allegheny Mountains and has a decidedly Appalachian feel. Yellow-crowned night heron and Carolina chickadee nest here, and long-tailed salamander and hellbender are largely restricted to this ecoregion.

Forests and Woodlands. Various forest and woodland types make up the matrix of the ecoregion, occurring on sloping or flat uplands. Dry-mesic upland forests, mesic upland forests, and flatwoods are common, on substrates ranging from alkaline to acidic. Except in particularly dry or acidic areas, the herbaceous understory is usually rich with twinleaf (*Jeffersonia diphylla*), bloodroot (*Sanguinaria canadensis*), drooping trillium (*Trillium flexipes*), and other wildflowers, forming a carpet of color in the spring; sedges (*Carex*) and ferns persist through the growing season.

In the easternmost portion (generally the eastern half of Ohio), terrain is hilly to the north, becoming rugged with steep hills and valleys to the south; this area is part of the

Allegheny Plateau, and forests characteristically harbor plant species with an affinity to the Appalachian Mountains. Similar topography reaches into the Shawnee Hills of southern Indiana and southern Illinois. Mixed mesophytic forests dominate, with fewer areas of dry-mesic and dry upland forests. Frequent dominant trees in more mesic settings include yellow buckeye (*Aesculus flava*) and basswood (*Tilia americana*); other common trees include sugar maple (*Acer saccharum*), American beech (*Fagus grandifolia*), and eastern hemlock (*Tsuga canadensis*). Hickories (*Carya*) and oaks (*Quercus*) characterize dry-mesic sites. Scarlet oak (*Q. coccinea*), chestnut oak (*Q. montana*), and post oak (*Q. stellata*) are common on the driest sites, with various sedges and heath family shrubs in the understory.

An extension of the higher elevation area surrounding the Central Basin of Tennessee extends into southern Indiana. This area is characterized by calcareous substrates and karst features (caves and sinkholes). Calciphiles such as blue ash (*Fraxinus quadrangulata*) and chinkapin oak (*Quercus muehlenbergii*) are common; Virginia pine (*Pinus virginiana*) and chestnut oak (*Q. montana*), frequent in surrounding regions with more acidic substrates, are absent.

Flatwoods and mesic to dry-mesic upland forests are common in till plains on flat to rolling topography, with cliffs and ravines restricted to areas along major river valleys and glacial drainage features. Tulip poplar (*Liriodendron tulipifera*), white oak (*Quercus alba*), and red oak (*Q. rubra*) are common, with flatwoods characterized by bur oak (*Q. macrocarpa*), pin oak (*Q. palustris*), and Shumard's oak (*Q. shumardii*), among others. Flatwoods are relatively flat with shallow depressional areas, yielding a mix of plant species that tolerate wet and moist to dry areas growing in close proximity.

TOP American beech (*Fagus grandifolia*) is a major component in mesic and dry-mesic forests of the Eastern Forests ecoregion. *MAH*

BOTTOM On relatively flat terrain, shallow depressions in flatwoods hold water from fall through spring, creating conditions for both wetland and upland plants; these pools are often important breeding grounds for amphibians. *SN*

More dissected topography with generally flat plateaus and deep valleys is present in areas impacted by the Illinois glaciation event but not the Wisconsin glaciation event. Dry to mesic forests are on plateaus and mesic forests are on slopes. Pignut hickory (*Carya glabra*), white oak (*Quercus alba*), and black oak (*Q. velutina*) characterize dry to dry-mesic forests, with shagbark hickory (*C. ovata*), chinkapin oak (*Q. muehlenbergii*), and red oak (*Q. rubra*) in mesic areas.

Floodplain forests along rivers, streams, and creeks vary substantially in composition depending on duration and frequency of flooding. Typical trees include silver maple (*Acer saccharinum*), green ash (*Fraxinus pennsylvanica*), and American elm (*Ulmus americana*), as well as trees common in mesic forests. Flooding also affects the herbaceous understory, where species present in mesic forests and swamps are common.

Glades and Barrens. Sandstone, siltstone, shale, and limestone glades and barrens are scattered primarily throughout unglaciated southern portions of the ecoregion. These prairie-like communities are situated on very shallow soil or in areas with extensive rock outcroppings, ranging in substrate chemistry from acidic to alkaline. Scattered trees including blackjack oak (*Quercus marilandica*) and post oak (*Q. stellata*) are somewhat stunted in more acidic settings, with poverty oat grass

(*Danthonia spicata*), little bluestem (*Schizachyrium scoparium*), and farkleberry (*Vaccinium arboreum*) characterizing the understory. Big bluestem (*Andropogon gerardii*), rattlesnake master (*Eryngium yuccifolium*), and Indian grass (*Sorghastrum nutans*) are frequent in species-rich alkaline glade and barren systems.

Prairies. Small, scattered prairies are a minor component of the ecoregion, particularly in glaciated regions. Unlike glades and barrens, prairies have a more substantial soil layer. Various prairie types are present, including hill prairies (sometimes with gravelly substrate) and dry to wet prairies. Dominant plant species are similar to those of prairies in the Tallgrass Prairie and Northern Lakes ecoregions.

Cliffs. The cliff community is well developed, primarily in unglaciated regions along drainages; cliffs are also local in the glaciated region. They are composed of sandstone or limestone and can be extremely treacherous, often vertical. Cliff faces range from very dry to having continuous seepage; this moisture gradient partially determines the specific flora that colonizes this community. Upper edges of cliffs characteristically have thin soil with dry or dry-mesic woodland or forest. Cliff faces support numerous ferns, among them walking fern (*Asplenium rhizophyllum*), bulblet fern (*Cystopteris bulbifera*), and marginal shield fern (*Dryopteris marginalis*). Wild hydrangea (*Hydrangea arborescens*) finds just enough soil in crevices on the cliffs. Forbs are infrequent but often include common alumroot (*Heuchera americana*). Sandstone cliffs sometimes

are associated with large, spectacular rock overhangs; these rockhouses, as they are known, can harbor very rare species such as Appalachian filmy fern (*Vandenboschia boschiana*).

Wetlands. Wetlands are well represented, primarily consisting of flatwoods, swamps, circumneutral seeps, and sinkhole swamps and ponds. These communities are aggregated in the Missouri Bootheel, southern Illinois, southwestern and southeastern Indiana, and the glaciated till plains of Indiana and Ohio. Many harbor species with an affinity to the Gulf of Mexico and Atlantic coastal plains.

Swamps are extensive in the southwestern portion of the ecoregion, where the Mississippi Embayment extends into the Missouri Bootheel and along major rivers into southern Illinois and southwestern Indiana. A rich tree flora is present in these swamps. Dominant species include water hickory (*Carya aquatica*), water locust (*Gleditsia aquatica*), and, in the understory, such shrubs and small trees as swamp privet (*Forestiera acuminata*) and possumhaw (*Ilex decidua*). The herbaceous stratum is variable, depending on water depth.

Circumneutral seeps (with approximately neutral pH) in this ecoregion are primarily in the Ohio and Indiana till plains, often situated on slopes. They are groundwater-fed wetlands with organic soils and range from open to having sparse tree cover. Black ash (*Fraxinus nigra*) and nannyberry (*Viburnum lentago*) are common; marsh marigold (*Caltha palustris*) and skunk cabbage (*Symplocarpus foetidus*), among others, typify the herbaceous stratum.

Sinkhole swamps and ponds occur primarily in unglaciated karst regions of the ecoregion. Dominant trees in sinkhole swamps include red maple (*Acer rubrum*), sweet gum (*Liquidambar styraciflua*), and swamp cottonwood (*Populus heterophylla*). Various sedges, buttonbush (*Cephalanthus occidentalis*), and

TOP Limestone glades in the Eastern Forests are often situated on south- or west-facing slopes. *MAH*

BOTTOM A vertical cliff dominated by mosses, lichens, and ferns crops out along a drainage in a forested landscape. *SN*

Circumneutral seeps are often sloping with mucky substrate and are wet year-round; skunk cabbage (*Symplocarpus foetidus*) is nearly always a dominant though ephemeral component of the understory. *LC*

other swamp herbaceous species are common. Sinkhole ponds occur in sinkholes with deeper water, resulting in a pond surrounded by marsh; spatterdock (*Nuphar advena*), softstem bulrush (*Schoenoplectus tabernaemontani*), and broad-leaved cattail (*Typha latifolia*) are common dominants.

NORTHERN LAKES

The Northern Lakes ecoregion consists of relatively newly exposed land. The coldest winter temperatures and greatest average snowfall amounts in the Midwest occur in this area, which was entirely covered by ice sheets of the Wisconsin glaciation. Until they retreated, the glaciers blanketed the majority of the region with sedimentary deposits (over 1200 feet deep in places!), but local areas of exposed bedrock (such as around Lake Superior and parts of Lake Michigan) form cliffs, balds, and lakeshores. Most of the Northern Lakes is forested, with open woodlands and savannas and smaller areas of open prairie concentrated in the western portion. Lakes, streams, and wetlands are also common. With a generally cool climate and mix of coniferous and deciduous trees, this region has a boreal (northern) quality, and some of its plant species are found in northern regions around the world (circumboreal). Wilson's snipe and chestnut-sided warbler nest in this ecoregion, and gray wolf, Blanding's turtle, and northern pike are found here.

Forests and Woodlands. Forests ranging from deciduous at the southern extent to mixed in northern reaches historically dominated the Northern Lakes ecoregion. Forest structure and composition is determined by climate, landscape position, and nature of the substrate. Dry forest (often with acidic substrate) occurs on coarse, well-drained soils, especially in the southern part of the ecoregion. Dry-mesic and mesic forests are prevalent, ranging in soil chemistry from acidic to basic. Dry-mesic sites are often on level ground or south- or west-facing slopes, whereas more mesic sites tend to be on north- or east-facing slopes, gently rolling till plains, and outwash plains, but the line between these two common communities is often blurred. Wet forests are present in depressions and in proximity to water bodies.

In the southern portion of this ecoregion, dry forests are typically dominated by oak. White oak (*Quercus alba*), black oak (*Q. velutina*), and sweet pignut hickory (*Carya ovalis*) characterize dry-mesic forests, often mixed with some "mesic" trees. Sugar maple (*Acer saccharum*), basswood (*Tilia americana*), and, in the east, American beech (*Fagus grandifolia*) make up the mesic forest matrix. Moving north, yellow birch (*Betula alleghaniensis*) is a common canopy component in mesic sites, and conifers such as eastern white pine (*Pinus strobus*) and eastern hemlock (*Tsuga canadensis*) become more frequent throughout, with jack pine (*P. banksiana*) dominating the driest sites. Boreal forests consisting of balsam fir (*Abies balsamea*) and white spruce (*Picea glauca*) in association with paper birch (*Betula papyrifera*) and balsam poplar (*Populus balsamifera*) occur in cooler microclimates along and near the Upper Great Lakes.

In mesic sites especially, wildflowers are abundant and provide an amazing show for a few weeks each spring. Among these are yellow trout lily (*Erythronium americanum*),

large-flowered trillium (*Trillium grandiflorum*), and Canada violet (*Viola canadensis*). Ferns and sedges are also common in the understory. In drier situations more susceptible to fire, trees are more widely spaced, comprising woodland communities that sometimes contain small, open prairie. The understory in dry forests is often comprised of shrubs in the heath family, and the herbaceous stratum is less ephemeral-dominated, with Pennsylvania oak sedge (*Carex pensylvanica*) and members of the aster family frequent. Southward, the pea family is also well represented in dry forests, and near the Great Lakes dry forests contain understory species more typical of dry-mesic or even mesic sites. Species richness generally decreases in dry forests from south to north.

Floodplain forests, present along rivers and streams, range from wet to mesic. Dominant species are variable. In frequently (and often deeply) flooded sites, typical trees include sycamore (*Platanus occidentalis*), eastern cottonwood (*Populus deltoides*), and American elm (*Ulmus americana*). Infrequently flooded zones support many trees typical of mesic forests. In addition to floodplain specialists, a wide variety of herbaceous species otherwise typical of open to forested wetlands or upland forests also occur, resulting in exceptional species richness in the best-developed examples.

Flatwoods, mostly restricted to the southeastern portion of the ecoregion, occur on relatively level topography where soils are underlain by a layer of relatively impermeable clay. Wetland trees occur where water is perched in the spring or after heavy precipitation events; mounds, which may be barely elevated above wet depressions, support upland trees. Trees are generally shallowly rooted, commonly resulting in tree tip-ups. Red maple (*Acer rubrum*), shellbark hickory (*Carya laciniosa*), and white oak (*Quercus alba*) are typical canopy components. The herbaceous stratum comprises plants representative

of both the swamp and mesic forest communities, and such ferns as royal fern (*Osmunda spectabilis*) and cinnamon fern (*Osmundastrum cinnamomeum*) are common.

Lakes and Wetlands. Unsurprisingly, lakes and wetlands are a defining characteristic of the Northern Lakes ecoregion. Glacial lakes, ranging from oligotrophic (nutrient-poor) to eutrophic (nutrient-rich), are often situated within otherwise forested rolling topography. A variety of wetland types are present in low-lying areas and in poorly drained soils.

Lakes are unvegetated or have submerged or floating vegetation. Shallow lake borders often support submergent and emergent marsh communities. Submergent marshes are characterized by milfoils (*Myriophyllum*), pondweeds (*Potamogeton*), and watermeals (*Wolffia*), among others. In the shallower emergent zone, these associate with pickerelweed (*Pontederia cordata*), common arrowhead (*Sagittaria latifolia*), and broad-leaved cattail (*Typha latifolia*). Floating muck flats and seasonally exposed substrates may support species with an affinity to the Atlantic and/ or Gulf of Mexico coastal plain. These coastal plain disjuncts are concentrated in southwestern Michigan and northwestern Indiana.

Swamps are most frequently associated with lakes and low-order streams southward and relatively level lake and outwash deposits northward. Groundwater seepage is often present in swamps of the Northern Lakes region. In the southern portion of the region, deciduous trees such as bur oak (*Quercus macrocarpa*) and American elm (*Ulmus americana*) are characteristic; buttonbush (*Cephalanthus occidentalis*) is present in the wettest portions. The herbaceous stratum often includes false nettle (*Boehmeria cylindrica*), cardinal flower (*Lobelia cardinalis*), and various sedges. Northward, swamps are typically dominated by conifers including tamarack (*Larix laricina*), black spruce (*Picea mariana*), and eastern white pine (*Pinus strobus*), with an herbaceous stratum of sedges, ferns, lichens, and mosses, particularly sphagnum moss. Gray alder (*Alnus incana*) and winterberry (*Ilex verticillata*) occur in areas that experience seasonal inundation, such as along streams.

Bogs are typically situated within glacial lake basins. Species richness is typically low, particularly where the rooting zone is isolated from mineralized groundwater. Thick deposits of sphagnum moss form the matrix of this community, and the living surface typically forms a characteristic hummock-hollow microtopography, with the relatively dry hummocks dominated by low shrubs in the heath family such as leatherleaf (*Chamaedaphne calyculata*) and cranberries (*Vaccinium macrocarpon, V. oxycoccos*). Older, elevated bog zones often support trees such as tamarack (*Larix laricina*) and black spruce (*Picea mariana*). Saturated or inundated hollows and open sphagnum "lawns," these sometimes floating over the water body, support several characteristic herbaceous species such as few-seeded sedge (*Carex oligosperma*), rose pogonia (*Pogonia ophioglossoides*), and carnivorous plants, such as purple pitcherplant (*Sarracenia purpurea*), which have adapted to the low-nutrient conditions by developing various ways to trap and digest insects. Muskegs (aka blanket bogs) are vegetatively similar to bogs but occur on expansive flat glacial outwash plains or glacial lake plains rather than in concave basins. These occur almost exclusively in the northern portion of the region.

TOP Mesic upland forests, characterized by having an abundance of colorful wildflowers including large-flowered trillium (*Trillium grandiflorum*), are a welcome sight after a long, cold winter. *BS*

BOTTOM The lakes of the Northern Lakes ecoregion often have concentric zones of vegetation, with marsh transitioning to shrub and forested zones around the perimeter. *PS*

Standing water is present, at least during the spring, in swamps of the Northern Lakes ecoregion, which are commonly vegetated by ferns and sedges, such as fringed sedge (*Carex crinita*). *LC*

Fen is a term applied variously to groundwater-fed wetlands that occur on muck or peat soils derived from poorly decomposed vegetation. Specific fen types include poor fens (acidic), rich fens (alkaline), prairie fens, circumneutral seeps, and marl beaches. In poor fens, calciphiles are absent or localized, and plant species typically associated with bogs but that grow under minerotrophic conditions are often present or even dominant; examples include bog birch (*Betula pumila*), star sedge (*Carex echinata*), and bog goldenrod (*Solidago uliginosa*). Neutral to calcareous fens and depressions in some patterned fens are characterized by bluejoint grass (*Calamagrostis canadensis*), shrubby cinquefoil (*Dasiphora fruticosa*), and Kalm's lobelia (*Lobelia kalmii*). Along and at the base of hillslopes in the southern portion of the Northern Lakes region, sparsely to densely forested circumneutral seeps are characterized by black ash (*Fraxinus nigra*) and skunk cabbage (*Symplocarpus foetidus*), with other species similar to those found in swamps of the region. Prairie fen, a type of rich fen occurring in the southern part of the region where oak savannas were once prevalent, includes a diverse mix of typical fen species and mesic prairie species. Marl beach is a community that occurs around lakeshores and on former lake beds where marl is a major component of the substrate. These areas are characterized by conservative calciphiles such as wicket spikerush (*Eleocharis rostellata*) and often are somewhat species-poor.

Along the shores of the Great Lakes (and rarely inland) are primary communities where little to no soil development has occurred. Instead, these communities occur on raw sand, gravel, cobble, or, especially along the northern Great Lakes shorelines, level bedrock or cliffs. These primary communities are vegetated by grasses, shrubs, and stunted trees, as well as lichens and mosses. Vegetation cover can be sparse depending on the texture and type of exposed substrate, and characteristic plant species varies by substrate type. Several plant species that are disjunct from the western United States, as well as species that are disjunct from the arctic-alpine region, can be found in the more northern primary communities. In some places, sand dunes and swales have formed as a result of repeated glacial advances and retreats. Where sand dunes have formed, the rhizomes of marram

TOP Bogs are commonly dominated by leatherleaf (*Chamaedaphne calyculata*), purple pitcherplant (*Sarracenia purpurea*), and sphagnum moss, with vegetation forming concentric zones around an open water "eye." *BS*

BOTTOM Rich, marly fens characteristically have patches of gray-colored soil with no vegetation surrounded by an abundance of sedges and a diverse mix of shrubs, grasses, and various goldenrods and other forbs. *LC*

grass (*Ammophila breviligulata*) provide stability that begins the process of primary succession. In these sandy soils, pannes, with calcareous substrates and a flora similar to that of rich fens, have formed between dunes where wind has blown sand away to the water table. Extensive marshes replace primary communities along the Great Lakes where water is shallow on broad, gentle slopes. These marshes have a composition similar to those found around glacial lakes, but they often include calciphiles and species typical of pannes such as twig rush (*Cladium mariscoides*) and low calamint (*Clinopodium arkansanum*).

Prairies. Although Midwestern landscape-scale prairies are restricted to the Tallgrass Prairie ecoregion, small, scattered wet,

A sand dune formation borders Lake Michigan, with marram grass (*Ammophila breviligulata*) providing much of the herbaceous vegetation cover. *BS*

mesic, and dry prairies were part of the pre-settlement Northern Lakes ecoregion. These prairies occur on a variety of landforms, including glacial lake plains, outwash plains, moraines, and loess hills. Plant species similar to those found in the Tallgrass Prairie are common. These relatively small, isolated prairies were historically maintained by drought (or periodic flooding in wet prairies), poor or shallow soils, and/or fire, and they often grade into partially treed savannas and forests. Northward, oak-dominated savannas give way to pine barrens dominated by jack pine (*Pinus banksiana*).

OZARK HIGHLANDS

The Ozark Highlands is an unglaciated landscape characterized by rolling hills, rock outcrops, glades, and a complex of forest types dominated by oak-hickory-pine forest. Karst features—including thousands of caves and springs nationally recognized for their size—are common. Numerous clear-flowing large rivers and streams dissect and carve deep valleys around the flanks of the plateau, which contains some of the oldest exposed bedrock in the country. The ecoregion has the greatest topographic relief of any area in the southern Midwest, namely, the St. Francois Mountains. It also has the highest summer maximum temperature averages (85–90°F) and the greatest amount of annual precipitation (40–50 inches) in the Midwest.

While a significant portion of the natural communities and vegetation of the Ozark Highlands has an eastern affinity, certain parts have a distinctly southern and/or western flavor, exemplified by such plants as southern maidenhair fern (*Adiantum capillus-veneris*), Texas greeneyes (*Berlandiera texana*), and soapweed yucca (*Yucca glauca*). There are also animals with these same affinities, among them the pygmy rattlesnake, greater roadrunner, and Texas brown tarantula.

Forests and Woodlands. Several types of forests and woodlands occur in the Ozarks. Most are dry and dry-mesic with localized areas of mesic forests. They are typically found in the hilliest terrain, often intersected by rivers in deep valleys, and occur on substrates that are acidic (soils developed on chert, sandstone, or igneous rock) or slightly alkaline (dolomite or limestone).

Trees of the acidic forests include mockernut hickory (*Carya tomentosa*), shortleaf pine (*Pinus echinata*), and white oak (*Quercus alba*). A few characteristic understory species are dittany (*Cunila origanoides*), bearded panic grass (*Dichanthelium boscii*), and wild comfrey (*Cynoglossum virginianum*). There is some overlap in dominants on alkaline sites, but in the latter there is also a high occurrence of sugar maple (*Acer saccharum*), blue ash (*Fraxinus quadrangulata*), and chinkapin oak (*Q. muehlenbergii*). Herbs in the higher pH sites include bulblet fern (*Cystopteris bulbifera*), goldenseal (*Hydrastis canadensis*), and bloodroot (*Sanguinaria canadensis*).

Substrates of woodlands are the same as those of forests but are mostly on the drier end of the moisture continuum and have a greater susceptibility to fire. On acidic sites black hickory (*Carya texana*), mockernut hickory (*C. tomentosa*), and shortleaf pine (*Pinus echinata*) are typical dominants. Typical herbaceous plants include Indian physic (*Gillenia stipulata*) and naked tick-trefoil (*Hylodesmum nudiflorum*). On limestone or dolomite substrates, dominants of woodlands may include white ash (*Fraxinus americana*), chinkapin oak (*Quercus muehlenbergii*), and local occurrences of blue ash (*F. quadrangulata*). Understory species may include purple coneflower (*Echinacea purpurea*), crested coral-root (*Hexalectris spicata*), and yellow pimpernel (*Taenidia integerrima*).

Floodplain Forests. Floodplain (or bottomland) forests are most developed along larger rivers. The tree dominants on smaller drainages are mesic or dry-mesic with sugar maple (*Acer saccharum*), bitternut hickory (*Carya cordiformis*), and black walnut (*Juglans nigra*), for example, whereas in larger, wetter floodplains there is also box elder (*A. negundo*), silver maple (*A. saccharinum*), and hackberry (*Celtis occidentalis*). In the understory one may find wood nettle (*Laportea canadensis*), cutleaf coneflower (*Rudbeckia laciniata*), and poison ivy (*Toxicodendron radicans*).

LEFT The rugged hills of the Ozark Highlands are dominated by oak-hickory-pine forest. *CG*

RIGHT Alkaline glades are open, grass-dominated communities punctuated by showy wildflowers, such as the Missouri evening primrose (*Oenothera macrocarpa*) shown here. *CG*

Prairies and Savannas. Prairies and savannas occur in patches on rolling or level uplands, but both communities are now considerably reduced in size due to fire suppression and agriculture. Typically, these are dominated by big bluestem (*Andropogon gerardii*), little bluestem (*Schizachyrium scoparium*), and Indian grass (*Sorghastrum nutans*) with an assortment of sun-loving forbs. In limestone/dolomite savannas chinkapin oak (*Quercus muehlenbergii*) and post oak (*Q. stellata*) are the dominant trees, whereas in the more acidic sandstone/ shale and chert types, blackjack oak (*Q. marilandica*) and black oak (*Q. velutina*) dominate. Many of the species also occur in glades.

Glades. There are over 182,000 acres of glades in the Ozark Highlands, the greatest area of them in the United States. These open, rocky communities typically occur on steep, upper slopes of south- or west-facing hills and are generally classified as either alkaline or acidic. Alkaline glades of dolomite or calcareous limestone have little bluestem (*Schizachyrium scoparium*) as a dominant along with other grasses such as side-oats grama (*Bouteloua curtipendula*) and poverty dropseed (*Sporobolus vaginiflorus*). Forbs include such colorful species as Indian paintbrush (*Castilleja coccinea*), purple prairie clover (*Dalea purpurea*), and yellow coneflower (*Echinacea paradoxa*).

Igneous, chert, and sandstone glades are acidic communities that usually harbor drought-tolerant grasses and forbs adapted to harsh, poor soils, such as poverty oat grass (*Danthonia spicata*), panic grasses (*Dichanthelium*), and flowering spurge (*Euphorbia corollata*).

Cliffs. Impressively large cliffs, a few over 500 feet in height, are particularly prevalent along river courses. South-facing cliffs are generally hot and dry; north-facing ones are cooler and moister. The chemical makeup of the predominant rock types influences species composition. Acidic cliffs may support various ferns including Bradley's spleenwort (*Asplenium bradleyi*), lobed spleenwort (*A. pinnatifidum*), and marginal shield fern (*Dryopteris marginalis*); vines such as Virginia creeper (*Parthenocissus quinquefolia*); shrubs such as wild hydrangea (*Hydrangea arborescens*); and stunted trees including red cedar (*Juniperus virginiana*) and shortleaf pine (*Pinus echinata*). Alkaline cliffs share some of the same species (minus the pine) but may add southern maidenhair fern (*Adiantum capillus-veneris*), wall-rue spleenwort (*Asplenium ruta-muraria*), and bulblet fern (*Cystopteris bulbifera*).

Springs/Streams/Rivers. Springs are common in the Ozarks, some exceptionally grand, with the largest ones discharging over 200 million gallons of water per day. They are most prominent in areas of limestone and dolomite. Borders of spring runs may harbor heart-leaved plantain (*Plantago cordata*), a species relatively common in the Ozarks but very rare elsewhere. Many Ozark stream and river borders

are composed of gravel- to boulder-sized rocks that are subject to intense flooding; characteristic plants in such gravel wash or river scour communities include vernal witch hazel (*Hamamelis vernalis*), sand grape (*Vitis rupestris*), and Carolina willow (*Salix caroliniana*). Herbaceous species that tolerate this dynamic environment include water willow (*Justicia americana*), cardinal flower (*Lobelia cardinalis*), and clammyweed (*Polanisia dodecandra*).

Wetlands. Wetlands are generally small and widely dispersed in the Ozarks. Deeper ones are formed in natural sinkholes, creating ponds and their associated marshes and swamps. Some harbor species that have affinities to Coastal Plain wetlands, such as cypress-knee sedge (*Carex decomposita*), water tupelo (*Nyssa aquatica*), and marsh millet

CLOCKWISE FROM TOP LEFT
Sandstone glades are acidic bedrock communities with shallow or no soil. *PN*

Borders of Ozark streams are often open due to scouring by floodwater. Water willow (*Justicia americana*) and other species that tolerate this dynamic environment line the water's edge. *PN*

Rock outcrops such as this large dolomite cliff are not uncommon in the Ozarks. They are particularly favorable habitats for a variety of rock-loving species, including several different ferns. *PN*

(*Zizaniopsis miliacea*). Ozark fens and seeps are fed by groundwater with a slow and diffuse rate of flow; they may be forested or graminoid. The flora may include several species quite unusual for the southern Midwest, such as tuberous grass pink (*Calopogon tuberosus*), marsh bellflower (*Campanula aparinoides*), and queen-of-the-prairie (*Filipendula rubra*).

TALLGRASS PRAIRIE

The Tallgrass Prairie ecoregion is characterized by mostly treeless terrain dominated by native grasses and a high diversity of forbs, especially those of the aster family. The ecoregion is a major eastern extension of grasslands connected to the Great Plains. This Prairie Peninsula, extending into an otherwise forested landscape, is thought to have been formed when the area experienced extreme summer drought and high temperatures during the Hypsithermal or Mid-Holocene Climatic Optimum, an extended warming period between 5000 and 10,000 years ago. Even today at least part of the ecoregion has the least average annual accumulation of precipitation in the Midwest, some areas with as little as 15–20 inches. These climatic conditions, in combination with historical fires and grazing and browsing by bison, elk, and other large herbivores, were conducive to creating and maintaining grasslands.

Almost all original prairie is gone, its millions of acres converted to agriculture or development. Its former extent is beyond most people's imagination, as today considerably less than 1 percent remains. Even in Illinois, the "Prairie State," it is very difficult to find native prairie vegetation. Fortunately, at least a few precious remnants remain, mostly in areas less conducive to agriculture, such as steep hillsides; land with soil too poor, rocky, or wet to plow; railroad rights-of-way; and pioneer cemeteries. Tallgrass prairie is clearly one of the most endangered natural community types in North America.

Prairies and Savannas. Prairies in the Midwest can be categorized in several ways, one of which is soil moisture. Mesic prairie is the classic prairie type in the Midwest. Much of it developed into highly organic "black soil," almost all of which is now in row-crop agriculture. Mesic prairie is usually found on level or rolling terrain that is evenly moist. Three grasses dominate, namely big bluestem (*Andropogon gerardii*), little bluestem (*Schizachyrium scoparium*), and Indian grass (*Sorghastrum nutans*). Other important species include rattlesnake master (*Eryngium yuccifolium*), prairie blazing star (*Liatris pycnostachya*), rosinweed (*Silphium integrifolium*), prairie dropseed (*Sporobolus heterolepis*), and gray-headed coneflower (*Ratibida pinnata*).

Wet prairie occurs where drainage is poor, such as in floodplains or in shallow, somewhat impervious swales or former lake basins on relatively level terrain. Species may include bluejoint grass (*Calamagrostis canadensis*), sawtooth sunflower (*Helianthus grosseserratus*), and cup plant (*Silphium perfoliatum*).

Dry prairie usually exists on steep slopes or substrates of coarse sand or gravel—places that assure quick and thorough drainage. Dry sand prairies occur where there is a significant deposit of sand, such as dunes. On the latter, porcupine grass (*Hesperostipa spartea*), June grass (*Koeleria macrantha*), and eastern prickly pear (*Opuntia cespitosa*) are characteristic species. Hill prairies (aka goat prairies) are often composed of loess or gravel. Some of their species include side-oats grama (*Bouteloua curtipendula*), downy painted-cup (*Castilleja sessiliflora*), and Great Plains ladies'-tresses (*Spiranthes magnicamporum*). On the western edge of the ecoregion, especially on loess hill prairies, occur several Great Plains species near the eastern edge of their natural range, including Engelmann's milkweed (*Asclepias engelmanniana*) and silver buffaloberry (*Shepherdia argentea*).

TOP Prairie vegetation once covered millions of acres in the Midwest. This rare remnant in Missouri exemplifies its former grandeur. *CG*

BOTTOM Midwestern tallgrass prairies are rich in wildflowers, such as the prairie coreopsis (*Coreopsis palmata*) and pale purple coneflower (*Echinacea pallida*) shown here. *BG*

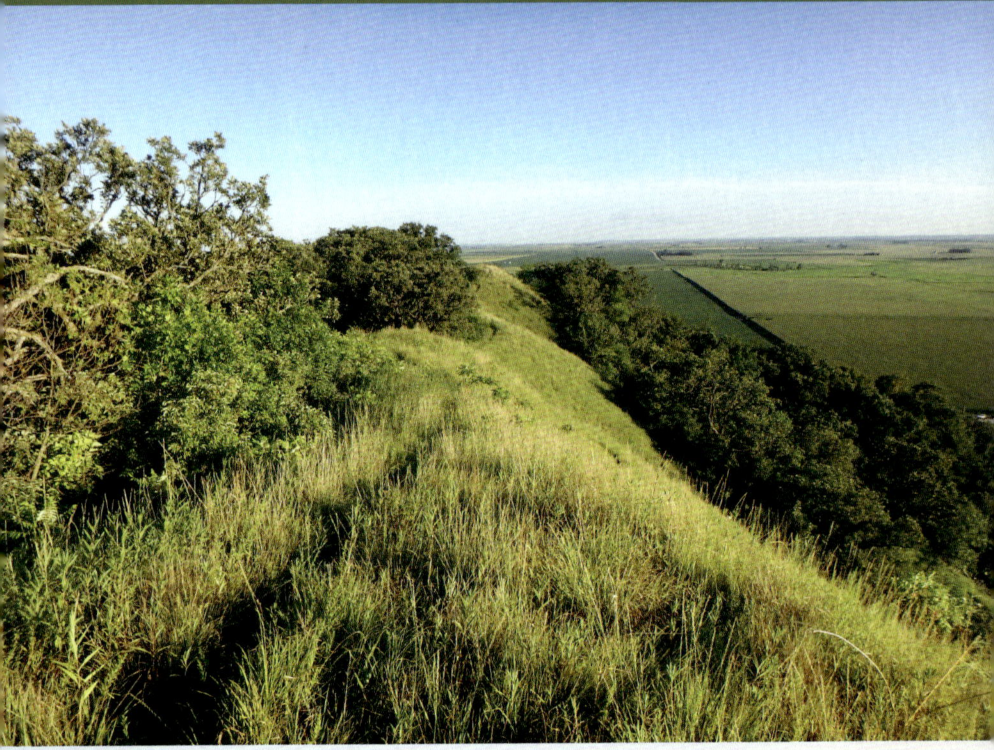

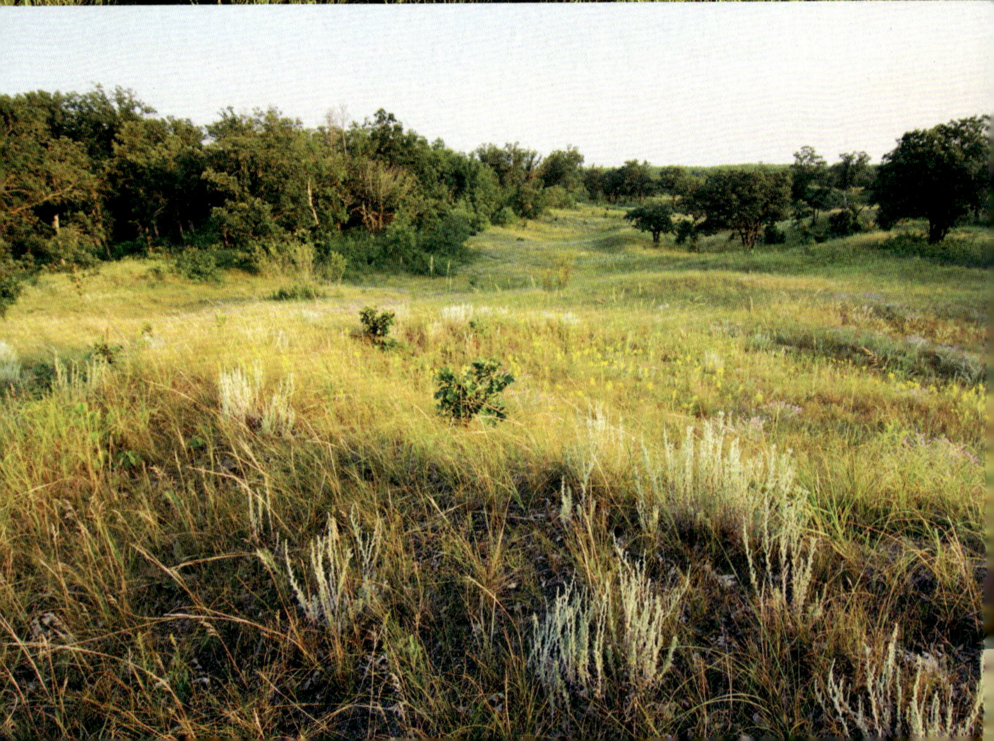

In Midwestern savannas the trees are usually oaks, particularly white oak (*Quercus alba*), bur oak (*Q. macrocarpa*), and black oak (*Q. velutina*). These may exist as full-size open-grown trees or "grubs" (sprouts from root balls). Hazelnut (*Corylus americana*) and prairie willow (*Salix humilis*) are characteristic shrubs. Many of the wildflower species found in savanna also occur in prairie. Before settlement, savannas were particularly prevalent in the hillier portions of the region or along larger water courses.

Forests and Woodlands. These natural communities, especially forests, were historically less common in this ecoregion. Forests chiefly occurred in areas that offered some protection from fire, such as in wet floodplains, leeward river corridors, and deep, shaded and/or cliff-lined ravines. Generally, the diversity of forest species declines going westward within the ecoregion. In woodlands, periodic fire creates open understories that allow for species dependent on higher light levels. The greater light allows for a mix of prairie and forest species to thrive.

Marshes, Fens, and Lakes. Impressively large marshes and various other wetlands are known in the ecoregion, one of which, the Grand Kankakee Marsh in Indiana, has been referred to as the "Everglades of the North." Unfortunately, most of its nearly 500,000 acres have been drained and cleared. In addition to species associated with wet prairies, these wetlands may also harbor swamp milkweed (*Asclepias incarnata*), southern blueflag (*Iris virginica*), and broad-leaved cattail (*Typha latifolia*). Groundwater seepage in a prairie landscape commonly forms a prairie fen. It is identified by the occurrence of both prairie and fen species, such as big bluestem (*Andropogon gerardii*), fen grass-of-Parnassus (*Parnassia glauca*), and prairie dropseed (*Sporobolus heterolepis*). Some natural bodies of water exist in the ecoregion, such as river oxbow lakes and a few glacial lakes. In western Minnesota and northcentral Iowa, shallow and often ephemeral bodies of water, commonly referred to as prairie potholes, are home to a variety of marsh and aquatic plants.

TOP Due to their steepness and unsuitability for agriculture, hill prairies provide sanctuary for prairie species. Note the complete conversion to row crops beyond this loess hill prairie in Iowa. *JP*

BOTTOM Savannas are a matrix of grasslands and widely scattered or copses of trees, such as can be seen in this Minnesota example with bur oak (*Quercus macrocarpa*) and quaking aspen (*Populus tremuloides*). *TRK*

As they are in savannas, oaks are the principal trees of Tallgrass Prairie woodlands. Black oak (*Quercus velutina*) is shown here. CG

PLANT FAMILIES

IN THE 18TH CENTURY, Carl Linnaeus developed our present-day system of taxonomy, or the classification of living things. This system involves grouping organisms together into increasing levels of similarity using shared characteristics. There are eight ranks within this hierarchy: domain, kingdom, phylum, class, order, family, genus, and species. Family is the highest classification level that is used regularly with plants.

Although it is possible to learn plants fairly well by simply flipping through the pages of a photographic guide, a much greater level of understanding comes from learning what characteristics unite plants into a given family. In fact, once you have a better understanding of family characteristics, you can travel beyond your floral area of comfort and still substantially narrow down the choices for an unknown plant by determining the family to which it belongs, as many plant families are cosmopolitan. For example, the family Fabaceae is well represented worldwide, and all species within the family have the same set of characteristics. No matter if the identification challenge before you is *Tephrosia virginiana* in the Midwest or *Cochlianthus gracilis* in China, both will have characteristics of the family Fabaceae.

The following accounts provide characteristics for all the plant families (and their genera) treated in this book. We hope that you refer back to this section regularly and eventually "see" the commonalities among our Midwestern wildflowers.

Acanthaceae—acanthus family. A family of herbaceous plants, shrubs, and trees. Leaves opposite, simple, usually entire. Flowers perfect, irregular and two-lipped or sometimes appearing nearly regular, tubular with four or five corolla lobes, two or four stamens, and a pistil with a superior ovary. Fruit a two-valved capsule. Genera: *Justicia*, *Ruellia*.

Acoraceae—sweet-flag family. A monotypic family of herbaceous wetland plants, formerly included in Araceae. Leaves long and sword-shaped with parallel venation, without a stalk, basally disposed, emitting a sweet scent when bruised. Flowers inconspicuous, perfect, regular with six tepals, six stamens, and a pistil, very numerous on nearly cylindrical spadix but lacking spathe. Fruit a berry. Genus: *Acorus*.

Agavaceae—century plant family. A family of herbaceous perennial plants, shrubs, and trees, formerly included in Liliaceae. Leaves mostly in rosettes, simple, stalkless, sometimes succulent. Flowers regular, perfect or unisexual, with six tepals in two whorls of three, six stamens with somewhat succulent filaments, and single pistil, in large terminal or axillary clusters. Fruit a capsule. Genera: *Manfreda*, *Yucca*.

Alismataceae—water plantain family. A small family of herbaceous annual to perennial wetland plants that grow submerged, in mudflats, or rooted in water but with emergent stalks. Leaves simple, entire, long-stalked in basal rosette with leafless stem from middle, containing milky sap; in some species submerged leaves take on ribbon-like appearance reminiscent of unrelated aquatic plants. Flowers regular with three white petals in terminal inflorescence with whorled branching pattern. Fruit an achene. Genera: *Alisma, Echinodorus, Sagittaria.*

Alliaceae—onion family. A small family of herbaceous perennial plants from bulbs, formerly included in Liliaceae. Leaves mostly in basal rosettes, sometimes alternate, simple, usually with onion or garlic odor when bruised. Flowers regular, perfect, with six tepals, six stamens, and single pistil (or flowers sometimes replaced by bulblets), single or in cluster forming an umbel at top of plant. Fruit a capsule. Genera: *Allium, Nothoscordum.*

Amaranthaceae—amaranth family. A family of monoecious, dioecious, or perfect-flowered herbaceous annual to perennial plants and subshrubs. Leaves opposite or alternate, simple. Flowers regular, small, individually inconspicuous, with zero to five perianth parts, one to five stamens, and single pistil, sometimes subtended by up to five bracts, in dense terminal or axillary clusters. Fruit an achene or utricle. Genera: *Amaranthus, Chenopodium, Cycloloma, Froelichia, Kali.*

Amaryllidaceae—amaryllis family. A family of perennial herbaceous plants usually from bulbs, formerly included in Liliaceae. Leaves basal, simple. Flowers regular, perfect, with six tepals, a tubular corona (crown), six stamens, and single pistil with inferior ovary, single or in cluster forming an umbel at top of naked stem, with conspicuous bracts. Fruit a capsule or fleshy berry. Genera: *Hymenocallis, Narcissus.*

Apiaceae—carrot family. A family of annual to perennial herbaceous plants, shrubs, and trees. Leaves mostly alternate, often with sheathing bases, usually deeply lobed to compound, rarely simple. Flowers small, regular or nearly so, perfect or unisexual, with five sepals, five petals, five stamens, and single pistil with inferior ovary, in clusters forming simple or compound umbels, with bracts (involucre) and bractlets (involucel) present or absent. Fruit a schizocarp splitting into two one-seeded mericarps. Genera: *Angelica, Chaerophyllum, Cicuta, Conium, Cryptotaenia, Daucus, Erigenia, Eryngium, Heracleum, Osmorhiza, Oxypolis, Pastinaca, Perideridia, Polytaenia, Ptilimnium, Sanicula, Sium, Spermolepis, Taenidia, Thaspium, Torilis, Zizia.*

Apocynaceae—dogbane family. A family of perennial herbaceous plants, vines, shrubs, and trees. Leaves opposite, alternate, or whorled, simple, entire, usually containing milky sap. Flowers regular, perfect, with five sepals, five petals, five stamens, and single pistil with superior ovary, one or two or in umbel-like or raceme-like axillary or terminal clusters; members of former Asclepiadaceae also often with five tubular hoods, each with horn protruding or included. Fruit usually a follicle, rarely a berry or drupe, seeds often with hairs at apex. Genera: *Amsonia, Apocynum, Asclepias, Cynanchum, Matelea, Vinca, Vincetoxicum.*

Araceae—arum family. A family of perennial herbaceous plants, shrubs, and vines. Leaves simple or compound, basal or alternate, containing sharp calcium oxalate crystals that can cause damage to mouth and throat if eaten. Flowers inconspicuous and lacking petals or with four or six inconspicuous perianth parts, with up to six stamens and a pistil, numerous and unisexual, perfect, or with male above and female below in spike-like inflorescence on a spadix surrounded by hood-like, often showy spathe. Fruit a berry. Genera: *Arisaema, Calla, Peltandra, Symplocarpus.*

Araliaceae—ginseng family. A family of perennial herbaceous plants, shrubs, trees, and woody vines. Leaves mostly alternate, simple or compound. Flowers small, regular, often perfect, with five sepals, five petals, five stamens, and single pistil with inferior ovary, in umbel or stalked clusters of umbels. Fruit a berry or drupe. Genera: *Aralia, Hydrocotyle, Panax.*

Aristolochiaceae—pipevine family. A small family of perennial herbaceous plants, woody vines, and shrubs. Leaves simple, basal or alternate (sometimes appearing opposite), generally cordate, entire. Flowers perfect, regular or irregular, with three petal-like sepals, six or 12 stamens, and a pistil with ovary inferior or nearly so, solitary in leaf axil or terminal. Fruit a capsule. Genera: *Asarum, Endodeca.*

Asparagaceae—asparagus family. A small family of perennial herbaceous plants, shrubs, and vines with leaf-like branchlets that function as leaves; formerly included in Liliaceae. True leaves thin and scale-like, alternate. Flowers regular, perfect or unisexual, with six perianth parts, six stamens, and single pistil with superior ovary, solitary or in racemes or umbels. Fruit a berry. Genus: *Asparagus.*

Asteraceae—aster family. One of the two largest plant families worldwide and the largest in the Midwest (nearly 600 species in our eight-state region). Leaves basal, opposite, alternate, or rarely whorled, simple or compound. Flowers in heads that resemble individual flowers, heads solitary or in many-flowered arrays of various arrangements, with ray flowers only, disk flowers only, or with ray flowers surrounding disk flowers, heads subtended by phyllaries (aka involucral bracts) in one to several series that comprise the involucre. Ray flowers irregular with small tube topped by flattened, spreading or ascending, strap-like ray, usually female or sterile. Disk flowers regular, tubular with four or five lobes at top, perfect or occasionally male. Ovary often topped by pappus that surrounds corolla and consists of bristles, scales, and/or awns. Fruit nearly always an achene. Genera: *Achillea, Ageratina, Ambrosia, Anaphalis, Antennaria, Anthemis, Arctium, Arnoglossum, Artemisia, Bidens, Boltonia, Brickellia, Carduus, Centaurea, Cichorium, Cirsium, Conoclinium, Coreopsis, Cyclachaena, Doellingeria, Dyssodia, Echinacea, Eclipta, Elephantopus, Erechtites, Erigeron, Eupatorium, Eurybia, Euthamia, Eutrochium, Galinsoga, Grindelia, Hasteola, Helenium, Helianthus, Heliopsis, Heterotheca, Hieracium, Hypochaeris, Inula, Ionactis, Iva, Krigia, Lactuca, Lapsana, Leucanthemum, Liatris, Lygodesmia, Matricaria, Nabalus, Nothocalais, Packera, Palafoxia, Parthenium, Petasites, Pluchea, Polymnia, Pseudognaphalium, Pyrrhopappus, Ratibida, Rudbeckia, Senecio, Sericocarpus, Silphium, Smallanthus, Solidago, Sonchus, Symphyotrichum, Tanacetum, Taraxacum, Tragopogon, Tussilago, Verbesina, Vernonia, Xanthium.*

Balsaminaceae—touch-me-not family. A family of annual and perennial herbaceous plants and subshrubs with fleshy stems. Leaves simple, usually alternate. Flowers perfect, irregular, with three (sometimes five) sepals (one usually with nectar spur), three (sometimes five) petals, five stamens, and pistil with a superior ovary. Fruit usually an explosive capsule, rarely a berry. Genus: *Impatiens.*

Berberidaceae—barberry family. A family of perennial herbaceous plants, shrubs, and trees. Leaves usually basal or alternate, simple or compound. Flowers regular, perfect, with up to 18 sepals, usually six petals (or petals absent), six to 12 stamens, and single pistil with superior ovary, one to many in various inflorescence types. Fruit a berry, capsule, achene, or follicle. Genera: *Caulophyllum, Jeffersonia, Podophyllum*.

Bignoniaceae—trumpet creeper family. A family of shrubs, trees, and woody vines. Leaves usually opposite or whorled, simple, up to three times pinnately compound, or palmately compound. Flowers perfect, irregular and usually two-lipped, tubular with five corolla lobes, four stamens and a shorter staminode, and a pistil with a superior ovary. Fruit a two-parted capsule, often elongated but sometimes nearly round. Genus: *Campsis*.

Boraginaceae—borage family. A family of annual to perennial herbaceous plants, shrubs, and small trees, often rough- or bristly-hairy. Leaves simple, basal and/or alternate. Flowers regular, perfect, tubular or with spreading corolla lobes, usually with five-lobed fused (at least basally) calyx and five-lobed fused (at least basally) corolla, five stamens, and single pistil usually with superior ovary, stalked in clusters that are often coiled initially but unfurling when flowering. Fruit a capsule or nutlet. Genera: *Buglossoides, Cynoglossum, Echium, Ellisia, Hackelia, Heliotropium, Hydrophyllum, Lappula, Lithospermum, Mertensia, Myosotis, Onosmodium, Phacelia*.

Brassicaceae—mustard family. A large family of annual to perennial herbaceous plants and shrubs. Leaves alternate and/or basal, usually simple but often deeply pinnately lobed. Flowers regular, perfect, with four sepals, four often yellow or white (less frequently purple) petals (sometimes absent), usually six stamens (four long, two short), and single pistil with superior ovary, usually in racemes. Fruit a two-parted capsule, either a silique (when at least three times as long as wide) or a silicle (when less than three times as long as wide). Genera: *Alliaria, Arabidopsis, Arabis, Barbarea, Berteroa, Boechera, Brassica, Camelina, Capsella, Cardamine, Descurainia, Draba, Erysimum, Hesperis, Iodanthus, Leavenworthia, Lepidium, Nasturtium, Planodes, Rorippa, Sisymbrium, Thlaspi, Turritis*.

Cabombaceae—watershield family. A small family of perennial aquatic plants. Leaves submerged and opposite or whorled or floating with peltate attachment to stalk. Flowers perfect, regular, with three (rarely four) sepals and petals and few to numerous stamens, solitary on long axillary stalks, on or just above water surface. Fruit an achene or follicle. Genus: *Brasenia*.

Cactaceae—cactus family. A family of perennial herbaceous plants, shrubs, and trees. Stems succulent, cylindrical to spherical, functioning as leaves. Leaves usually reduced or absent. Areoles present on pads; spines and often glochids present on areoles. Flowers regular, usually perfect, with usually numerous tepals and stamens, solitary and sessile. Fruit dry to fleshy, usually with many seeds. Genus: *Opuntia*.

Campanulaceae—bellflower family. A family of annual to perennial herbaceous plants (rarely trees). Leaves usually simple and alternate, often exuding milky sap when bruised. Flowers perfect, regular or irregular, with five sepals, fused corolla with five lobes sometimes in two lips. Fruit a capsule. Genera: *Campanula, Campanulastrum, Lobelia, Triodanis*.

Cannabaceae—hemp family. A small family of trees, shrubs, and perennial herbaceous plants, sometimes vines. Leaves opposite or alternate. Flowers regular, perfect or unisexual (then plants dioecious), small, with four to six perianth parts and stamens, in terminal or axillary clusters. Fruit an achene or drupe. Genera: *Cannabis, Humulus*.

Caprifoliaceae—honeysuckle family. A family of subshrubs, shrubs, small trees, perennial herbaceous plants, and vines. Leaves opposite, usually simple. Flowers perfect, usually irregular and tubular, calyx and corolla with five lobes, five stamens, and single pistil with inferior ovary. Fruit a berry or drupe. Genera: *Lonicera, Symphoricarpos, Triosteum*.

Caryophyllaceae—pink family. A family of annual and perennial herbaceous plants. Leaves usually opposite, simple, entire. Flowers regular, usually perfect, usually with five sepals and petals (rarely four or petals absent), petals sometimes lobed, often with five or ten stamens, one to many in branched inflorescences. Fruit a capsule. Genera: *Arenaria, Cerastium, Dianthus, Gypsophila, Holosteum, Minuartia, Moehringia, Paronychia, Saponaria, Scleranthus, Silene, Stellaria*.

Celastraceae—bittersweet family. A family of shrubs (sometimes vine-like) and trees. Leaves opposite or alternate, simple. Flowers perfect, regular, with four or five sepals and petals (petals rarely absent) and four or five stamens. Fruit a capsule, achene, berry, drupe, or nutlet; seeds often covered by a brightly colored fleshy appendage (aril). Genus: *Euonymus*.

Cistaceae—rockrose family. A family of annual to perennial herbaceous plants and shrubs. Leaves simple, alternate or opposite (rarely whorled). Flowers regular or nearly so, usually perfect, with three or five sepals (usually of two widths) and four or five petals (rarely three or absent), solitary or in branched clusters. Fruit a capsule. Genera: *Crocanthemum, Lechea*.

Cleomaceae—spiderflower family. A family of annual to perennial herbaceous plants and shrubs. Leaves alternate, palmately compound with up to seven leaflets. Flowers perfect, regular or irregular, usually with four sepals, four petals, six stamens, and single pistil with superior ovary, solitary or stalked in a raceme or head. Fruit a capsule or nutlet. Genus: *Polanisia*.

Commelinaceae—spiderwort family. A family of herbaceous annual to perennial plants. Leaves simple, entire, alternate, sheathing or clasping stem. Flowers perfect, regular or irregular, with three sepals, three petals, six stamens, and a pistil with a superior ovary, in branched or umbel-like cluster, often subtended by a bract. Fruit a capsule. Genera: *Commelina, Tradescantia*.

Convallariaceae—lily-of-the-valley family. A family of perennial herbaceous plants, formerly included in Liliaceae. Leaves basal, alternate, or whorled, simple, entire. Flowers regular, perfect, with four, six, or eight tepals and stamens and single pistil, solitary and axillary or stalked and sometimes branched in terminal or axillary clusters. Fruit a berry or capsule. Genera: *Clintonia, Convallaria, Maianthemum, Medeola, Polygonatum, Streptopus, Uvularia*.

Convolvulaceae—morning-glory family. A family of annual to perennial herbaceous plants and subshrubs, usually vining or trailing. Leaves absent or simple and alternate, often lobed. Flowers perfect, regular, with five (sometimes four) sepals subtending fused, bell-shaped corolla with five often shallow lobes, five stamens, and a pistil with a superior ovary, solitary in leaf axils or in clusters. Fruit usually a capsule. Genera: *Calystegia*, *Convolvulus*, *Cuscuta*, *Ipomoea*.

Cornaceae—dogwood family. A small family of herbaceous plants, shrubs, and trees. Leaves simple, usually opposite and entire, and usually with arching veins. Flowers regular, usually perfect with four-lobed calyx and four or five petals subtending four or five stamens and single pistil with inferior ovary, stalked in umbel- or head-like clusters. Fruit a drupe (rarely a berry). Genus: *Cornus*.

Crassulaceae—stonecrop family. A family of annual to perennial herbaceous plants and shrubs (sometimes tree-like or climbing). Leaves succulent, usually simple, alternate, opposite, or basal. Flowers perfect, regular, with three to five sepals and petals (sometimes fused at base), several to numerous stamens, and three to five pistils. Fruit a cluster of follicles. Genus: *Sedum*.

Cucurbitaceae—gourd family. A family of annual to perennial herbaceous plants, often coarsely bristly, usually trailing or climbing by branched, coiled tendrils. Leaves usually simple, alternate, and palmately lobed. Flowers usually unisexual (plant monoecious), regular, with five sepals (sometimes absent) subtending corolla with usually five lobes; male flowers with three to five stamens, in small clusters at nodes, female flowers with pistil with inferior ovary, usually solitary at nodes. Fruit usually a berry or capsule, often with a tough outer coating and often prickly. Genus: *Echinocystis*.

Diervillaceae—bush honeysuckle family. A small family of shrubs. Leaves opposite, simple, toothed. Flowers perfect, irregular, calyx and corolla tubular and five-lobed, with five

stamens and single pistil with inferior ovary. Fruit a capsule. Genus: *Diervilla*.

Dioscoreaceae—yam family. A family of perennial herbaceous vines. Leaves alternate, opposite, or whorled basally, usually simple and cordate, with arching veins. Flowers usually unisexual with male and female on separate plants (dioecious) but rarely monoecious or with perfect flowers, regular, with six tepals fused at least basally, male flowers with six stamens, female flowers with three pistils, solitary or in branched axillary clusters. Fruit a capsule, often winged. Genus: *Dioscorea*.

Dipsacaceae—teasel family. A small family of annual, biennial, or perennial herbaceous plants (rarely shrubs), sometimes prickly. Leaves simple, usually in basal rosettes and opposite and fused across stem. Flowers perfect, irregular or nearly so, with funnel-shaped four- or five-lobed corolla, usually four stamens, and a pistil with inferior ovary, numerous and densely packed in spherical to cylindrical terminal, long-stalked head. Head subtended by involucre, each flower subtended by involucel. Fruit an achene. Genus: *Dipsacus*.

Droseraceae—sundew family. A small family of carnivorous annual and perennial herbaceous plants (rarely subshrubs). Leaves usually in basal rosettes, with top surface covered in sticky, gland-tipped hairs that trap and digest insects. Flowers regular, perfect, usually with five sepals and petals, few to numerous stamens, and single pistil with a superior ovary, solitary or several stalked in terminal cluster. Fruit a capsule. Genus: *Drosera*.

Ericaceae—heath family. A family of perennial herbaceous plants, shrubs, and trees, usually growing in acidic substrates. Leaves usually simple or essentially absent, alternate or opposite (rarely whorled). Flowers perfect, regular, generally bell-shaped, urn-shaped, or cylindrical, usually with four or five sepals, four or five separate or fused petals, two to ten stamens, and single pistil. Fruit a berry, drupe, or capsule. Genera: *Andromeda, Arctostaphylos, Chamaedaphne, Chimaphila, Epigaea, Gaultheria, Hypopitys, Kalmia, Moneses, Monotropa, Orthilia, Pyrola, Rhododendron, Vaccinium.*

Euphorbiaceae—spurge family. A large family of annual to perennial herbaceous plants, shrubs, trees, vines, and succulents, with clear or milky sap. Leaves opposite, alternate, or whorled, usually simple. Flowers regular, male or female on monoecious or dioecious plants, with up to six sepals, usually lacking petals (sometimes five), with superior ovary, solitary or in terminal or axillary clusters, often subtended by bracts. Fruit usually a capsule. Genera: *Acalypha, Croton, Euphorbia, Phyllanthus.*

Fabaceae—pea family. A large family of annual to perennial herbaceous plants, vines, shrubs, and trees. Leaves usually alternate and compound, entire, often with stipules at base. Flowers usually perfect and irregular, with five sepals (often fused at base), and five petals often arranged into a large upper petal (the banner or standard), two lateral petals (wings), and two lower petals that are often fused (keel), with a superior ovary. Fruit a legume. Genera: *Amorpha, Amphicarpaea, Apios, Astragalus, Baptisia, Chamaecrista, Clitoria, Crotalaria, Dalea, Desmanthus, Desmodium, Galactia, Glycyrrhiza, Hylodesmum, Kummerowia, Lathyrus, Lespedeza, Lotus, Lupinus, Medicago, Melilotus, Mimosa, Orbexilum, Oxytropis, Pediomelum, Phaseolus, Pueraria, Securigera, Senna, Strophostyles, Stylosanthes, Tephrosia, Trifolium, Vicia.*

Gentianaceae—gentian family. A family of annual to perennial herbaceous plants, shrubs, and trees. Leaves basal, opposite, or whorled, simple, entire. Flowers regular, perfect, with four- or five-lobed calyx and corolla, four stamens, and single pistil with superior ovary, solitary or in clusters. Fruit a capsule. Genera: *Bartonia, Centaurium, Frasera, Gentiana, Gentianella, Gentianopsis, Halenia, Obolaria, Sabatia.*

Geraniaceae—geranium family. A family of mostly annual and perennial herbaceous plants. Leaves opposite, alternate, or basal, simple or compound. Flowers usually regular and perfect, with five sepals, usually five petals, five, ten, or 15 stamens, and single pistil, few to numerous in clusters sometimes resembling umbels. Fruit capsule-like with long beak, separating into five parts. Genus: *Geranium.*

Haloragaceae—water-milfoil family. A small family of annual to perennial herbaceous plants (rarely shrubs), usually aquatic or partially terrestrial. Leaves opposite, alternate, or whorled, submersed pinnately divided into thread-like segments, emergent simple and entire to divided. Flowers male or female on monoecious or dioecious plants, regular, calyx with two to four lobes, corolla with two to four petals (or absent), with four or eight stamens and single style with inferior ovary, solitary or in clusters. Fruit fleshy or nut-like. Genus: *Proserpinaca.*

Hemerocallidaceae—hemerocallis family. A family of perennial herbaceous plants, formerly included in Liliaceae. Leaves basal, simple, linear, elongated. Flowers regular, perfect, with six tepals, six stamens, and single pistil with superior ovary, in branched cluster at top of stem. Fruit a capsule. Genus: *Hemerocallis.*

Hyacinthaceae—hyacinth family. A family of perennial herbaceous plants usually from bulbs, formerly included in Liliaceae. Leaves basal, simple, linear, elongated, usually fleshy.

Flowers usually regular, perfect, with six tepals, six stamens, and single pistil with superior ovary, on stalks in racemes at top of stem. Fruit a capsule. Genera: *Camassia, Muscari, Ornithogalum, Othocallis*.

Hypericaceae—St. John's-wort family. A family of annual to perennial herbaceous plants, shrubs, and trees. Leaves opposite or whorled, simple. Flowers regular, perfect, with five sepals often fused at base, usually five separate petals, several to many stamens, and single pistil with superior ovary, solitary or in branched terminal or axillary clusters. Fruit a capsule. Genus: *Hypericum*.

Hypoxidaceae—star grass family. A small family of perennial herbaceous plants, formerly included in Liliaceae. Leaves basal, simple, linear, elongated. Flowers regular, perfect, with six tepals, six stamens, and single pistil with inferior ovary, in branched cluster at top of naked stem. Fruit a capsule or berry. Genus: *Hypoxis*.

Iridaceae—iris family. A family of annual and perennial herbaceous plants and shrubs. Leaves mostly basal, some alternate, simple, sword-shaped, with fan-like arrangement. Flowers regular or nearly so, usually perfect, with six tepals (or three sepals and three petals), three stamens, and single pistil, solitary or in cluster, sometimes subtended by bracts. Fruit a capsule. Genera: *Belamcanda, Iris, Sisyrinchium*.

Juncaginaceae—arrowgrass family. A small family of annual and perennial herbaceous plants. Leaves basal and alternate, simple, flat or cylindric. Flowers regular, perfect, usually with six tepals (rarely three or four), up to six stamens, and single pistil, in a spike or raceme at top of plant. Fruit an achene. Genus: *Triglochin*.

Lamiaceae—mint family. A large family of herbaceous plants, shrubs, trees, and vines, often aromatic. Stems square in cross section. Leaves opposite, simple, occasionally deeply lobed. Flowers usually perfect, irregular (sometimes nearly regular), calyx tubular and usually five-lobed, corolla tubular and usually two-lipped with five lobes, usually with four stamens and a pistil with a superior ovary. Fruit a cluster of four nutlets. Genera: *Agastache, Blephilia, Clinopodium, Collinsonia, Cunila, Glechoma, Hedeoma, Lamium, Leonurus, Lycopus, Marrubium, Mentha, Monarda, Nepeta, Perilla, Physostegia, Prunella, Pycnanthemum, Salvia, Scutellaria, Stachys, Synandra, Teucrium, Trichostema*.

Lemnaceae—duckweed family. A family of small, perennial, aquatic herbaceous plants. Stems absent. Leaves simple, entire, sometimes in clusters, sometimes bearing roots below. Flowers minuscule and inconspicuous, lacking calyx and corolla, perfect, with one or two stamens and single pistil with superior ovary, in floral cavity or sac-like spathe within lateral pouch. Fruit bladder-like. Genus: *Lemna*.

Lentibulariaceae—bladderwort family. A small family of carnivorous annual and perennial herbaceous plants, usually aquatic or in saturated soils. Leaves simple in basal rosette, or simple or dissected and alternate or whorled and bearing sac-like bladders. Flowers perfect, irregular, calyx and corolla two-lipped, corolla with spur at base, with two stamens and single style with superior ovary, solitary or stalked in raceme at top of stem. Fruit a capsule. Genera: *Pinguicula, Utricularia*.

Liliaceae—lily family. A family of perennial herbaceous plants, sometimes from bulbs. Leaves basal, alternate, opposite, or whorled, simple. Flowers regular, perfect, with six tepals, six stamens, and single pistil, solitary, paired, or in clusters. Fruit a capsule or berry. Genera: *Erythronium*, *Lilium*.

Limnanthaceae—false mermaid family. A small family of annual, ephemeral herbs. Leaves alternate, pinnately compound. Flowers tiny, regular, perfect, with three petals and three sepals. Fruit a schizocarp. Genus: *Floerkea*.

Linaceae—flax family. A family of mostly herbaceous annuals and perennials. Leaves simple, opposite or alternate, entire. Flowers regular, perfect, mostly five-petaled, yellow or blue, occurring in terminal branching clusters. Fruit a capsule. Genus: *Linum*.

Linnaeaceae—twinflower family. A family of subshrubs or shrubs, creeping or erect, mostly evergreen. Leaves opposite, mostly simple. Flowers tubular, perfect, mostly regular, with five lobes. Fruit a single-seeded achene or capsule. Genus: *Linnaea*.

Loganiaceae—logania family. A mostly tropical family of herbaceous and woody plants. Leaves opposite, simple, entire. Inflorescences terminal or axillary one-sided racemes, often coiled at the tip. Flowers regular, perfect, corolla and calyx five-lobed with five stamens and single pistil. Fruit a capsule. Genus: *Spigelia*.

Lythraceae—loosestrife family. A family of annual to perennial herbs and shrubs. Leaves mostly opposite and entire. Flowers mostly regular, perfect, with a hypanthium (floral cup) of four to six petals that are commonly pink to purplish and crinkled. Fruit a capsule. Genera: *Ammannia*, *Cuphea*, *Decodon*, *Lythrum*, *Rotala*.

Malvaceae—mallow family. A family of annual to perennial herbs and shrubs. Leaves alternate, petiolate, simple, generally palmately lobed or toothed and with stellate hairs. Flowers regular, perfect, with five petals and five sepals, stamens united forming a tube around the style. Fruit a capsule. Genera: *Abutilon*, *Callirhoe*, *Hibiscus*, *Malva*, *Malvastrum*, *Napaea*, *Sida*.

Melanthiaceae—bunchflower family. A family of perennial herbs, formerly included in Liliaceae. Leaves linear to broadly elliptic, mostly basal, with alternate stem leaves. Flowers regular, perfect, usually small but several, with six petal-like tepals. Fruit a capsule. Genera: *Aletris*, *Anticlea*, *Melanthium*, *Triantha*, *Veratrum*.

Melastomataceae—melostome family. A large, mostly tropical family of herbs, shrubs, and trees. Leaves mostly opposite, entire and usually sessile with three to seven deeply ribbed and arching parallel veins. Flowers regular, perfect, with four rounded petals attached to vase-shaped hypanthium (floral cup). Fruit a capsule or berry. Genus: *Rhexia*.

Menispermaceae—moonseed family. A family of mostly twining woody and herbaceous vines, with alternate, simple, palmately veined leaves. Flowers small, white to green with six petals and six sepals and of separate sexes on separate plants (dioecious). Fruit a drupe. Genus: *Menispermum*.

Menyanthaceae—buckbean family. A cosmopolitan family of mostly aquatic or semi-aquatic perennials with alternate or basal, simple or trifoliolate leaves. Flowers mostly perfect, regular, five-parted, tubular or funnel-shaped, with five stamens alternating with the lobes. Fruit a capsule. Genus: *Menyanthes*.

Molluginaceae—carpetweed family. A family of mostly low-growing or creeping, slightly succulent annual and perennial herbs. Leaves entire, opposite, alternate, or whorled. Flowers small, regular, and perfect, petals absent or with five petal-like sepals. Fruit an achene or capsule. Genus: *Mollugo*.

Montiaceae—miner's lettuce family. A family of somewhat succulent annual, biennial, and perennial herbs, formerly included in Portulacaceae. Leaves simple, entire, and mostly opposite. Flowers regular and perfect, generally with two sepals and two to 19 petals. Fruit a capsule, with seeds typically bearing elaiosomes (food bodies). Genus: *Claytonia*.

Myrsinaceae—myrsine family. A family of mostly annual and perennial herbs. Leaves simple, opposite or whorled, some dotted or streaked with dark resin canals. Flowers perfect, regular, four- to nine-parted, tubular, and funnel or bell-shaped; sometimes dotted or streaked. Fruit a capsule or drupe. Genera: *Anagalis, Lysimachia, Trientalis*.

Nelumbonaceae—water lotus family. A monotypic family of aquatic perennials. Leaves large, round, simple, entire, peltate (umbrella-like), arising from a rhizome buried in mud below the water's surface. Flowers regular, perfect, with numerous tepals, held well above the water's surface on long stalks. Fruit acorn-like, held loosely within a broad, hard, pitted receptacle. Genus: *Nelumbo*.

Nyctaginaceae—four-o'clock family. A family of primarily annuals, perennials, and vining shrubs. Leaves mostly simple and entire, opposite, the pairs unequal or equal in size. Flowers regular, perfect, the petals absent but with five petal-like sepals, often congested above sepal-like bracts. Fruit an achene. Genus: *Mirabilis*.

Nymphaeaceae—water lily family. A family of aquatic perennials. Leaves floating to somewhat emergent, bases usually two-lobed, arising from often large rhizomes buried in mud below the water's surface. Flowers commonly large and showy, regular, perfect, with numerous petals and sepals, the latter sometimes petal-like. Fruit berry-like and spongy. Genera: *Nuphar, Nymphaea*.

Onagraceae—evening primrose family. A family of annuals to perennials. Leaves opposite, alternate or whorled, entire or toothed. Flowers often showy, normally four-parted, regular and perfect, the floral tube elongated and sometimes extending beyond the ovary. Fruit a capsule or nut-like. Genera: *Chamaenerion, Circaea, Epilobium, Gaura, Ludwigia, Oenothera*.

Orchidaceae—orchid family. A large family of perennial herbs. Leaves (if present) alternate, opposite, or whorled, mostly sessile, and mostly simple and entire. Flowers with three petals, one of which (the lip) is typically larger, dissimilar in shape, and usually positioned lowermost. Sepals three. Male and female organs fused on a column. Fruit a capsule. Genera: *Aplectrum, Arethusa, Calopogon, Calypso, Coeloglossum, Corallorhiza, Cypripedium, Epipactis, Galearis, Goodyera, Hexalectris, Isotria, Liparis, Malaxis, Neottia, Platanthera, Pogonia, Spiranthes, Tipularia, Triphora*.

Orobanchaceae—broomrape family. A large family of root-parasitic (or partially so) annuals to perennials, possessing chlorophyll or not. Leaves simple or lobed, alternate, some reduced to scales. Flowers irregular, perfect, two-lipped and five-lobed. Fruit a capsule. Several members formerly in Scrophulariaceae. Genera: *Agalinis, Aureolaria, Buchnera, Castilleja, Conopholis, Dasistoma, Epifagus, Melampyrum, Orobanche, Pedicularis*.

Oxalidaceae—wood sorrel family. A family of annual to perennial herbs. Leaves mostly palmately compound, with three rounded leaflets notched at the tips, basal or alternate on stem, petiolate (leaflets generally sessile). Flowers regular, perfect, with five petals and sepals. Fruit a capsule or berry. Genus: *Oxalis*.

Papaveraceae—poppy family. A family of annual to perennial herbs, sometimes woody. Leaves simple, entire or compound, basal or alternate on stem. Flowers regular or irregular, perfect, with petals mostly four or more. Fruit a capsule. Genera: *Adlumia, Capnoides, Corydalis, Dicentra, Sanguinaria, Stylophorum*.

Parnassiaceae—grass of parnassus family. A small family of perennials. Leaves somewhat waxy, rounded, basal, simple and entire. Flowers regular, perfect, with five petals and sepals, solitary on long stalks. Fruit a capsule. Genus: *Parnassia*.

Passifloraceae—passionflower family. A family of mostly perennial vines, woody or herbaceous, some with axillary tendrils. Leaves alternate and mostly simple but commonly lobed. Flowers disk-shaped and complex, commonly bearing a fringe of hair-like filaments, regular, perfect, with five petals, five sepals, and three club-shaped styles. Fruit a berry or capsule. Genus: *Passiflora*.

Penthoraceae—ditch stonecrop family. A monotypic family of wetland perennials from rhizomes. Leaves simple, alternate, and mostly sessile. Flowers regular, perfect, with five inconspicuous petals (or lacking) and five evident sepals. Fruit a capsule. Genus: *Penthorum*.

Phrymaceae—lopseed family. A family of annual to perennial herbs and shrubs. Leaves simple, opposite, entire or toothed. Flowers generally perfect, irregular, tubular, two-lipped, the upper lip two-lobed, the lower three-lobed. Fruit a capsule or achene. Genera: *Mimulus, Phryma*.

Phytolaccaceae—pokeweed family. A family of annual to perennial herbs and woody plants. Leaves simple, entire, alternate or opposite, petiolate, often with wavy margins. Flowers small, regular, generally perfect, lacking petals but with four to eight persistent petal-like sepals. Fruit a berry or achene. Genus: *Phytolacca*.

Plantaginaceae—plantain family. A family of annual to perennial herbs and woody plants. Leaves basal, alternate or opposite, entire or toothed. Flowers regular or irregular, unisexual or perfect, and generally four- or five-parted. Fruit typically a capsule. Many members formerly in Scrophulariaceae. Genera: *Besseya, Callitriche, Chaenorhinum, Chelone, Collinsia, Gratiola, Leucospora, Linaria, Lindernia, Nuttallanthus, Penstemon, Plantago, Veronica, Veronicastrum*.

Polemoniaceae—phlox family. A family of annual to perennial herbs, subshrubs, and vines. Leaves simple or compound, alternate or opposite, basal or on stems. Flowers regular or irregular, perfect, trumpet- or bell-shaped, five-lobed. Fruit a capsule. Genera: *Collomia, Phlox, Polemonium*.

Polygalaceae—milkwort family. A family of annual to perennial herbs, shrubs, and trees. Leaves simple, entire, mostly alternate. Flowers generally irregular, perfect, with five sepals, the lateral two (wings) are larger and petal-like, true petals small, three or five, fused to stamen tube. Fruit a capsule. Genus: *Polygala*.

Polygonaceae—buckwheat family. A family of annual to perennial herbs, vines, shrubs, and trees. Leaves generally simple, entire, alternate. Most species bear papery sheaths (ocreae) that surround stem nodes. Flowers generally small, regular, perfect, with two to six tepals, often in two whorls. Fruit an achene. Genera: *Fallopia, Persicaria, Polygonum, Reynoutria, Rumex*.

Pontederiaceae—pickerelweed family. A family of annual to perennial aquatic or wetland herbs. Leaves simple, alternate or whorled, and linear, round, or sagittate (arrowhead-shaped). Flowers regular or irregular, perfect, tubular, with six tepals. Fruit a capsule or dry and one-seeded. Genera: *Heteranthera, Pontederia*.

Portulacaceae—purslane family. A family of annual and perennial herbs. Leaves fleshy, simple, entire, opposite or alternate, linear or obovate. Flowers regular, perfect, with two sepals and five petals, the latter fused at base, forming a ring. Fruit a capsule. Genera: *Phemeranthus, Portulaca*.

Potamogetonaceae—pondweed family. A family of aquatic herbaceous plants. Leaves floating and/or submersed, simple, alternate or subopposite, linear to elliptic-ovate, petiolate, sheathing and stipuled. Flowers tiny on cylindrical spikes generally held above water, regular, perfect, with four tepals. Fruit a nutlet or drupe. Genus: *Potamogeton*.

Primulaceae—primrose family. A family of annual to perennial herbs. Leaves generally basal and mostly simple, entire or finely toothed. Flowers regular, perfect, with four or five petals and sepals, solitary or on long-stalked umbels. Fruit a capsule. Genera: *Androsace, Dodecatheon, Primula, Samolus*.

Ranunculaceae—buttercup family. A large and highly variable family of mostly annual and perennial herbs. Leaves both basal and along stem, alternate or opposite, petiolate, and simple or compound. Flowers generally regular and perfect, sepals three to six with some petal-like, petals absent to several. Fruit an achene, berry, capsule, or follicle. Genera: *Actaea, Anemone, Aquilegia, Caltha, Clematis, Coptis, Delphinium, Enemion, Ficaria, Hepatica, Hydrastis, Myosurus, Pulsatilla, Ranunculus, Thalictrum*.

Rhamnaceae—buckthorn family. A family of primarily shrubs and trees. Leaves simple, mostly alternate, petiolate, entire or toothed, and pinnately veined. Flowers small, regular, perfect, with four or five sepals and petals. Fruit a drupe, capsule, schizocarp, or samara. Genus: *Ceanothus*.

Rosaceae—rose family. A well-known family of annual to perennial herbs, shrubs, and trees, many ornamental and/or edible. Leaves simple or compound, mostly toothed, alternate. Flowers usually regular and perfect, with five petals and sepals. Fruit an achene, drupe, pome, or capsule, among others. Genera: *Agrimonia, Aruncus, Comarum, Dasiphora, Drymocallis, Filipendula, Fragaria, Geum, Gillenia, Potentilla, Rosa, Rubus, Sanguisorba, Sibbaldia, Spiraea, Waldsteinia*.

Rubiaceae—coffee or madder family. A family of annual to perennial herbs, shrubs, and trees. Leaves mostly entire, opposite, and petiolate, with stipules on stems between bases of opposing petioles. Flowers generally regular, perfect, and tubular with four- or five-lobed petals and sepals. Fruit a drupe, berry, or capsule. Genera: *Cruciata, Diodia, Galium, Houstonia, Mitchella, Spermacoce*.

Santalaceae—sandalwood family. A family of woody to herbaceous perennials, many of which are partially parasitic on roots of other plants. Leaves mostly simple and entire, alternate or opposite. Flowers regular, perfect or unisexual on same individual or different individuals (monoecious or dioecious, respectively), with four or five tepals. Fruit a nut or drupe. Genus: *Comandra*.

Sarraceniaceae—pitcher plant family. An amazing family of carnivorous wetland perennials. Leaves modified as tubular, water-filled traps conducive to harvesting insects. Flowers regular, perfect, nodding, with five sepals and five dangling petals surrounding a five-lobed disk-like style. Fruit a capsule. Genus: *Sarracenia*.

Saururaceae—lizard's tail family. A family of wetland perennial herbs. Leaves alternate, simple, entire, and petiolate, with fused stipules. Flowers small, perfect, without sepals or petals but with two to eight showy stamens. Fruit a capsule or dry fruit splitting into separate segments. Genus: *Saururus*.

Saxifragaceae—saxifrage family. A family of mostly perennial herbs. Leaves usually simple, basal or alternate on stem, entire or toothed. Flowers mostly regular, perfect, with five sepals and five petals commonly narrowed at their bases. Fruit a capsule. Genera: *Chrysosplenium, Heuchera, Micranthes, Mitella, Sullivantia, Tiarella*.

Scheuchzeriaceae—scheuchzeria family. A family with only one species, podgrass (*Scheuchzeria palustris*), a perennial herb essentially restricted to acid bogs. Leaves alternate,

entire, linear, sessile and sheathing. Flowers regular, perfect, with six tepals. Fruit a capsule. Genus: *Scheuchzeria*.

Scrophulariaceae—figwort family. A family of annual to perennial herbs and woody plants. Leaves alternate or opposite, entire or toothed. Flowers regular or irregular, generally perfect, with four- or five-parted calyx and corolla. Fruit a capsule. Many genera formerly here now placed in other families. Genera: *Scrophularia, Verbascum*.

Smilacaceae—catbrier family. A family of mostly prickly, climbing woody vines with tendrils, but also herbaceous, non-prickly and non-climbing perennials. Leaves often somewhat thick and leathery, simple, opposite or alternate, and strongly parallel-veined, with smaller interconnecting veins between the three to five large veins. Flowers regular, unisexual, male and female on separate plants (dioecious), with six tepals. Fruit a berry. Genus: *Smilax*.

Solanaceae—nightshade family. A family of annuals, perennials, and woody plants. Leaves simple, generally alternate, entire or toothed. Flowers generally regular, perfect, with five sepals and five petals, these with lobes united at the base. Fruit a berry or capsule. Genera: *Datura, Leucophysalis, Physalis, Solanum*.

Trilliaceae—trillium family. A family of perennial herbs, formerly included in Liliaceae. Large leafy bracts (generally viewed as leaves) occur mostly in whorls of three and are simple, entire, ovate or obovate to elliptical. True leaves underground, alternate, scale-like along horizontal stem. The solitary flower is regular, perfect, with three petals, three sepals and six stamens. Fruit a berry. Genus: *Trillium*.

Typhaceae—cat-tail family. A family of strongly colonial wetland herbaceous perennials. Leaves basal, alternate on stems, two-ranked, narrowly linear and strap-shaped, flat or triangular in cross section with sheathing bases. Flowers minute and densely packed, without petals or sepals, male and female separate on same plant (monoecious). Fruit a follicle or achene-like drupe. Genera: *Sparganium*, *Typha*.

Urticaceae—nettle family. A family of annual or perennial herbs and woody plants, often with stinging hairs. Leaves simple and opposite or alternate. Flowers minute, petals lacking, sepals four or five, male and female separate on same plant (monoecious). Fruit an achene. Genera: *Boehmeria*, *Laportea*, *Parietaria*, *Pilea*, *Urtica*.

Valerianaceae—valerian family. A family of annual to perennial herbs. Leaves opposite, simple, or pinnately lobed or divided. Flowers regular or irregular, perfect or unisexual, tubular, and five-lobed. Fruit an achene. Genera: *Valeriana*, *Valerianella*.

Verbenaceae—vervain family. A family of annual to perennial herbs, shrubs, and trees, many with four-angled stems. Leaves simple, opposite, with margins entire, toothed, or deeply lobed or divided. Flowers regular to irregular, mostly perfect, tubular with five lobes, often in long narrow spikes. Fruit a schizocarp or drupelike. Genera: *Glandularia*, *Phyla*, *Verbena*.

Violaceae—violet family. A family of annuals, perennials, and shrubs. Leaves basal or alternate on stem, simple, entire to toothed or lobed, and stipulate. Flowers irregular, perfect, with five sepals and five petals, the lowest petal often the largest and prolonged into a spur. Fruit a capsule. Genera: *Hybanthus*, *Viola*.

Xyridaceae—yellow-eyed grass family. A family of annual to perennial herbs. Leaves basal, long and linear (grass-like), sheathing the naked flowering stem. Flowers regular, perfect, sepals three with one dissimilar to the others, and three petals in cone-like heads on stems that may be twisted. Fruit a capsule. Genus: *Xyris*.

Zygophyllaceae—caltrop family. A family of annual to perennial herbs, shrubs, and trees, often with angled or swollen nodes. Leaves usually opposite, mostly compound, often with spiny stipules. Flowers regular, perfect, with five sepals and five petals. Fruit a capsule or schizocarp. Genus: *Tribulus*.

HOW TO USE THIS BOOK

WITHOUT QUESTION, the best way to identify a plant is to use the dichotomous keys in technical references, but this method is not for everyone. With this in mind, we wrote *Wildflowers of the Midwest* primarily for keen amateurs and beginner wildflower enthusiasts. Included in this field guide are native and non-native herbaceous plants, non-woody vines, subshrubs, and some shrubs—over 1000 plant species and photographs; similar species (and how to distinguish them from the subject species) are discussed when appropriate. Even so, we had to make some very excruciating decisions about which species to include and which to leave out; we generally chose those that are the most widespread in the eight-state Midwest region, but some were included if they are common within a particular ecoregion or if they are particularly showy. Keep this in mind if you find a wildflower in the field that doesn't quite match one of those in this book, and check other references as necessary.

At least a rudimentary understanding of the parts of a plant are necessary to use this book properly. Don't let that scare you if you have no experience identifying plants! The line drawings inside the front and back covers of the book are there to help you—and here are a few basics to get you started. The parts of the plant that you should know, moving from the bottom to the top, are the roots (underground portion of the plant that lacks leaves and provides stability for the plant), the stem (portion of the plant that has leaves, can be below ground as rhizomes, creeping along the ground, ascending, vining, or erect), the leaves (the usually expanded, photosynthetic organs of the plant), and the flower (the reproductive portion of the plant).

The flower is the focus of this book. The flower includes stamens (the male reproductive organs), pistils (the female reproductive organs), or both. A stamen has two parts: the anther, which produces pollen, and the filament, which is a stalk connecting the anther to the rest of the flower. A pistil has three parts: a stigma, which receives pollen, an ovary, which is where reproduction happens (the mature ovary being the fruit, which contains seeds), and a style, which connects the stigma to the ovary. A flower also usually includes a perianth made up of a calyx (or sepals, when separate from the base) or both a calyx and a corolla (petals, when separate from the base). When the calyx and corolla are indistinguishable from one another, they are often referred to as tepals.

When a flower includes both male and female parts, it is said to be perfect (or bisexual); it is considered imperfect (or unisexual) when it includes only male or female parts. Male flowers are staminate; female flowers are pistillate. When imperfect flowers are present, a plant can either be monoecious (both male and female flowers on the same plant) or dioecious (male and female flowers on separate plants).

The symmetry of a flower is also important. A flower is said to be regular when, looking head on, it can be bisected in more than one plane and still result in mirror images;

it is irregular when there is only one plane in which it can be bisected and still result in mirror images. We recommend that you review the previous chapter to learn about some of the plant families with very distinctive flower types: the arum family (Araceae), the aster family (Asteraceae), the mustard family (Brassicaceae), the heath family (Ericaceae), the pea family (Fabaceae), the orchid family (Orchidaceae), the milkwort family (Polygalaceae), and the violet family (Violaceae).

This book is first arranged by typical flower color, the main color you see. This is generally the color of the corolla, but when a corolla is not present, it may be the color of the calyx or the showy modified leaf associated with a flower (aka a bract or, in Araceae, a spathe). The general flower colors are shown on the page margins, making finding the section of the book in which a certain species is located fairly straightforward. But note: flower color is often debatable (especially blue vs. purple and pale pink vs. pale lavender) and can shift in different light or as flowers age. In addition, some species have flowers that can range in color, even within a given population. For example, moth mullein (*Verbascum blattaria*) can have either white or yellow flowers, but it is treated only in the White Flowers section; in such situations, the range of flower colors is included in the species description. When a plant has bicolored flowers, it is included in the section of the dominant color; for example, wild columbine (*Aquilegia canadensis*) has flowers with red exteriors and yellow interiors, but they "read" as red, so it is included in the Pink to Red Flowers section. And when several species in a genus are covered, especially those with similarly colored flowers, an attempt has been made to keep them together for easier comparison; for example, all *Nabalus* species are in the White Flowers section. The upshot?

If you are unable to find a plant in one flower color section, try looking in another.

Within each flower color section of this book, species are arranged alphabetically first by family, and within each plant family by genus and then specific epithet. As such, it is important to develop a good understanding of characteristics of each plant family (again, review the previous chapter). Until then, the best way to use this field guide is to go to the section with the flower color of the plant you are trying to identify and to leaf through that section looking at the photographs until you find your plant or something similar. Then read the description of that plant. If you know the name of a plant and want to see what it looks like, use the index, which includes both common and botanical names.

The following information is included in each species account: botanical and family names, common name(s), photograph(s), habitat, bloom time, a description of the plant, and the Midwest geographical distribution.

BOTANICAL AND FAMILY NAMES

Although the boundaries are not often clean, a species is defined as individuals that share physical and genetic characteristics and can interbreed and produce fertile offspring. Species that are related are grouped together into categories called genera (singular = genus). Every plant species has a botanical (or scientific) name that consists of two parts: the first part is the genus, which is capitalized, and the second part is the specific epithet, which is lowercase. Botanical names can give us insight into the species. In large-flowered trillium (*Trillium grandiflorum*), for example, the genus, *Trillium*, references the three-parted nature of this lily-like species, and the specific epithet, *grandiflorum*, is a reference to the plant's large flowers. Each species account in *Wildflowers of the Midwest* begins with the botanical name; botanical names are typically italicized or underlined in formal writing. In some instances, we have included subspecies (subsp.) or varieties (var.) within the species description; these are classifications of somewhat distinct entities below the species level. The botanical name is followed immediately by the family name, which begins with a capital letter and ends in -aceae. See the previous chapter for the characteristics of all the families encountered in this book.

When it comes to botanical names, even the experts sometimes disagree as to the most appropriate classification and, with continued research into evolutionary relationships, names are constantly changing. In general, we use the most current botanical names and family names, but in some situations we disagree with a reclassification and have chosen to stick with an older name (see starflower, *Trientalis borealis*, for example). Occasionally, synonyms (syns.) of botanical names are included.

COMMON NAMES

One or several common names is included for each species treated in this book. Many plants have been assigned numerous common names over time, especially within different regions of the country, and some common names can apply to multiple species. That is why it is always best to use a botanical name to identify a plant.

PHOTOGRAPHS AND PHOTOGRAPHERS

The photographs in *Wildflowers of the Midwest* will probably be the most useful tool in helping you to identify a plant. In most cases, each species treated in this book is represented by a single photograph.

Unfortunately, it is impossible to show all the important characteristics for identification of a plant in a single image; in addition, many plant species are somewhat variable, so one photograph cannot display all the variation in a single species. To make this guide as useful as possible, we have selected photographs that best represent the plant and illustrate the key characters for identifying the plant (such as foliage and habit), with emphasis on flowers. In some cases, an inset is included that illustrates either a similar species or another part of the plant that can be useful for identification. Although we have taken many of the photographs included in this book, numerous other photographers have also generously contributed photographs. An abbreviation recognizing each photographer (their initials) is included at the end of each photo caption and/or species account; see page 579 for a key to the full names of all photographers.

HABITAT

Plants do not grow just anywhere, randomly. Rather, every plant typically favors a specific set of environmental conditions. For example, blue vervain (*Verbena hastata*) occurs in moist to wet soils, whereas creeping vervain (*V. bracteata*) grows in dry soils. Under natural conditions (those similar to what would have been in the Midwest prior to European settlement), species that can tolerate similar growing conditions grow together; these natural communities, and even more degraded habitat types that don't represent natural communities, can be useful in helping to narrow down a plant identification, so we have included the habitat(s) in which each plant typically occurs in each species account. Descriptions of these habitats can be found in the Ecoregions and Natural Communities chapter.

BLOOM TIME

Following the habitat(s) in each species account is the seasonal bloom time: spring (March to May), summer (June to August), fall (September to November), and winter (December to February). We have tried to give the typical bloom time for a given species throughout the entire region, but latitude matters. Pinkweed (*Persicaria pensylvanica*), for example, begins blooming in the southern portion of the Midwest in late spring but doesn't begin blooming in the northern portion of our region until summer—yet it continues blooming into the fall, especially northward; to accommodate this range, we have reported the bloom time as spring, summer, and fall. But there are always exceptions (particularly warm springs can bring on early bloom, for example); such anomalies are ignored.

DESCRIPTION

This section follows the same general sequence for each species entry, starting with the overall aspect of the plant—your first impression when you see a plant. This includes its height or size (we have usually provided the maximum height typically seen under natural conditions) and whether the plant is an annual (germinating, flowering, producing fruit and seeds, and dying within one year), biennial (germinating and producing typically a rosette of leaves the first year, then bolting and producing flowers, fruit, and seeds, and dying during the second year), or perennial (living more than two years, some for many years, others short-lived, but not necessarily flowering every year), or whether it grows as a vine (twining or growing on structures or other plants with or without the use of tendrils) or a shrub (a low woody plant that maintains its stem from year to year).

Next, a plant's leaves and flowers are described. The description of the leaves includes their arrangement on the plant (opposite, alternate, or whorled along the stem, or basal or in rosettes), as well as their dimensions (in inches), margins (entire, toothed, wavy), whether or not they are on stalks, that is, petiolate (and petiole length), and often shape (ovate, lanceolate, cordate). In some instances we have also mentioned if the leaves are hairy or glabrous (smooth).

For species with multi-colored flowers, or with flowers that can range from one color to another, flower color is included in the description; but in most cases, flower color is omitted, as it matches the color on the margin of the page. Aspects that are consistently discussed include floral symmetry (regular or irregular), dimensions, shape (tubular, urn-shaped, bell-shaped), and other characters that can be helpful such as number of stamens and if they are exserted from the flower or included within the flower. How the flowers are arranged in the inflorescence (spike, raceme, panicle) is also usually included.

When the fruit of a given species is an important character for identification, we have included a brief statement concerning it. This often includes fruit type (berry, capsule, achene) and size, and sometimes includes mention of a notable characteristic of the seeds.

MIDWEST DISTRIBUTION

At the end of each species account is the geographical distribution of the species within the Midwest. States in which the species is known to occur (or have occurred if the species is considered extirpated from the state and no longer present there) are abbreviated: Iowa (IA), Illinois (IL), Indiana (IN), Michigan (MI), Minnesota (MN), Missouri (MO), Ohio (OH), and Wisconsin (WI). If a plant is found under natural conditions in only one or a few states, or in the Midwest and nowhere else in the world, it is considered endemic to that state or region, and this is noted. A plant can be assumed to be native at least somewhere in the Midwest unless "introduced" follows the distribution information; "introduced" means that although the species is now found in the Midwest, it was not a part of the flora of the Midwest prior to European settlement in the region and has found a way, either intentionally or unintentionally, into our present-day flora. Distribution information concludes with a summary of the species' abundance and/or frequency (using terms such as common, rare, or widespread) and the ecoregions within which it occurs (see map on page 000).

FINAL THOUGHTS

Identifying a plant correctly takes careful and patient observation. Before even opening this book to look through the pages for a wildflower you are trying to identify, we suggest you first observe everything you can about the plant and its parts—leaf arrangement, petal count, overall hairiness (or lack of hairs)—anything that stands out. Once you've taken the time to do that, open up *Wildflowers of the Midwest*. Our hope is that through a combination of the features of each species account, you will be able to accurately identify most wildflowers that you encounter in the eight-state Midwest region. You should be able to start with the flower color and readily flip through that section of the guide to find a likely match for your plant. This may be a slow process at first, but with experience and especially as you become more familiar with the characteristics of plant families, you will be able to narrow down your search. After finding a likely option, always compare your plant, the habitat within which it is growing, bloom time, and where you found it geographically with the information in the species account for a more confident identification. It is also useful to check other references (books or websites) to verify an identification. If your plant doesn't match well with information in the species account, it could just be an odd individual (maybe you've heard the saying, "Plants don't read books"), or you may want to continue searching in this book for another likely option; or, since we had to limit the scope of this book to the most common wildflowers in the Midwest, it is possible that you have found a species that is not included in this book. In some cases, reading the description may lead you to similar species, and you can check other resources to see if your plant matches one of those. Once you have correctly identified your plant, you will feel a great sense of accomplishment—but don't stop there. Teach others about that plant; take a photograph; sketch the plant. All these activities will help you to remember the plant the next time you see it, and teaching and sharing what you've learned truly is one of the best ways to increase your knowledge and appreciation of our flora!

WHITE FLOWERS

Justicia americana
ACANTHACEAE

water willow

Shallow and/or quiet water in streams, rivers, lakes. Summer.

An erect perennial, unbranched, glabrous, 1½–2½ ft. tall, colonial. Leaves opposite, usually sessile, linear-lanceolate, 3–6 in. long × ⅓–1 in. wide. Flowers irregular, ¾–1 in. wide, tubular, 2-lipped, the upper with a notched tip, the lower with 3 spreading lobes, the larger middle one with dark purple markings, occurring in compact heads on upright axillary stalks. Water willow is a common host of dodder (*Cuscuta*). Contrary to what its common name suggests, this species is not a true willow (*Salix*). IA, IL, IN, MI, MO, OH. Occurs primarily in the lower Midwest. Common. *SN*

Yucca flaccida
AGAVACEAE

Adam's needle

Cemeteries, roadsides, fields, old homesites, sand prairies. Summer.

A semi-woody perennial greater than 5 ft. tall. Leaves evergreen, several in a basal rosette, linear to narrowly lanceolate and somewhat flexible with sharp tips, margins frayed into curling fibrous strands, 1½–2½ ft. long × 1–3 in. wide. Flowers regular, 2–3 in. long, bell-shaped and nodding, bearing 6 thick and fleshy tepals, numerous on a terminal, spreading, and densely hairy panicle. Fruit a thick, cylindrical capsule. Our only native yucca, soapweed yucca (*Y. glauca*), reaches its easternmost natural range in western IA and southern MO. IL, IN, MI, MO, OH, WI. Introduced from farther south. Scattered in the Midwest. *DT*

Alisma subcordatum
ALISMATACEAE

common water plantain

Mudflats, swamps, marshes, lake and pond edges, ditches. Summer.

A glabrous perennial 1–2 ft. tall. Leaves basal, petiolate, ovate to somewhat subcordate, 5–12 in. long × 1–6 in. wide, petioles commonly longer than the blades. Flowers (inset, left) regular, about ⅛ in. wide, with 3 petals and sepals about ¹⁄₁₆ in. long, floral array highly branched with numerous flowers. The similar large-flowered water plantain (*A. triviale*) is more common in the northern part of our region; its flowers are larger, about ¼ in. wide, with petals longer than sepals (inset, right). IA, IL, IN, MI, MN, MO, OH, WI. Common and widespread throughout. *MAH/SN*

Echinodorus berteroi
ALISMATACEAE

upright burhead

Marshes, ponds, wet open areas. Summer, fall.

An upright annual or perennial to 2⅓ ft. tall. Leaves basal, long-petiolate, rounded to cordate at base, entire, glabrous, to 6 in. long and wide. Flowers regular, to ½ inch wide, with 3 sepals and 3 petals, in whorls along branches, creating open, conical arrangements along stem. Fruit forms bur-like heads. Creeping burhead (*E. cordifolius*) has stems that arch or creep along the ground and root at their tips. May be confused with *Alisma* or *Sagittaria*. IA, IL, IN, MO, OH, WI. Mostly in southern portion of the Tallgrass Prairie, with few occurrences in the Ozark Highlands and Eastern Forests. *SN*

Sagittaria calycina
ALISMATACEAE

hooded arrowhead

Mudflats, especially along rivers and oxbow ponds. Summer.

An erect to lax glabrous annual, 6–24 in. tall. Leaves all basal, the petioles long and often rather spongy, blades sagittate with 2 spreading lower lobes, up to 1 ft. long and about as wide. Flowers regular, up to 1 in. wide, with 3 petals and 3 sepals, the latter closely appressed to fruiting head, on thick downwardly curving pedicels. Inflorescence erect, consisting of 3–10 whorls. Fruit an achene with beak projecting horizontally. Syn. *S. montevidensis*. **IA, IL, IN, MI, MN, MO, OH, WI. Occurs mostly in the lower Midwest.** *SN*

Sagittaria cuneata
ALISMATACEAE

arumleaf arrowhead

Marshes, mudflats, lakes, streams. Summer, fall.

A glabrous perennial to 2 ft. tall, monoecious. Leaves basal, entire, variable, those on submerged plants ribbon-like, also with long-petiolate floating sagittate leaves (or lacking basal lobes), when emergent with short-lobed sagittate leaves (sometimes unlobed) to 8 in. long × 4 in. wide, on stout petioles to 1 ft. long. Flowers regular, to 1 in. wide, male (above) and female (below), with 3 spreading petals, stalked, in whorls of 3 in terminal inflorescence to 8 in. long. Fruit an achene with minute erect beak (see inset), numerous in globular heads. **IA, IL, IN, MI, MN, OH, WI. More frequent northward, scattered southward.** *SN*

Sagittaria graminea
ALISMATACEAE

grass-leaved arrowhead

Marshes, shores, lakes. Summer, fall.

A glabrous perennial to 3 ft. tall, monoecious. Leaves basal, entire, variable, those of submerged plants in rosettes, stiff and sessile with parallel margins, when emergent with parallel margins or narrowly ovate, to 7 in. long × 1½ in. wide, on petioles to 7 in. long. Flowers regular, to 1 in. wide, male (above) and female (below), with 3 spreading petals, stalked, in whorls of 3 in terminal inflorescence to 8 in. long. Fruit an achene, broader above middle, with minute erect beak on one side, numerous in globular heads. Crested arrowhead (*S. cristata*) is more northern and has slightly longer beaks on fruit. IA, IL, IN, MI, MN, MO, OH, WI. Scattered. *SN*

Sagittaria latifolia
ALISMATACEAE

common arrowhead

Marshes, shallow portions of lakes and ponds, swamps, streams, ditches. Summer.

An erect, glabrous, usually monoecious perennial, tuberous, 1–3 ft. tall. Leaves basal, long-petiolate, blades variable but usually sagittate with 2 spreading lower lobes, up to 1 ft. long × 6 in. wide. Flowers regular, 1–1½ in. wide, with 3 petals and 3 sepals, the latter reflexed in fruit, on slender pedicels, in 2–10 whorls on flowering stem. Leafy bracts below each whorl are ¼–⅓ in. long and boat-shaped. Fruit an achene with a horizontal beak (see inset). In shortbeak arrowhead (*S. brevirostra*), the bracts are ⅓–⅔ in. long and long-tapering. IA, IL, IN, MI, MN, MO, OH, WI. Common, possibly in every county. *SN*

Sagittaria rigida
ALISMATACEAE

stiff arrowhead, sessilefruit arrowhead

Marshes, lakes, ponds, rivers. Summer, fall.

A glabrous perennial to 3 ft. tall, monoecious. Leaves basal, entire, variable, those on submerged plants in rosettes, stiff and sessile with parallel margins, when emergent with parallel margins or more often ovate, rarely sagittate with small basal lobes, to 6 in. long × 4 in. wide, on long petioles. Flowers regular, to 1 in. wide, stalked male (above) and sessile female (below), with 3 spreading petals, in whorls of 3 in terminal inflorescence to 6 in. long. Fruit an achene, broader above middle, with ascending beak on one side, numerous in sessile globular heads. IA, IL, IN, MI, MN, MO, OH, WI. Scattered, more frequent westward. *PD*

Allium burdickii
ALLIACEAE

narrow-leaved wild leek

Dry-mesic and mesic upland woods. Summer.

A glabrous perennial herb from a bulb, commonly occurring in clumps, 6–8 in. tall. Leaves mostly present only in spring, flat, narrowly lanceolate, pale green throughout, 4–10 in. long × 1–2 in. wide, gradually tapering to the ground without a distinct leaf stalk (see inset). Has onion smell. Flowers regular, about ⅛–¼ in. long, somewhat bell-shaped with 6 tepals. Inflorescence of 10–20 flowers in an umbel atop an 8-in.-tall leafless stalk. Leaves of the similar wild leek (*A. tricoccum*) are wider with reddish petioles. IA, IL, IN, MI, MN, MO, OH, WI. Widespread, more common than many maps indicate. *PR/CB*

Allium tricoccum
ALLIACEAE

wild leek, ramps

Rich, moist upland woods, ravines.
Summer.

A glabrous perennial herb from a bulb,
unbranched, 8–12 in. tall, commonly occurring in large colonies. Leaves flat, oval to
broadly elliptic, bright green, 4–12 in. long ×
2–4 in. wide, tapering abruptly to a distinct
reddish sheath and petiole (see inset), mostly
present only in spring, smelling of onion when
bruised, somewhat resembling ornamental
tulip leaves. Flowers regular, about ¼ in. wide,
with 6 tepals. Inflorescence usually with
18–50 flowers in a rounded umbel atop a leafless stalk. Shiny black seeds evident in late
summer and fall. IA, IL, IN, MI, MN, MO,
OH, WI. Widespread and locally common.
MAH/CR

Nothoscordum bivalve
ALLIACEAE

false garlic

Prairies, glades, exposed bluffs,
woodlands. Spring.

An erect perennial, unbranched except
in inflorescence, glabrous, 4–12 in. tall.
Leaves all basal from a bulb, linear, flat,
non-odoriferous, 4–12 in. tall × ¹⁄₁₆–⅛ in.
wide. Flowers regular, about ½ in. wide, with 6
spreading tepals, each with a greenish-yellow
base, 5–8 in a terminal umbel. This species
looks remarkably like an onion (*Allium*) but
lacks the onion odor. A good way to make a
quick distinction is simply to bruise a leaf and
smell it. IL, IN, MO, OH. Occurs mostly in
the Midwest's lower third, being particularly
common in the Ozark Highlands. SN

Froelichia floridana
AMARANTHACEAE

prairie cottonweed, large cottonweed, plains snakecotton

Sparsely vegetated sand, savannas, gravel bars, roadsides, railroads. Spring, summer, fall.

A hairy, mostly unbranched annual to 3 ft. tall. Leaves opposite, entire, to 4 in. long × ¾ in. wide, mostly on lower third of stem. Flowers regular, to ⅓ in. wide, with white-woolly calyx sometimes pink at 5-lobed tips, lacking petals, in 5-ranked spikes to 4 in. long in upper part of plant, terminal spike solitary, lateral spikes opposite and well spaced along stem. Slender cottonweed (*F. gracilis*) is shorter and more branched with leaves to ½ in. wide and flowering spikes to 1½ in. long. IA, IL, IN, MI, MN, MO, OH, WI. Primarily in the Mississippi and Kankakee drainages with scattered occurrences elsewhere. *MJH*

Hymenocallis occidentalis
AMARYLLIDACEAE

spiderlily

Floodplain forests, swamps, wet flatwoods, moist slopes. Summer.

An erect, glabrous perennial from a bulb, 2–3 ft. tall. Leaves basal, somewhat fleshy, linear and strap-shaped, glaucous, 1–2 ft. long × 1–2½ in. wide. Flowers regular, up to 6 in. wide composed of 6 spreading tepals and a shallow, funnel-shaped tube about 2 in. wide bearing 6 upcurved stamens. Inflorescence of 2–6 fragrant flowers atop a naked peduncle. Fruit a broadly triangular capsule. This spectacular plant blooms during the hottest part of the summer, typically late July into August. IL, IN, MO. Restricted primarily to the southwestern counties of the Eastern Forests. *MAH*

Angelica atropurpurea
APIACEAE

great angelica, purplestem angelica

Marshes, fens, seeps, sedge meadows, streambanks, swamps, ditches. Spring, summer.

A perennial to 10 ft. tall, unbranched except toward top. Leaves alternate on long petioles with inflated sheathing bases connecting to purple, glaucous, hollow stem; lower leaves to 2 ft. long and wide, 2–4 times compound with toothed leaflets to 4 in. long × 2 in. wide; upper leaves smaller. Flowers regular, with 5 petals, to ¼ in. wide, numerous, forming nearly spherical compound umbels to 9 in. wide; several inflorescences at top of plant. IA, IL, IN, MI, MN, OH, WI. Primarily in the Northern Lakes and northern portion of the Eastern Forests, extending to some extent into northern portion of the Tallgrass Prairie. *SN*

Angelica venenosa
APIACEAE

hairy angelica, wood angelica

Rocky forests, savannas, prairies, streambanks. Spring, summer.

A perennial to 5 ft. tall. Leaves alternate and basal, petiolate with slightly inflated sheathing bases connecting to purplish stem; lower leaves to 10 in. long, 2–4 times compound with toothed leaflets to 2 in. long × ½ in. wide; upper leaves smaller. Flowers regular, small, with 5 petals, forming several flat to hemispherical compound umbels to 6 in. wide at top of plant. Fruit short-hairy. Lovage (*Ligusticum canadense*), mostly restricted to the Ozark Highlands, has ridged, glabrous fruit. IL, IN, MI, MO, OH. Southeastern corner of MI; throughout the Ozark Highlands and eastern portion of the Eastern Forests, scattered elsewhere in the Eastern Forests. *AG*

Chaerophyllum procumbens
APIACEAE

wild chervil

Floodplain forests, stream terraces, moist woodlands. Spring.

A weakly ascending annual, branching, glabrous to sparsely hairy, 6–18 in. long. Leaves alternate and also basal, broadly ovate in outline, finely divided, 2–4 times compound, to 4 in. long and about as wide. Flowers regular, about ⅛ in. wide, 5-petaled, in compound umbels. Fruit elliptic, about ¼ in. long. This dainty spring ephemeral blooms early, sets seed, and disappears before summer's arrival. The uncommon var. *shortii* differs from typical var. *procumbens* by the presence of hairs on the fruit. Southern wild chervil (*C. tainturieri*) has lanceolate fruit. IA, IL, IN, MI, MO, OH, WI. **Mostly common except absent from much of the Northern Lakes.** *MAH*

Cicuta bulbifera
APIACEAE

bulblet-bearing water hemlock

Marshes, bogs, swamps, lake and stream margins. Summer, fall.

A spindly, glaucous perennial to 4 ft. tall. Leaves alternate, to 12 in. long × 8 in. wide, smaller above, petiolate, pinnately compound with linear, toothed leaflets to ¼ in. wide, with veins ending in sinuses of teeth. Flowers regular, to ⅛ in. wide with 5 shallowly lobed petals, forming flat, compound umbels to 3 in. wide at ends of branches and on stalks from upper leaf axils. The bulblets that develop in upper part of plant are a means of vegetative reproduction. Extremely poisonous if ingested. **IA, IL, IN, MI, MN, OH, WI. Mostly in the Northern Lakes and adjacent portions of the Tallgrass Prairie and Eastern Forests.** *PD*

Cicuta maculata
APIACEAE

water hemlock

Floodplain forests, swamps, fens, marshes, shores. Spring, summer.

A perennial with hollow stems, erect and commonly ascending, branched, glabrous, pale but sometimes mottled with purple at base. Leaves alternate and basal, petiolate to sessile, broadly ovate, 2 or 3 times pinnately compound with toothed margins, up to 1½ ft. long × 10 in. wide. Flowers regular, about ⅛ in. wide, 5 obovate petals, in compound umbels with about 20+ flowers per umbellet. Water hemlock is reputed to be the most poisonous native plant in all North America. Ingestion by livestock or humans may lead to death in as little as 15 minutes. IA, IL, IN, MI, MN, MO, OH, WI. **Widespread and common throughout.** *SN*

Conium maculatum
APIACEAE

poison hemlock

Disturbed areas, roadsides, ditches, pastures. Spring, summer.

A biennial, erect, stems hollow, branching above, glabrous and often glaucous, spotted with purple, 4–8+ ft. tall. Leaves alternate and basal, 3 or 4 times pinnately compound and highly dissected, lower ones long-petiolate, upper nearly sessile, broadly triangular, with toothed margins, up to 1½ ft. long × 1 ft. wide. Flowers regular, about ⅛ in. across, with 5 obovate petals, in compound umbels, about 20+ per umbellet. Malodorous. The sap is reported to have been the poison that killed Greek philosopher Socrates. IA, IL, IN, MI, MN, MO, OH, WI. **Introduced from Eurasia. Widespread and common but less so in northern counties.** *MAH*

Cryptotaenia canadensis
APIACEAE

honewort

Moist upland forests, floodplain forests, stream terraces, seeps. Spring, summer.

An erect perennial, branching, glabrous, 1–2½ ft. tall. Leaves alternate, petiolate or sessile upward, once compound with 3 leaflets, these lanceolate to ovate, some divided, with sharply toothed margins, 2–4 in. long × 1–4 in. wide. Flowers regular, about ¹⁄₁₆ in. wide, with 5 obovate petals, in compound umbels, up to 10+ or so in each umbellet. Honewort possibly gets its name for having been used to treat hones (a former name for tumors or swellings of the cheek). **IA, IL, IN, MI, MN, MO, OH, WI. Common and widespread, possibly in every county.** *SN*

Daucus carota
APIACEAE

Queen Anne's lace, wild carrot

Roadsides, old fields, meadows, ditch banks, waste areas. Spring, summer, fall.

An erect biennial with some branching, commonly rough-hairy, 2–4 ft. tall. Leaves alternate, lowest petiolate, oblong, 2–4 times pinnately compound, smallest segments linear to lanceolate, up to 9 in. long × 2–6 in. wide. Flowers regular, ¹⁄₁₆–⅛ in. wide, with 5 obovate petals, the middle flower commonly dark purple in a 3- to 5-in.-wide flat-topped compound umbel with large, divided bracts beneath. Queen Anne's lace is the ancestor of the cultivated carrot. The similar caraway (*Carum carvi*) lacks large bracts beneath the umbels. **IA, IL, IN, MI, MN, MO, OH, WI. Introduced from Eurasia. Common throughout.** *SN*

Erigenia bulbosa

APIACEAE

harbinger of spring, pepper and salt

Moist and floodplain forests. Spring.

A glabrous, sprawling perennial to 8 in. tall. Leaves basal or alternate, to 5 in. long × 3 in. wide, 2 or 3 times ternately compound. Flowers regular, to ⅕ in. wide, with 5 petals that contrast with dark purple-black anthers, in tight clusters within flat compound umbels at top of plant. Often with reddish-brown stems that contrast sharply with bright green foliage. One of our earliest-blooming native wildflowers and thus a true harbinger of spring. IL, IN, MI, MO, OH, WI. **Widespread in the Eastern Forests, Ozark Highlands, and southeastern portion of the Northern Lakes; less common in southeastern portion of the Tallgrass Prairie.** *SN*

Eryngium yuccifolium

APIACEAE

rattlesnake master

Wet to dry prairies, savannas, barrens, glades. Summer.

A glabrous, gray-green perennial to 5 ft. tall, unbranched except in inflorescence. Leaves alternate, clasping, mostly near base of plant, to 2½ ft. long × 1½ in. wide, much smaller above, strap-like with parallel veins and sparse spine-tipped marginal teeth. Flowers regular, tiny, with 5 shallowly lobed petals, forming several conspicuous spiny round clusters to 1 in. wide at top of plant. The specific epithet is a reference to yucca-like foliage, and the common name alludes to use of roots by Native Americans to treat rattlesnake bites. IA, IL, IN, MI, MN, MO, OH, WI. **Widespread in the Ozark Highlands and southern portion of the Tallgrass Prairie, less frequent elsewhere.** *EH*

Heracleum maximum
APIACEAE

cow parsnip

Floodplains, wet meadows, forest openings, thickets. Spring, summer.

A bristly-hairy perennial to 8 ft. tall. Leaves basal and alternate, compound with 3 large, deeply and sharply lobed, toothed leaflets, to 20 in. long and wide, largest at base, smaller, simple and 3-lobed above, with conspicuously inflated sheaths at base of petioles. Flowers irregular, with 5 uneven, notched petals, to ¼ in. wide, numerous, forming flat, compound umbels to 8 in. wide, with up to 50 umbellets per umbel; several umbels in upper part of plant. Giant hogweed (*H. mantegazzianum*), introduced from Europe, is larger with larger leaves and has 70+ umbellets per umbel. IA, IL, IN, MI, MN, MO, OH, WI. Throughout, but rare in the Ozark Highlands. *SN*

Osmorhiza longistylis
APIACEAE

anise root

Rich upland forests, moist woodlands. Spring.

An erect perennial, branched, glabrous or hairy, 1½–3 ft. tall. Leaves alternate, petiolate, 2 or 3 times pinnately compound, 2–9 in. long × 5–9 in. wide, triangular with toothed leaflets, emits anise scent when bruised. Flowers regular, about ⅛ in. wide, with 5 oblanceolate petals, in a compound umbel with 10+ flowers per umbellet, leafy bracts present beneath umbellets, styles ⅛ in. long, persistent in fruit (see inset). The wide-ranging sweet cicely (*O. claytonii*) has shorter styles and lacks strong anise scent. Mountain sweet cicely (*O. berteroi*) of the upper Great Lakes lacks leafy bracts beneath umbellets. IA, IL, IN, MI, MN, MO, OH, WI. Common throughout. *MAH/SN*

Oxypolis rigidior

APIACEAE

cowbane, stiff cowbane

Marshes, fens, wet prairies, sedge meadows, swamps, seeps, floodplains, ditches. Summer, fall.

A glaucous perennial to 5 ft. tall. Leaves alternate, to 1½ ft. long, smaller above, petiolate with inflated, papery sheath, pinnately compound with up to 11 entire to irregularly sparsely toothed leaflets to 4 in. long × 1 in. wide, with veins ending in tips of teeth and leaflet margins rolled under. Flowers regular, to ⅛ in. wide, with 5 shallowly lobed petals, forming domed, compound umbels to 6 in. wide in upper part of plant. Water parsnip (*Sium suave*) has abundant, regularly spaced teeth on leaflets (or leaves deeply dissected). IA, IL, IN, MI, MN, MO, OH, WI. Scattered to widespread but absent in northernmost regions. *SN*

Perideridia americana

APIACEAE

wild dill, eastern yampah

Moist floodplain forests, dry upland forests, prairies, glades. Spring, summer.

An erect perennial, glabrous, 2–4 ft. tall. Leaves alternate, petiolate, oblong-ovate to triangular, 1–3 times pinnately compound with very narrow and entire leaf segments, 1–6 in. long × 3–4 in. wide. Flowers regular, about ⅛ in. wide, with 5 obovate and notched petals, in round-topped compound umbels of 10+ per umbellet. Shortly after flowering and fruiting in early summer, plants go into senescence. One common name is a reference to its similarity to garden dill (*Anethum graveolens*), as both have very narrow leaf segments. IL, IN, MO, OH. Most occurrences are in northern IL and the Ozark Highlands. *ST*

Ptilimnium nuttallii
APIACEAE

laceflower, Ozark mock bishop's-weed

Wet prairies, wet spots in glades, wet areas along railroads. Summer.

A glabrous annual to 2 ft. tall. Leaves alternate, to 2 in. long and wide, short-petiolate, pinnately compound to twice compound with filiform leaflets. Flowers regular, to ⅛ in. wide with 5 shallowly lobed petals, forming slightly domed compound umbels to 2½ in. wide at ends of branches at top of plant, umbels subtended by 3–15 filiform bracts. Fruit smooth and spherical. Big mock bishop's-weed (*P. costatum*) is a taller plant with longer leaves and larger umbels. **IL, MO. Scattered in the Ozark Highlands and surrounding portions of the Eastern Forests and the Tallgrass Prairie.** *DT*

Sanicula canadensis
APIACEAE

black snakeroot

Mesic and dry-mesic upland forests. Spring, summer.

An erect biennial, unbranched, glabrous, 6–24 in. tall. Leaves alternate, petiolate, 3–5 times palmately divided with toothed margins, up to 5 in. long × 4–5 in. wide. Flowers regular, with 5 obscure greenish-white petals, both male and perfect present, the latter bristly, with straight styles that are concealed by bristles, in small compact umbels of 2 or 3 perfect flowers and up to 7 male ones per umbellet. Fruits rounded and bristly. Clustered black snakeroot (*S. odorata*) has longer styles and yellow flowers and usually blooms earlier than black snakeroot. **IA, IL, IN, MI, MN, MO, OH, WI. Common and widespread, less so in the Northern Lakes.** *KC*

Sanicula trifoliata
APIACEAE

beaked snakeroot

Mesic upland forests, stream terraces.
Spring, summer.

An erect biennial, glabrous, 1–2½ ft. tall.
Leaves alternate, petiolate, palmately compound with 3 leaflets, coarsely toothed, up
to 4 in. long × 5–7 in. wide. Flowers regular,
5-petaled, greenish-white and obscure, both
male and perfect present, the latter bristly,
with styles straight, shorter than and concealed by bristles, calyx lobes equal to the bristles and forming a small beak atop the fruit,
in compound umbels of 2 or 3 perfect flowers
and up to 5 male ones per umbellet. Fruit ¼
in. long and bristly, looking like a miniature
pineapple. IA, IL, IN, MI, MN, OH, WI. **Most
common in Eastern Forests and southern
Northern Lakes.** *SN*

Sium suave
APIACEAE

water parsnip

Swamps, floodplain forests, vernal pools,
marshes, wet prairies. Summer.

An erect perennial, glabrous, 2–5 ft. tall.
Leaves alternate, petiolate, narrowly oblong,
pinnately compound, 1–12 in. long and up to 8
in. wide. Leaflets opposite and sessile (except
for terminal leaflet), ranging from linear to
ovate-lanceolate, with saw-tooth margins,
the teeth evenly spaced. Submerged leaves
are finely dissected (see inset). Flowers regular, about ⅛ in. across, petals 5, elliptic and
shallowly notched, in somewhat flat-topped
to rounded compound umbels of 20+ flowers
per umbellet. Water parsnip may grow in relatively deep water. IA, IL, IN, MI, MN, MO,
OH, WI. **Common but less so in unglaciated
hill country.** *ST*

Spermolepis inermis
APIACEAE

spreading scaleseed

Prairies, glades, fields, roadsides, railroads. Spring, summer.

A glabrous annual to 2 ft. tall. Leaves alternate, to 2 in. long and wide, short-petiolate to sessile, 2–4 times pinnately compound with filiform leaflets. Flowers regular, to ⅛ in. wide, with 5 petals, forming slightly domed compound umbels to 2 in. wide at ends of stems and from leaf axils, umbels lacking bracts, umbellets with 1–3 bractlets. Fruit usually roughened with small warty bumps (rarely smooth), broadly ovate. Hooked scaleseed (*S. echinata*) has fruit covered in hooked bristles. IA, IL, IN, MI, MN, MO. Scattered to rare, mostly in the Ozark Highlands and the Tallgrass Prairie. *SN*

Torilis arvensis
APIACEAE

hedge parsley

Roadsides, railroads, waste areas. Summer.

An ascending to erect annual, branched, with appressed hairs, 1–3 ft. tall. Leaves alternate, petiolate, 1–3 times pinnately compound, overall outline of blades ovate to triangular, 1–6 in. long × ½–4 in. wide. Flowers regular, about ⅛ in. across, with 5 petals notched and unequal in size, 5–15+ flowers per umbellet, spreading in several somewhat flat-topped umbels with no (or 1) narrow bract immediately beneath. Fruit bristly. The less-common Japanese hedge parsley (*T. japonica*) has 2 to many elongate bracts below the primary umbel. IA, IL, IN, MI, MN, MO, OH, WI. Introduced from Europe. Common throughout the lower Midwest. *MJH*

Apocynum cannabinum

APOCYNACEAE

common dogbane, Indian hemp

Prairies, thickets, savannas, streambanks, shores, fields, roadsides. Summer.

A colony-forming perennial to 6+ ft. tall. Leaves opposite, entire, to 6 in. long × 2½ in. wide, short-petiolate, spreading or ascending, containing milky sap. Flowers regular, bell-shaped with 5 erect to spreading lobes, to ¼ in. long, stalked in clusters in upper part of plant. Fruit a long, skinny, dangling bean-like follicle that dries and splits to release seeds with long white silky hairs at one end. Plants with sessile or very short-petiolate leaves with cordate bases are sometimes elevated to species status as *A. sibiricum*. Fibers in stem used to make cordage. IA, IL, IN, MI, MN, MO, OH, WI. Widespread. *MJH*

Asclepias exaltata

APOCYNACEAE

poke milkweed

Upland forests, open woodlands. Spring, summer.

A perennial to 5 ft. tall. Leaves opposite, entire, shiny, glabrous, elliptic-ovate, to 8 in. long × 3 in. wide, on petioles to 1 in., containing milky sap. Flowers regular, to ½ in. long with 5 reflexed pale green corolla lobes and 5 erect white to pale pink corona hoods, each with protruding horn, on stalks to 2 in. long, drooping in few-flowered umbels from leaf axils. Fruit a lanceolate, hairy follicle to 6 in. with seeds bearing tuft of hairs. Common name inspired by the resemblance of leaves in color, texture, and size to leaves of pokeweed (*Phytolacca americana*), which are alternate. IA, IL, IN, MI, MN, MO, OH, WI. Widespread except in IA and MO. *SN*

Asclepias hirtella
APOCYNACEAE

prairie milkweed, tall green milkweed

Prairies, glades, fields. Spring, summer.

A perennial to 3 ft. tall. Leaves numerous, alternate, linear, entire, to 6 in. long × ½ in. wide, short-petiolate, containing milky sap. Flowers regular, to ⅓ in. long, white to greenish-white, with 5 reflexed-spreading corolla lobes (often with purplish tips) and 5 erect corona hoods (often with purplish base), without horns, up to 100 in 3-in. round umbels in upper leaf axils. Fruit a narrowly lanceolate, hairy follicle to 5½ in. long with seeds bearing tuft of hairs. Sometimes treated as a variety of *A. longifolia*. Slimleaf milkweed (*A. stenophylla*) has narrower leaves and smaller, fewer-flowered umbels. IA, IL, IN, MI, MN, MO, OH, WI. Occurs in much of the Tallgrass Prairie, scattered elsewhere. *SN*

Asclepias lanuginosa
APOCYNACEAE

woolly milkweed

Prairies, open woods. Summer.

A hairy perennial to 1 ft. tall. Leaves opposite, entire, 4 in. long × 1¼ in. wide, sessile or short-petiolate, containing milky sap. Flowers regular, to ⅓ in. long with 5 declined to reflexed-spreading pale green to cream-colored corolla lobes with purplish tips and 5 erect white corona hoods, without horns, on stalks in solitary rounded umbel. Fruit a lanceolate, hairy follicle with seeds bearing tuft of hairs. Mead's milkweed (*A. meadii*) is glabrous with larger flowers with horns. *Asclepias* species produce pollen sacs (pollinia) in slits near anthers; these attach to insect feet and are transported to another flower during pollination. IA, IL, MN, WI. Rare in northern portion of the Tallgrass Prairie and adjacent Northern Lakes. *CB*

Asclepias ovalifolia
APOCYNACEAE

dwarf milkweed, oval-leaved milkweed

Prairies, savannas, open woods. Spring, summer.

A finely hairy perennial to 2 ft. tall. Leaves opposite, entire, oval to ovate, to 3 in. long × 1½ in. wide, short-petiolate, containing milky sap. Flowers regular, to ½ in. long with 5 reflexed-spreading white corolla lobes with pink tips and 5 erect white corona hoods with incurved horns, on stalks in rounded umbels to 2 in. wide at top of plant and from upper 1 or 2 leaf axils. Fruit a lanceolate, finely hairy follicle to 3 in. long with seeds bearing tuft of hairs. IA, IL, MI, MN, WI. Uncommon in western portion of the Northern Lakes and northern portion of the Tallgrass Prairie. *SN*

Asclepias quadrifolia
APOCYNACEAE

fourleaf milkweed

Rocky moist to dry upland forests, commonly with calcareous soils. Spring, summer.

An erect, unbranched perennial, glabrous to short-hairy, 8–18 in. tall, with milky sap. Leaves opposite or in whorls of 4, short-petiolate, lanceolate, entire with hairy margins and midveins, 1–5 in. long × ½–2½ in. wide. Flowers regular, about ¼ in. wide with 5 reflexed white to pale pink corolla lobes and 5 white hoods that are longer than incurved horns. Inflorescence of terminal and axillary umbels with up to 35 flowers. Fruit a follicle with flattened seeds each bearing a tuft of long hairs. IA, IL, IN, MO, OH. Widespread in a patchy distribution. *EH*

Asclepias variegata
APOCYNACEAE

white milkweed, red-ring milkweed

Dry upland woodlands, barrens, savannas, roadsides. Spring, summer.

An erect, unbranched perennial, glabrous to short-hairy, 1–3 ft. tall, with milky sap. Leaves opposite, petiolate, entire, oblong-elliptic, short-hairy underneath, 1–6 in. long × ½–3½ in. wide. Flowers regular, about ⅓ in. wide with 5 reflexed white corolla lobes and 5 white hoods ringed with reddish-purple at their bases, longer than the incurved horns, up to 40 in compact, terminal umbels. Fruit a smooth follicle with flat seeds each bearing a tuft of long hairs. White milkweed is principally a species of the U.S. Southeast. IL, IN, MO, OH. Found mostly in our far southern counties. *CB*

Asclepias verticillata
APOCYNACEAE

whorled milkweed

Prairies, glades, savannas, woodland openings, fields, roadsides. Summer.

A slender perennial to 2 ft. tall. Leaves numerous, mostly in whorls of 3–6 (sometimes alternate), to 3 in. long × ⅛ in. wide, margins rolled under, sessile or short-petiolate, containing milky sap. Flowers regular, to ½ in. long, with 5 reflexed-spreading white to greenish-white corolla lobes (often with pale purplish tips) and 5 erect white corona hoods, each with an ascending incurved horn, in several small convex umbels. Fruit a narrowly lanceolate, smooth follicle to 4 in. long with seeds bearing tuft of hairs. IA, IL, IN, MI, MN, MO, OH, WI. Widespread in the Tallgrass Prairie and Ozark Highlands; common in adjacent portions of the Northern Lakes and Eastern Forests. *TT*

Cynanchum laeve
APOCYNACEAE

blue vine

Fencerows, forest edges, gardens, thickets. Summer, fall.

A climbing perennial herbaceous vine, sparsely hairy, growing to 15 ft. Leaves opposite, petiolate, ovate, cordate, 2–6 in. long × 1–4 in. wide. Flowers regular, ⅛–¼ in. long, with 5 lanceolate and petal-like segments, calyx 5-lobed. Inflorescence in axillary umbels of up to 40 flowers, fragrant. Fruit a smooth follicle. Blue vine has clear sap. Monarch larvae, which usually feed upon milkweed (*Asclepias*) leaves, sometimes also feed upon blue vine. Its flowers are touted as a good source of nectar for honeybees. Syns. *Ampelamus albidus, A. laevis.* IA, IL, IN, MI, MO, OH. **Widespread but mostly absent in the Northern Lakes.** *MAH*

Calla palustris
ARACEAE

wild calla, water arum, bog arum

Bogs, swamps, lakes, rivers, usually in areas that are at least slightly acidic. Spring, summer.

A colony-forming perennial to 16 in. tall. Leaves basal, mostly upright, to 7 in. long × 5 in. wide, simple, cordate with canoe-shaped tip, entire, long-petiolate, with midvein from which numerous lateral veins arise and arch toward tip. Flowers regular, inconspicuous, lacking petals, numerous on thick cylindrical spadix to 1¼ in. long within conspicuous white, oval, leaf-like spathe to 4 in. long × 2 in. wide; spathe and spadix on stalk to 12 in. above water. Fruit a bright red berry, clustered. IL, IN, MI, MN, OH, WI. **Almost exclusively in the Northern Lakes, especially northward.** *SN*

Aralia hispida
ARALIACEAE

bristly sarsaparilla

Dry sandy to rocky savannas,
open woods, cliffs, swamps. Summer.

A perennial to 4 ft. tall, woody at base. Stems
glabrous above, bristly below, often becoming
reddish. Leaves alternate, twice compound,
short-petiolate, terminating in coarsely toothed
shiny leaflets to 4 in. long × 1½ in. wide. Flowers regular, ⅛ in. wide, with 5 recurved white
to green petals and long-exserted stamens, on
stalks, forming many-flowered hemispherical
umbels to 1¼ in. wide (see inset); numerous
umbels on long stalks in upper portion of
plant. Fruit a bluish-black spherical berry to
⅓ in. wide. Increases with disturbance and
intolerant of competition. IL, IN, MI, MN, OH,
WI. Primarily in the Northern Lakes, rare
elsewhere. SN

Aralia nudicaulis
ARALIACEAE

wild sarsaparilla

Forests, swamps, dunes. Spring, summer.

A colony-forming perennial to 2 ft. tall. Single compound leaf arises from underground
stem, its petiole to 15 in. long; 3 parts each
dissected into pinnately arranged glabrous
leaflets to 5 in. long × 2 in. wide, toothed and
with abruptly pointed tips. Early leaves often
copper-colored. Flowers regular, to ⅛ in. wide
with 5 recurved white to green petals, stalked,
forming usually 3 many-flowered stalked,
spherical umbels to 2 in. wide on leafless
stem to 9 in. tall beneath leaves. Fruit a ¼-in.
bluish-black berry. IA, IL, IN, MI, MN, MO,
OH, WI. Widespread in the Northern Lakes
and northern portion of the Tallgrass Prairie;
common in eastern portion of the Eastern
Forests. SN

Aralia racemosa
ARALIACEAE

spikenard

Rich, moist forests, deep ravines,
north-facing slopes, springy areas, moist
savannas. Summer.

A branched perennial herb with
spreading-hairy to glabrous stems up to 5 ft.
tall. Leaves alternate, glabrous, with large
3-branched compound leaves further divided
pinnately into ovate or somewhat cordate
toothed leaflets up to 8 in. long × 5 in. wide.
Flowers regular, less than ¼ in. wide, with
5 petals. Inflorescence composed of small
umbels, each with about 10 flowers on terminal and/or axillary panicles. Because of its
stoutness and huge compound leaves, one may
get the mistaken impression that spikenard
is a shrub. IA, IL, IN, MI, MN, MO, OH, WI.
Widespread but local. *MJH*

Panax trifolius
ARALIACEAE

dwarf ginseng

Moist to swampy forests, floodplains.
Spring, summer.

A perennial to 8 in. tall. Leaves 3 (rarely 4 or
5), in single whorl, on petioles to 2 in. long,
palmately compound with 3–5 stalkless (or
nearly so) leaflets to 1½ in. long × ½ in. wide,
with toothed margins. Flowers regular, to ⅛
in. wide with 5 spreading to reflexed petals,
several to numerous on white stalks radiating
from center and forming domed to spherical
cluster to ¾ in. long and wide on erect stalk to
3 in. above leaves. Fruit a pyramidal green to
yellowish drupe. IN, MI, MN, OH, WI. Mostly
in the Northern Lakes, also in adjacent
Tallgrass Prairie and spotty in the Eastern
Forests. *MAH*

Achillea millefolium
ASTERACEAE

yarrow

Fields, roadsides, grasslands. Summer.

An erect, unbranched, woolly perennial, 1–2½ ft. tall. Leaves alternate, narrowly lanceolate, soft and finely dissected, feathery or fern-like, aromatic, 2–6 in. long and up to 1 in. wide. Flowers in composite heads ¼–½ in. wide, ray flowers 5, disk flowers 5–20, white to rarely pink, arranged in a flat-topped cluster of up to 50 heads. Some consider yarrow a naturalized introduction from Europe; others, a native North American species. Several yarrow cultivars are used in landscaping, some with quite colorful flowers. IA, IL, IN, MI, MN, MO, OH, WI. **Common throughout.** *SN*

Ageratina altissima
ASTERACEAE

white snakeroot

Dry to moist woodlands, edges, thickets. Summer, fall.

An erect perennial, branched above, mostly glabrous, 2–3 ft. tall. Leaves opposite, simple, triangular to ovate, sharply toothed, 4–6 in. long × 2–4 in. wide, on petioles 1–2 in. long. Flowers in composite heads about ¼ in. wide, ray flowers lacking, 10–20 disk flowers with each flower ⅛ in. long and tubular with 5 lobes, inflorescence an array spanning 4–6 in. wide. White snakeroot caused milk sickness in early settlers who drank milk of cows that ate the plant. Abraham Lincoln's mother is thought to have died from the affliction. IA, IL, IN, MI, MN, MO, OH, WI. **Abundant, widespread throughout.** *SN*

Anaphalis margaritacea
ASTERACEAE

pearly everlasting

Fields, roadsides, forest openings, shores; usually in dry soil but occasionally in moist areas. Summer, fall.

A usually dioecious perennial to 3 ft. tall, typically densely white-woolly. Leaves alternate, entire, sessile, to 5 in. long × ¾ in. wide. Flowers in composite heads; numerous tiny disk flowers in conspicuous ¼-in. roundish heads with spreading white phyllaries and yellow-brown tips; several to many heads in generally flat-topped array at top of plant. Similar in appearance to plants in *Gnaphalium* and *Pseudognaphalium*, which have perfect flowers, but crushed inflorescences of the latter emit a pleasant balsam scent. IA, IL, IN, MI, MN, MO, OH, WI. Primarily in the Northern Lakes, with sparse occurrences elsewhere. *ST*

Antennaria neglecta
ASTERACEAE

field pussytoes

Prairies, open woodlands, savannas, old fields, roadsides. Spring.

A colony-forming, stoloniferous, dioecious, white-woolly perennial to 1 ft. tall. Leaves entire, mostly basal, to 2½ in. long × ¾ in. wide, broadest above middle with single vein running length of blade, with sparse spiderweb-like hair on top surface (can become smooth with age), densely woolly beneath; stem leaves few, alternate, sessile, smaller. Flowers in composite heads; numerous tiny disk flowers in ⅓-in.-wide heads; up to 8 heads in flat to rounded array at top of plant. Small pussytoes (*A. howellii*) has basal leaves green and glabrous. IA, IL, IN, MI, MN, MO, OH, WI. Throughout, less frequent in northern MI, southeastern MO, and western OH. *MJH*

Antennaria parlinii
ASTERACEAE

Parlin's pussytoes

Prairies, open woodlands, savannas, forests, rock outcrops, fields, roadsides. Spring.

A colony-forming, stoloniferous, dioecious, white-woolly perennial to 1½ ft. tall. Leaves entire, mostly basal, 3¾ in. long × 1¾ in. wide, broadest near middle with 3–5 parallel veins, densely woolly beneath, smooth on top (subsp. *parlinii*) or woolly in young plants (subsp. *fallax*); stem leaves few, alternate, sessile, much smaller. Flowers in composite heads; numerous tiny disk flowers in ¾-in-wide heads, involucres of female heads to ½ in. long; up to 12 heads in flat to rounded array at top of plant. Plantain-leaved pussytoes (*A. plantaginifolia*) has smaller female involucres. IA, IL, IN, MI, MN, MO, OH, WI. **Throughout, less frequent in western MN and western OH.** *CB*

Anthemis cotula
ASTERACEAE

dog-fennel, stinking chamomile

Gardens, lawns, fields, parking lots, roadsides, waste areas. Summer.

A branching, upright annual to 2 ft. tall. Leaves to 2½ in. long × 1¼ in. wide, alternate, sessile or nearly so, once to (usually) twice pinnately lobed into narrow, linear, flat segments, emitting unpleasant odor when crushed. Flowers in composite heads; a yellow, dome-shaped disk is surrounded by up to 16 white, 3-lobed ray flowers, in numerous solitary heads to 1¼ in. wide at ends of branches. German chamomile (*Matricaria chamomilla*) has narrower leaf segments with margins rolled under and emits a pleasant scent when bruised. IA, IL, IN, MI, MN, MO, OH, WI. **Introduced from Eurasia. Scattered to widespread.** *PD*

Arnoglossum atriplicifolium
ASTERACEAE

pale Indian-plantain

Dry open woodlands, barrens, moist prairies, savannas, thickets. Summer, fall.

An erect, unbranched perennial with a glabrous, glaucous stem rounded in cross section, 3–7 ft. tall. Leaves alternate, petiolate, glabrous, cordate to triangular with palmate venation and large irregular lobes on the margins, glaucous beneath, the lower ones 6–14 in. long × 7–10 in. wide. Flowers in ¼-in.-wide composite heads consisting of 5 disk flowers, ray flowers lacking, the heads clustered together in a flat-topped inflorescence atop the main stem. Syn. *Cacalia atriplicifolia*. IA, IL, IN, MI, MO, OH, WI. Widespread and relatively common except in the far north. *MJH*

Arnoglossum plantagineum
ASTERACEAE

prairie Indian-plantain

Glades, prairies, sedge meadows, fens, seeps, savannas. Spring, summer.

A perennial to 5 ft. tall. Leaves glabrous, thick, rubbery, with entire or shallowly toothed margins and parallel veins, mostly basal and long-petiolate, to 7½ in. long × 3½ in. wide (see inset), stem leaves smaller, alternate, short-petiolate or sessile. Flowers in composite heads; disk ⅜ in. long, cylindrical, ray flowers absent, phyllaries green with white wings; many stalked heads in conspicuous flat-topped array to 10 in. wide at top of plant. Syn. *Cacalia plantaginea*. IA, IL, IN, MI, MN, MO, OH, WI. Primarily in southern portion of the Tallgrass Prairie and Ozark Highlands; uncommon in Northern Lakes and Eastern Forests. *SN/NP*

Arnoglossum reniforme
ASTERACEAE

great Indian-plantain

Rich woods, well-drained stream terraces, thickets. Summer, fall.

An erect, unbranched perennial with a glabrous and angled stem, 3–7 ft. tall. Leaves alternate, petiolate, glabrous, kidney-shaped, irregularly toothed on margin, green on both surfaces, 5–8 in. long × 8–12 in. wide. Flowers in ¼-in.-wide composite heads consisting of 5 disk flowers, ray flowers lacking, the heads clustered together in a flat-topped array atop the main stem. Syn. *Cacalia muhlenbergii*. IA, IL, IN, MN, MO, OH, WI. Widespread but local. *ST*

Artemisia ludoviciana
ASTERACEAE

white sage, western mugwort

Prairies, fields, roadsides, railroads. Summer, fall.

A conspicuous, densely silver-hairy perennial to 3 ft. tall. Leaves alternate, sometimes with a few coarse teeth or lobes at tips but usually simple, unlobed, entire, to 3½ in. long × 1 in. wide, sessile or nearly so, becoming smaller into inflorescence, aromatic when bruised. Flowers in composite heads; ray flowers absent, disk flowers inconspicuous, in numerous short-stalked, small, mostly ascending heads along branches of narrow panicle-like array at top of plant. IA, IL, IN, MI, MN, MO, OH, WI. Widespread in western portion of the Northern Lakes and northern portion of the Tallgrass Prairie, scattered to rare south and east. *CB*

Boltonia asteroides
ASTERACEAE

false aster, white doll's daisy

Riverbanks, borders of oxbow lakes, flatwoods, wet prairies. Summer, fall.

An erect perennial, upwardly branching with glabrous, non-winged stems. Leaves alternate, short-petiolate, entire, glabrous, linear to lance-elliptic, 2–5 in. long × ¼–¾ in. wide. Flowers in composite heads ¾ in. wide, ray flowers 30–50, disk flowers 2–3 times as many, on thin, widely spreading wiry stems. Variety *latisquama* has spatulate phyllaries; the more common and widespread var. *recognita* has linear phyllaries. In decurrent false aster (*B. decurrens*), a Midwest endemic confined to the floodplains of the Illinois and Mississippi rivers, the stems are winged. **IA, IL, IN, MI, MN, MO, OH, WI. Occurs mostly in the Tallgrass Prairie. MJH**

Brickellia eupatorioides
ASTERACEAE

false boneset

Prairies, savannas, glades, woodlands. Summer, fall.

An erect perennial, upwardly branched, hairy, 1–3 ft. tall. Leaves alternate but can appear opposite due to close spacing, sessile or short-petiolate, entire or with few teeth, linear to lanceolate, 1–4 in. long, to 1¾ in. wide. Flowers in composite heads about ¼ in. wide, disk flowers 6–20, ray flowers lacking, in clusters of 2 or 3 terminating branches in a somewhat flat-topped inflorescence. Four (of 6) named varieties, based mostly on differences in the flower heads, are known within our region. **IA, IL, IN, MI, MN, MO, OH, WI. Mostly confined to the Tallgrass Prairie and Ozark Highlands. SN**

Cirsium pitcheri
ASTERACEAE

Pitcher's thistle, dune thistle

Beaches, dunes. Spring, summer.

A densely silvery-hairy perennial to 4 ft. tall. Leaves alternate, to over 1 ft. long, very deeply lobed into linear segments terminating in short spines. Flowers in composite heads; numerous white to pale pinkish or bluish (very rarely purple) disk flowers in solitary heads to 2 in. long at ends of branches in upper part of plant; heads subtended by long, triangular phyllaries with spreading to appressed spine tips. Fruiting head fluffy from silky pappus. Exists in rosette stage for up to 7 years, then bolts, flowers, disperses seed, and dies. IL, IN, MI, WI. A Great Lakes endemic and federally threatened species, immediately bordering Lake Huron, Lake Superior, and Lake Michigan. *SN*

Doellingeria umbellata
ASTERACEAE

flat-topped white aster, parasol white-top

Sedge meadows, wet prairies, fens, bogs, swamps, moist forests, marshes, rocky shores. Summer, fall.

A perennial to 5 ft. tall. Leaves alternate, sessile or very short-petiolate, entire, to 5 in. long × 1 in. wide. Flowers in composite heads; a yellow disk is surrounded by up to 12 white ray flowers, in numerous flower heads in a flattish-topped array at top of plant; heads to ¾ in. wide. Leaves have distinctive impressed venation, and flower heads appear as though 1 or more ray flowers are missing. Syn. *Aster umbellatus*. IA, IL, IN, MI, MN, MO, OH, WI. Widespread in the Northern Lakes, adjacent portions of the Tallgrass Prairie, and eastern portion of the Eastern Forests. *SN*

Eclipta prostrata
ASTERACEAE

yerba de tajo, false daisy

Swamps, mudflats, cultivated fields, ditches. Summer, fall.

A branching annual, prostrate or ascending with appressed hairs, up to 2 ft. long. Leaves opposite and mostly sessile, lanceolate with small teeth, 1–4 in. long × ¼–1 in. wide. Flowers in composite heads up to ½ in. wide, ray flowers 20+ with an equal number or more disk flowers, arranged in axillary or terminal clusters. Yerba de tajo is a native weed with worldwide distribution that favors sites with disturbance and periodic flooding. IA, IL, IN, MI, MN, MO, OH, WI. Widespread and common in southern counties, rare in the north. *MAH*

Elephantopus carolinianus
ASTERACEAE

elephant's foot

Various forest types, from moist to relatively dry and rocky. Summer, fall.

An erect perennial with stem unbranched and somewhat zigzag, hairy to sparsely so, 1–3 ft. tall. Leaves alternate, sessile, ovate to lanceolate, with shallow teeth, moderately hairy, 2–8 in. long × 1–4 in. wide. Flowers in composite heads forming clusters about 1 in. wide, ray flowers lacking, disk flowers white to pale pink or lavender and deeply lobed, usually 4 in a head (see inset), clustered above 1–3 leafy bracts. IL, IN, MO, OH. Relatively common in southern third of the Midwest. *MAH*

Erigeron annuus
ASTERACEAE

daisy fleabane, annual fleabane

Prairies, fields, forest openings, roadsides, railroads, waste areas. Summer.

A spreading-hairy annual or biennial to 4 ft. tall. Basal and lower stem leaves ovate and coarsely toothed, to 5 in. long × 3 in. wide, long-petiolate, stem leaves alternate, abundant, becoming narrower, less toothed, and sessile above. Flowers in composite heads; a yellow ¼-in.-wide disk is surrounded by numerous white to pink ray flowers in several to many heads to ¾ in. wide at top of plant. Prairie fleabane (*E. strigosus*), a perennial more frequent in natural areas, has appressed stem hairs and fewer leaves that are less toothed. IA, IL, IN, MI, MN, MO, OH, WI. Abundant and widespread, with fewer occurrences in western MN. *SN*

Erigeron canadensis
ASTERACEAE

horseweed, marestail

Prairies, fields, roadsides, railroads, waste areas. Summer, fall.

A spreading-hairy annual or biennial to 7 ft. tall, unbranched except above. Leaves narrowly elliptic, sessile, alternate, entire or irregularly toothed, to 4 in. long × ½ in. wide, lowest often absent when flowering. Flowers in composite heads; a yellow disk is surrounded by many tiny white ray flowers in heads to ⅛ in. wide (see inset), many heads in large, branching cluster at top of plant. Syn. *Conyza canadensis*. Less common is dwarf horseweed (*E. divaricatus*), a much smaller and branched plant with narrower leaves and grayish cast. IA, IL, IN, MI, MN, MO, OH, WI. Abundant and widespread. *MAH*

Erigeron philadelphicus
ASTERACEAE

Philadelphia fleabane

Moist forests, streambanks, trail and
road corridors, meadows, prairies.
Spring, summer.

An erect annual, biennial, or short-lived
perennial, unbranched, hairy, 2–3 ft. tall.
Leaves alternate, elliptic to lanceolate, with
scattered teeth, bases rounded to cordate and
clasping stem, somewhat hairy, 2–3 in. long
× ½–1½ in. wide. Flowers in composite heads
about ¾ in. wide, ray flowers 100+, white to
pinkish, disk flowers 100+, yellow, on widely
spreading branches atop plant. Philadelphia
fleabane resembles asters (*Symphyotrichum*),
but its earlier blooming period and greater
number of ray flowers help to distinguish it
from them. IA, IL, IN, MI, MN, MO, OH, WI.
Widespread and common throughout. *SN*

Erigeron pulchellus
ASTERACEAE

robin's plantain

Dry upland woodlands, savannas, cliff
ledges. Spring.

An erect perennial, unbranched, hairy, to 1½
ft. tall. Leaves mostly basal, few along stem,
obovate to oblanceolate, entire to shallowly
toothed, 1–6 in. long × 1½–3 in. wide. Flowers
in composite heads about 1 in. wide, ray flowers 50–100, white to pink, disk flowers 100+,
yellow, on widely spreading branches atop
plant. Fruit a mostly glabrous achene. In our
region is the typical var. *pulchellus* and the very
local var. *tolsteadtii*, endemic to eastern MN.
The latter is restricted to rock ledges and has
hairy achenes. IA, IL, IN, MI, MN, MO, OH,
WI. Widespread and common except in the
north and west. *MAH*

Eupatorium altissimum

ASTERACEAE

tall boneset, tall thoroughwort

Prairies, glades, meadows, roadsides, disturbed sites. Summer, fall.

An erect perennial, unbranched, hairy, 2–5 ft. tall. Leaves opposite, narrowly lance-elliptic and tapering to base, sessile, conspicuously 3 nerved with marginal teeth toward tip, 1–6 in. long × ¼–1 in. wide, small leaves commonly in leaf axils oriented at 90° to larger pair. Flowers in composite heads ⅛–¼ in. wide, ray flowers absent, disk flowers usually 5, heads clustered in a terminal flat-topped array. Upland boneset (*E. sessilifolium*) lacks axillary leaves, and its leaves have rounded bases. IA, IL, IN, MI, MN, MO, OH, WI. Widespread and common except mostly absent in the Northern Lakes. *KC*

Eupatorium perfoliatum

ASTERACEAE

common boneset

Fens, swamps, marshes, wet meadows, floodplains. Summer, fall.

A spreading-hairy perennial to 4 ft. tall. Leaves opposite, simple, perfoliate-clasping, coarsely toothed, to 8 in. long × 2 in. wide, with impressed reticulate venation. Flowers in composite heads; ray flowers absent, disk flowers form heads to ¼ in. wide, heads numerous, forming flat or dome-topped arrays to 8 in. across at top of plant. Other *Eupatorium* species have petiolate or sessile leaves, but leaves are not clasping and connected to one another across stem as in *E. perfoliatum*. Common name a reference to fused leaves, with appearance similar to broken bones being reset. IA, IL, IN, MI, MN, MO, OH, WI. Widespread and common. *MAH*

Eupatorium serotinum

ASTERACEAE

late boneset, late thoroughwort

Moist to wet prairies, fields, roadsides, railroads, disturbed sites, waste areas. Summer, fall.

A colony-forming perennial to 6 ft. tall with appressed-hairy stems. Leaves opposite, simple, distinctly petiolate, coarsely toothed, to 7 in. long × 2½ in. wide, tapering from base to tip. Flowers in composite heads; ray flowers absent, disk flowers form heads to ¼ in. wide, heads numerous, forming flat-topped arrays in upper part of plant. Tall boneset (*E. altissimum*) lacks distinct petioles (leaves taper to base) and has leaves entire or toothed only toward tips. IA, IL, IN, MI, MN, MO, OH, WI. Widespread and common in IL, IN, and MO, scattered in surrounding areas, and rare to absent elsewhere. *DT*

Eupatorium sessilifolium

ASTERACEAE

upland boneset

Dry to dry-mesic woodland, rocky slopes, barrens, savannas. Summer, fall.

An erect perennial, unbranched, glabrous below inflorescence, 2–5 ft. tall. Leaves opposite, lanceolate, sessile, glabrous, gland-dotted with rounded base and 1 main nerve, 1–6 in. long × ½–2 in. wide. Flowers in composite heads about ¼ in. wide, ray flowers lacking, disk flowers white and usually 5, clustered in a terminal, somewhat flat-topped array. Upland boneset is a more "conservative" *Eupatorium* species, having a greater fidelity to natural communities and thus not as likely to appear in highly disturbed sites. IA, IL, IN, MI, MN, MO, OH, WI. Mostly in the lower Midwest. *SN*

Eurybia furcata
ASTERACEAE

forked aster

Rich mesic woods, seeps, moist limestone ledges, streambanks. Summer, fall.

An erect, colonial perennial with unbranched, somewhat zigzag stems, glabrous below, hairy above, 1–3 ft. tall. Leaves alternate, cordate to ovate, surfaces rough-hairy, margins toothed, non-glandular, 2–6 in. long × 1–4 in. wide, basal leaves normally absent at flowering. Flowers in composite heads about 1 in. wide, ray flowers 10–20, with up to twice as many yellow disk flowers, in a terminal, somewhat flat-topped array. Forked aster is a Midwest endemic. Syn. *Aster furcatus*. In eastern OH, a somewhat similar species, white wood aster (*E. divaricata*), has mostly glabrous or sparsely hairy leaves. IA, IL, IN, MI, MO, WI. Very local, mostly in northern IL and southeastern WI. *SN*

Eurybia macrophylla
ASTERACEAE

bigleaf aster

Moist and dry-mesic upland forests, dune forests. Summer, fall.

An erect perennial with somewhat zigzag stems, colonial, 1–3 ft. tall. Leaves alternate, lower ones ovate with cordate base on long petioles, thick and somewhat rough-hairy with toothed margins, 4–10 in. long × 2–6 in. wide. Flowers in composite heads 1–1½ in. wide, ray flowers lavender to rarely white, 10–20, with 20–40 yellow disk flowers, in a flat-topped array bearing gland-tipped hairs. Syn. *Aster macrophyllus*. The similar Schreber's aster (*E. schreberi*) occurs mostly in OH with widely scattered occurrences elsewhere; it has mostly white ray flowers and no glands in the inflorescence. IA, IL, IN, MI, MN, MO, OH, WI. Principally in the eastern and northern Midwest. *MJH*

Galinsoga quadriradiata
ASTERACEAE

quickweed, Peruvian daisy, shaggy-soldier, hairy galinsoga

Lawns, gardens, pavement cracks, alongside urban buildings, roadsides. Summer, fall.

A weak, spreading-hairy annual to 2 ft. tall. Leaves opposite, petiolate, coarsely toothed, to 3 in. long × 2 in. wide. Flowers in composite heads; a yellow disk is surrounded by up to 6 short, white, 3-lobed ray flowers, in several solitary ¼-in.-wide stalked heads. Small-flowered galinsoga (*G. parviflora*), less frequent, has shorter, appressed hairs on stem, and scales atop the tiny fruit lack an awn (present in quickweed). IA, IL, IN, MI, MN, MO, OH, WI. Introduced from Central and South America. Frequent, with greatest abundance in the Northern Lakes, Eastern Forests, and eastern portion of the Tallgrass Prairie. *SN*

Hasteola suaveolens
ASTERACEAE

false Indian-plantain

Wet meadows, fens, marshes, prairies, floodplain forests. Summer.

An erect perennial, unbranched, glabrous, 3–6 ft. tall. Leaves alternate, petiolate, hastate with widely diverging lobes (except for uppermost leaves) and serrated margins, 6–12 in. long × 4–6 in. wide. Flowers in composite heads about ¾ in. long × ¼ in. wide, ray flowers lacking, disk flowers 20+ arranged in a terminal, flat-topped inflorescence. The flower heads are reported to have a sweet fragrance. Syn. *Cacalia suaveolens*. IA, IL, IN, MN, MO, OH, WI. Rather local and uncommon throughout. *BS*

Leucanthemum vulgare
ASTERACEAE

oxeye daisy

Old fields, roadsides, waste areas.
Spring, summer.

An erect perennial, unbranched, mostly glabrous, 1–2 ft. tall, colonial. Leaves alternate,
petiolate to sessile, oblanceolate to linear,
some pinnately lobed or toothed, 3–5 in. long
× ½–¾ in. wide. Flowers in composite heads
about 2 in. wide, ray flowers 20–40, disk flowers 200+, borne singly at stem tips. Luther
Burbank used oxeye daisy as one of the species
to produce Shasta daisy (*L. ×superbum*), a
showy and commonly cultivated hybrid. IA,
IL, IN, MI, MN, MO, OH, WI. Introduced
from Eurasia. Common and widespread
throughout. *MAH*

Nabalus albus
ASTERACEAE

white lettuce, white rattlesnake-root

Forests, woodlands, floodplains, thickets.
Summer, fall.

A glabrous perennial to 5 ft. tall (usually
shorter). Leaves variable, simple, alternate,
8 in. long × 6 in. wide, 3- to 5-lobed, smaller
and unlobed above, coarsely toothed to entire,
short-petiolate, containing milky sap. Flowers
in numerous composite heads to ¾ in. long
× ½ in. wide, 8–15 ray flowers, disk flowers
absent, phyllaries appressed, in drooping
panicle-like array at top of plant (see inset).
Fruit an achene with cinnamon-colored pappus. Syn. *Prenanthes alba*. IA, IL, IN, MI, MN,
MO, OH, WI. Widespread and common in
the Northern Lakes and northern portion of
the Tallgrass Prairie, scattered in the Eastern
Forests. *SN*

Nabalus altissimus
ASTERACEAE

tall white lettuce

Moist upland forests, ravine slopes, savannas. Fall.

An erect perennial, glabrous, 2–5 ft. tall. Leaves alternate, highly variable in shape, from deeply palmately lobed to ovate to triangular, petiolate, entire or toothed, 3–8 in. long × 1–6 in. wide, milky sap present. Flowers in composite heads about ½ in. wide, ray flowers 4–6, greenish-yellow to nearly white, disk flowers lacking, phyllaries glabrous (see inset). Flowers dangle from tips of terminal panicles. Syn. *Prenanthes altissima*. The similar white lettuce (*N. albus*) has 8–15 ray flowers per head, and lion's foot (*N. serpentarius*) has phyllaries with at least a few bristly hairs. IL, IN, MI, MO, OH. Mostly in the Eastern Forests and Ozark Highlands. *MJH/MAH*

Nabalus asper
ASTERACEAE

rough white lettuce, rough rattlesnake-root

Prairies, glades, savannas, forest openings. Summer, fall.

A rough-hairy perennial to 5 ft. tall. Leaves simple, unlobed, alternate, to 4½ in. long × 2 in. wide, smaller and entire above, broadly toothed below, sessile, some clasping, containing milky sap. Flowers in composite heads; 10 or more white to cream-colored ray flowers in heads to ½ in. long × 1 in. wide, disk flowers absent, phyllaries appressed and hairy, heads numerous, upright to spreading, stalked in narrow, wand-like terminal array, also in upper leaf axils. Fruit an achene with light brown pappus. Syn. *Prenanthes aspera*. IA, IL, IN, MN, MO, OH, WI. Scattered in the Tallgrass Prairie, rare elsewhere. *BS*

Nabalus crepidineus
ASTERACEAE

nodding rattlesnake-root

Floodplain forests, mesic woods, ravines. Summer, fall.

An erect perennial, glabrous, 3–6 ft. tall, colonial. Leaves alternate, elliptic, triangular to sagittate, shallowly lobed or toothed, often with prominently winged petioles, milky sap present. Flowers in composite heads about ¾ in. wide, ray flowers 15–30, disk flowers lacking, phyllaries coarsely hairy, heads drooping in terminal flat-topped or spreading panicles from branch tips. Most plants produce basal leaves that die back in summer; very few bolt and produce flowers. Syn. *Prenanthes crepidinea.* Lion's foot (*N. serpentarius*) of southeastern OH typically has half the number of ray flowers as nodding rattlesnake-root. IL, IN, MI, MN, MO, OH, WI. **Widely scattered but uncommon to absent in the north.** *MAH*

Nabalus racemosus
ASTERACEAE

glaucous white lettuce

Wet prairies, fens, calcareous shores, interdunal wetlands. Summer, fall.

A glabrous (except inflorescence), glaucous perennial to 5 ft. tall. Leaves simple, alternate, to 10 in. long × 3 in. wide, long-petiolate, smaller and clasping above, mostly entire, containing milky sap. Flowers in composite heads; 9+ ray flowers, no disk flowers, phyllaries appressed, purple, densely hairy, heads ½ in. long and wide, upright to spreading, stalked in wand-like terminal array and upper leaf axils. Fruit an achene with creamy white pappus. Syn. *Prenanthes racemosa.* IA, IL, IN, MI, MN, MO, OH, WI. **Most frequent in transition zone between the Northern Lakes and the Tallgrass Prairie, scattered elsewhere.** *SN*

Parthenium integrifolium
ASTERACEAE

wild quinine, American feverfew

Prairies, fields, glades, open woods, savannas, barrens, roadsides. Spring, summer, fall.

A perennial to 3 ft. tall. Basal leaves simple, rough-textured, dark green, coarsely toothed and wavy-margined, ovate, to 10 in. long × 5 in. wide, tapering to long, winged petioles. Stem leaves similar but smaller and with shorter petioles, becoming sessile and clasping above. Flowers in composite heads; 5 or 6 inconspicuous, cupped ray flowers surround disk in heads to ⅓ in.; many stalked heads in conspicuous flat-topped array at top of plant. IA, IL, IN, MI, MN, MO, OH, WI. Widespread in the Ozark Highlands and eastern portion of the Tallgrass Prairie, scattered elsewhere. *KB*

Petasites frigidus
ASTERACEAE

sweet coltsfoot

Moist meadows and forests, swamps. Spring.

A colony-forming, mostly dioecious perennial to 2 ft. tall. Basal leaves round in outline, to 6 in. long, with 5–11 deep lobes that are lobed and toothed, long-petiolate, thick-textured with undersides covered in white felt-like hair. Stem leaves alternate, sessile, linear, pointed, scale-like above. Flowers in composite heads; several tiny disk flowers surrounded by few to many ray flowers in heads to ¾ in. wide, white to pinkish. Heads in terminal, round-topped cluster to 4 in. wide. Fruit with cottony pappus. Arrowleaf sweet coltsfoot (*P. sagittatus*) has larger, sagittate, unlobed leaves with reverse scallop-toothed margins. The two hybridize. MI, MN, WI. Widespread in northern portion of the Northern Lakes. *PD*

Polymnia canadensis
ASTERACEAE

leaf cup

Moist slopes, especially on calcareous substrates, cliffs, thickets. Summer, fall.

An erect perennial, branching, hairy or glandular-hairy to glabrous, 3–5 ft. tall. Leaves opposite, petioles stipulate and commonly with wings of leafy tissue along petiole, ovate, pinnately lobed with toothed margins, 5–10 in. long × 5–8 in. wide. Flowers in composite heads about ½ in. wide or wider, ray flowers typically minute or absent, 2–6, disk flowers 10–30+, arranged in small clusters at branch tips. Most populations have small ray flowers, but in forma *radiata* (see inset) they are comparatively large. IA, IL, IN, MI, MN, MO, OH, WI. **Common except in far northern and western counties.** *PD/MAH*

Pseudognaphalium obtusifolium
ASTERACEAE

sweet everlasting, rabbit-tobacco, old-field balsam, fragrant cudweed

Prairies, savannas, fields, pastures, forest openings, roadsides, disturbed sites. Summer, fall.

A white-woolly biennial to 2½ ft. tall. Leaves alternate, somewhat wavy-margined, sessile, to 3 in. long × ⅓ in. wide. Flowers in composite heads; tiny yellow-tipped perfect disk flowers in ¼-in.-long heads with appressed white phyllaries that spread in fruit; many heads in generally flat-topped array at top of plant. Crushed inflorescences and foliage emit pleasant balsam scent. Clammy cudweed (*P. macounii*) has leaf bases decurrent along stem. Low cudweed (*Gnaphalium uliginosum*) is bushy with smaller heads. IA, IL, IN, MI, MN, MO, OH, WI. **Wide-spread, except in western IA and MN.** *SN*

Sericocarpus asteroides
ASTERACEAE

toothed white-topped aster

Dry woodlands, thickets, barrens.
Summer.

An erect perennial, branching upward,
finely hairy, 8–24 in. tall. Basal leaves
oblanceolate, stem leaves alternate, sessile,
ovate, with toothed margins, 1–4 in. long
× ¼–1½ in. wide. Flowers in composite
heads about ½ in. wide, with 3–8 ray flowers
and 10–20 yellowish to white disk flowers, in a flat-topped array. Narrow-leaved
white-topped aster (*S. linifolius*), a species
restricted to rugged hills in southern IN and
OH, has narrow leaves that lack teeth. OH.
**Restricted to eastern half of OH, principally
in unglaciated territory.** *CR*

Solidago ptarmicoides
ASTERACEAE

**prairie goldenrod,
upland white goldenrod**

Prairies, glades, savannas, open
woodlands, pannes, fens, calcareous
outcrops and crevices. Summer, fall.

A perennial to 2 ft. tall. Leaves alternate, entire
to shallowly toothed, to 6 in. long × ⅓ in. wide,
smaller above, stiff, petiolate below, sessile
above. Flowers in composite heads; up to 25
white ray flowers surround creamy white disk
in heads to ¾ in. wide, each head subtended
by appressed phyllaries, stalked in branched,
flat-topped, terminal array to 6 in. wide. Syn.
Oligoneuron album. IA, IL, IN, MI, MN, MO,
OH, WI. **Widespread in the Ozark Highlands
and in transition zone between the Northern
Lakes and the Tallgrass Prairie, scattered to
rare elsewhere.** *ST*

Symphyotrichum boreale
ASTERACEAE

rush aster, northern bog aster

Wet sand prairies, conifer swamps, shores, fens, bogs, swales. Summer, fall.

A perennial to 3 ft. tall. Leaves alternate, simple, entire to sparsely shallowly toothed, somewhat rolled under along margins, to 6 in. long × ¼ in. wide, sessile, somewhat sheathing, scattered and to 1 in. long on inflorescence branches. Flowers in composite heads; up to 50 white to pale blue ray flowers surround yellow disk (turning purplish) in 1-in.-wide heads, with appressed to slightly spreading phyllaries, heads solitary at ends of branches in upper part of plant, 20 or fewer (often to 12), generally forming flattish cluster. IA, IL, IN, MI, MN, OH, WI. Widespread northward, scattered eastward. *SN*

Symphyotrichum dumosum
ASTERACEAE

rice button aster, bushy aster

Wet prairies, sedge meadows, savannas, fens, shores, sand flats, pannes. Summer, fall.

A perennial to 3 ft. tall. Leaves alternate, entire to sparsely shallowly toothed, somewhat rolled under along margins, to 3 in. long × ¼ in. wide, sessile; abundant ascending leafy bracts to ¼ in. long in inflorescence. Flowers in composite heads; up to 25 white to pale blue ray flowers surround yellow disk (turning purplish) in heads to ½ in. wide, with appressed phyllaries, several (often more than 20) upward-facing along branches in upper part of plant. Small-headed aster (*S. parviceps*) has up to 15 ray flowers in smaller heads subtended by phyllaries with awl-shaped tips. IA, IL, IN, MI, MO, OH, WI. Scattered, most frequent near the Great Lakes. *MAH*

Symphyotrichum ericoides
ASTERACEAE

white heath aster, wreath aster

Prairies, grassy barrens, open woodlands, roadsides. Summer, fall.

A branched perennial, hairy throughout, colonial from rhizomes, 1–3 ft. tall. Leaves alternate, entire, linear to narrowly oblanceolate, lower leaves up to 2½ in. long × ½ in. wide, becoming much smaller upward and within the inflorescence. Flowers in composite heads ¼–½ in. wide, ray flowers 5–20, slightly fewer disk flowers, phyllaries hairy with pointed tips and spreading to erect. Inflorescence a compact array of 100+ heads aligned 1-sided along branches. The specific epithet and one common name allude to its small, linear leaves, which are somewhat reminiscent of heath (*Erica*). IA, IL, IN, MI, MN, MO, OH, WI. **Principally in the Tallgrass Prairie.** *MJH*

Symphyotrichum lanceolatum
ASTERACEAE

panicled aster

Prairies, fields, marshes, wet forests, floodplains, shores, interdunal wetlands, fens, roadsides. Summer, fall.

A colony-forming perennial to 5 ft. tall. Leaves alternate, sparsely toothed, to 5 in. long × ¾ in. wide, sessile, smaller in inflorescence. Flowers in composite heads; up to 50 white ray flowers (rarely pale blue) surround yellow disk (turning reddish-orange) in heads to 1 in. wide, with appressed to slightly spreading phyllaries, numerous along branches (often more than 20 per branch) in panicle-like array. Ontario aster (*S. ontarionis*) has leaf undersides densely short-hairy. Small white aster (*S. racemosum*) has flower heads at ends of short-bracted branches. IA, IL, IN, MI, MN, MO, OH, WI. **Widespread and common.** *SN*

Symphyotrichum pilosum
ASTERACEAE

hairy aster

Prairies, old fields, roadsides, fencerows, pastures. Summer, fall.

An erect perennial, highly branched above, typically hairy, 2–4 ft. tall. Leaves alternate, entire, lanceolate to oblanceolate becoming elliptic and linear toward inflorescence, ½–4 in. long × ¹⁄₁₆–1 in. wide. Flowers in composite heads ½–¾ in. wide, ray flowers 10–30, disk flowers 15–40, involucre urn-shaped, phyllaries appressed or slightly spreading with outer surface glabrous, commonly inrolled and tapering to a spine-like, whitened point. Array widely spreading atop plant. Stems and leaves of the weedy var. *pilosum* are usually quite hairy; they are glabrous in var. *pringlei*. The latter is comparatively uncommon. IA, IL, IN, MI, MN, MO, OH, WI. **Abundant except in the northwest.** *DT*

Symphyotrichum urophyllum
ASTERACEAE

arrowleaf aster

Fields, glades, savannas, woodland edges, roadsides. Summer, fall.

A perennial to 4 ft. tall. Leaves basal and alternate, simple, cordate, shallowly toothed, to 5 in. long × 2 in. wide, with winged petiole to 5 in. long, becoming ovate to lanceolate just above base, then becoming smaller, narrow, and short-petiolate to sessile above. Flowers in composite heads; up to 12 white (rarely pale lavender) ray flowers surround yellow disk (turning purplish) in heads to ½ in. wide, with appressed to slightly spreading phyllaries, heads numerous in narrow, branched, panicle-like array at top of plant. IA, IL, IN, MI, MN, MO, OH, WI. **Scattered to widespread, less frequent westward.** *KC*

Verbesina virginica
ASTERACEAE

white crownbeard, frostweed

Open woodlands, stream terraces, thickets, roadsides. Summer, fall.

An erect perennial, stems hairy, winged, 3–8 ft. tall. Leaves alternate, broadly lanceolate to elliptic, finely toothed to entire, rough-hairy, 4–9 in. long × 1–4 in. wide. Flowers in composite heads ½–¾ in. wide, ray flowers 1–7, disk flowers 8–15, in 20–100+ heads in a somewhat flat-topped array atop the main stem and branches. One common name refers to the petal-like ribbons of frost "flowers" that form when sap leaks from the ruptured stem base during the first hard freezes of fall. IL, IN, MO, OH. Occurs mostly in the Ozark Highlands. DT

Jeffersonia diphylla
BERBERIDACEAE

twinleaf

Rich mesic and dry-mesic forests, often on rocky slopes. Spring.

An erect perennial, unbranched, glabrous, 6–10 in. tall. Leaves basal, on long petioles, blades in twin halves, each half obliquely ovate, with entire or wavy to shallowly lobed margins, 1–1½ in. long × ¾–1 in. wide when flowering, 4–7 in. long × 2–4 in. wide at maturity. Flowers regular, about 1 in. wide, usually with 8 petals, occurring singly atop a leafless stalk. Petals are short-lived. The genus name honors Thomas Jefferson. IA, IL, IN, MI, MN, OH, WI. Located mostly east of the Mississippi River and south of the Great Lakes. CR

Podophyllum peltatum

BERBERIDACEAE

mayapple

Moist upland forests, ravines, stream terraces, thickets. Spring.

An erect glabrous perennial, unbranched (except forked in fertile plants), 8–24 in. tall, colonial. Leaves round and umbrella-shaped, about 1 ft. wide, single-leaved and peltate when sterile, double-leaved when flowering, typically deeply 5- or 6-lobed, lobes oblanceolate and commonly further divided and toothed. Flowers regular, about 1½ in. wide, with 6–9 obovate petals, occurring singly and drooping from base of paired leaves, fragrant. To some people the opened ripe fruit has a scent reminiscent of passionflower fruit (*Passiflora incarnata*). IA, IL, IN, MI, MN, MO, OH, WI. **Common but local or absent in the upper Northern Lakes.** *SN*

Buglossoides arvensis

BORAGINACEAE

corn gromwell, field gromwell

Sandy fields, vacant lots, roadsides, railroads. Spring, summer.

A hairy annual to 2 ft. tall. Leaves alternate, entire, sessile, to 1½ in. long × ¼ in. wide, with strong midvein. Flowers regular with 5 spreading to ascending lobes, fused at base, to ¼ in. long and wide, solitary on short stalks in axils of upper leaves, condensed early but becoming spaced out later in season. Fruit in 4-nutlet clusters, gray-brown, stone-like, wrinkled and pitted, about ⅛ in. long. Syn. *Lithospermum arvense.* IA, IL, IN, MI, MN, MO, OH, WI. **Introduced from Eurasia. Scattered to widespread in southeastern quarter of the Midwest.** *ST*

Ellisia nyctelea
BORAGINACEAE

Aunt Lucy, false babyblueeyes

Moist forests, floodplains, disturbed
open areas with sparse vegetation.
Spring, summer.

A weak, branching annual to 16 in. tall (often
shorter). Stems purplish, fleshy, glaucous.
Leaves opposite below, alternate above, to 4 in.
long × 1 in. wide, deeply pinnately lobed into
toothed or lobed segments. Flowers regular,
with sepals longer than corolla, corolla tubular
with 5 spreading lobes, white to pale blue with
purplish spots, to ¼ in. wide, few in upper leaf
axils. Coville's phacelia (*Phacelia covillei*) and
oceanblue phacelia (*P. ranunculacea*) are rare
and have smaller, shorter-lobed leaves. IA,
IL, IN, MI, MN, MO, OH, WI. Widespread in
the Tallgrass Prairie and Ozark Highlands,
uncommon elsewhere. *SN*

Hackelia virginiana
BORAGINACEAE

stickseed

Upland forests, woodlands, thickets.
Summer, fall.

An erect biennial, branched above, hairy,
1–2 ft. tall. Stem leaves alternate, petiolate,
ovate-elliptic, entire, hairy, first-year leaves
all basal and commonly cordate, 2–12 in. long
× 1–4 in. wide. Flowers regular, ⅛–⅙ in. wide,
with short tube and 5 rounded and flared
lobes (see inset), in widely spreading pairs of
spike-like racemes. Fruit, a bristle-covered
cluster composed of 4 nutlets. This is a serious
sticktight! Nodding stickseed (*H. deflexa* var.
americana) of the northern Midwest is similar
but smaller with blue flowers and bristles limited to the nutlet margins. IA, IL, IN, MI, MN,
MO, OH, WI. Widespread but uncommon in
the far north. *SN/MAH*

Heliotropium tenellum
BORAGINACEAE

glade heliotrope, slender heliotrope

Calcareous glades, dry open and rocky
woodlands. Summer.

An erect, branched annual with silvery
appressed hairs, 4–10 in. tall. Leaves alternate, short-petiolate, linear with inwardly
rolled margins, ½–1¼ in. long × ¹⁄₁₆–⅛ in.
wide. Flowers regular, ⅛–¼ in. wide, slightly
funnel-shaped with 5 rounded lobes and a
glabrous, yellowish throat, occurring singly
at tips of widely spreading branches. Glade
heliotrope is almost exclusively confined in
our region to rocky, mostly treeless glades. An
unusual heliotrope in that its inflorescence
is not coiled. Syn. *Euploca tenella.* **IL, IN, MO,
OH. Occurs most commonly in the Ozark
Highlands.** *MAH*

Hydrophyllum canadense
BORAGINACEAE

Canada waterleaf, broadleaved waterleaf

Rich mesic forests, north-facing slopes,
ravines. Spring.

An erect perennial, glabrous to sparsely
hairy, rhizomatous, 8–18 in. tall. Leaves of
early spring are basal, pinnately lobed and
splotched with gray, the later flowering stem
leaves alternate, rounded and palmately
5-lobed and toothed, 4–6 in. long and wide.
Flowers regular, about ½ in. wide, bell-shaped
and 5-lobed, margins slightly wavy, with 5
narrow sepals that lack appendages between
them. Inflorescence a compact cluster typically situated at or below the leaves. **IL, IN,
MI, MO, OH. Occurs primarily in the eastern
half of the Midwest.** *MAH*

Hydrophyllum macrophyllum
BORAGINACEAE

largeleaf waterleaf

Moist forests. Spring.

A hairy perennial to 2 ft. tall. Leaves in basal rosette and a few alternate along stem, to 16 in. long × 5 in. wide, shorter up stem, deeply pinnately lobed or compound with approximately 9 lobes or leaflets, toothed, gray-green with silvery-white along midvein and extending into lobes, with additional random silvery-white spots. Flowers regular, to ½ in. wide, with 5 ascending lobes and long-exserted stamens, short-stalked and numerous in dense, short-branched clusters at top of stem. Vegetatively can be confused early in year with appendaged waterleaf (*H. appendiculatum*) but has more lateral lobes that are more regularly spaced. IL, IN, OH. Confined almost entirely to the Eastern Forests. *PR*

Hydrophyllum virginianum
BORAGINACEAE

Virginia waterleaf

Moist forests, savannas, thickets, floodplains, clearings. Spring, summer.

A perennial to 2 ft. tall. Leaves basal and alternate, to 6 in. long × 4 in. wide, deeply pinnately lobed or compound with 3–5 lobes or leaflets, gray-green with silvery-white mottled splotches, toothed, petiolate. Flowers regular, ½ in. wide, with 5 white to lavender ascending lobes and long-exserted stamens, short-stalked and numerous in short-branched rounded clusters to 2 in. across at top of stem. Compare with fernleaf phacelia (*Phacelia bipinnatifida*), which has smaller leaves with fewer lobes/leaflets and shorter stamens. IA, IL, IN, MI, MN, MO, OH, WI. Widespread and common, with fewer occurrences in extreme northern and southern parts of Midwest. *SN*

Myosotis verna
BORAGINACEAE

spring forget-me-not

Fields, pastures, open woodlands, rock outcrops, roadsides. Spring.

An erect annual, unbranched to branched, hairy, 5–15 in. tall. Leaves alternate, sessile to short-petiolate, oblanceolate to elliptic, entire, ½–2 in. long × ⅛–⅓ in. wide. Flowers regular, about ⅛ in. wide, funnel-shaped, 5-lobed, calyx 5-lobed with upper 3 lobes slightly shorter than the lower 2, fruiting stalks erect, in a terminal, elongated raceme. Bigseed forget-me-not (*M. macrosperma*) differs mainly by its overall larger size and the spreading orientation of its fruiting stalks. **IA, IL, IN, MI, MN, MO, OH, WI. Widespread but most common southward.** *ST*

Onosmodium occidentale
BORAGINACEAE

western marbleseed

Dry prairies, glades, bluffs. Summer.

An erect perennial, branched above, densely hairy, 1–2½ ft. tall. Leaves alternate, lanceolate, sessile, upper surface hairs mostly appressed and of similar length, leaves 2–4 in. long × ½–2 in. wide. Flowers regular, about ½ in. long, tubular with 5 slender, triangular lobes converging at their hairy tips, and a long-exserted style, in a 1-sided arching inflorescence atop the stem. The similar and widely scattered rough marbleseed (*O. hispidissimum*) has leaves with spreading hairs. Softhair marbleseed (*O. subsetosum*) of the Ozark Highlands has mostly glabrous lower stems. Some place our *Onosmodium* species in *Lithospermum*. **IA, IL, MN, MO, WI. Occurs primarily in the Tallgrass Prairie.** *SN*

Alliaria petiolata
BRASSICACEAE

garlic mustard

Rich upland and lowland forests, wooded
edges, thickets, roadsides. Spring.

A biennial having basal leaves only in first
season, then blooming in second season on
unbranched, glabrous stems 1–3 ft. tall. Basal
leaves somewhat rounded or kidney-shaped,
coarsely toothed, 2–4 in. wide (see inset);
flowering stalk leaves alternate, triangular,
coarsely toothed, 2–3 in. long and wide. All
leaves smell of garlic when crushed. Flowers
regular, ¼–½ in. wide, petals 4, in an elongated terminal raceme. Fruit a slender silique
2–3 in. long. Invasive and challenging to control. IA, IL, IN, MI, MN, MO, OH, WI. Introduced from Eurasia. Established over much
of the Midwest and spreading. *MAH/SN*

Arabidopsis lyrata
BRASSICACEAE

sand cress, lyre-leaved rock cress

Dry prairies, savannas, open woodlands,
sand dunes, cliffs on thin soils. Spring,
summer.

A biennial or short-lived perennial to 14 in.
tall. Leaves mostly basal, pinnately lobed with
terminal lobe largest, to 2 in. long, withering
before flowering; stem leaves few, alternate,
sessile, linear, entire, to 1½ in. long × ¼ in.
wide. Flowers regular, to ½ in. wide, with 4
spreading petals, in racemes at top of plant.
Fruit an ascending silique to 1¾ in. long. Syn.
Arabis lyrata. IA, IL, IN, MI, MN, MO, OH,
WI. Mostly in the Northern Lakes (somewhat
concentrated around the Great Lakes) and
adjacent parts of the Tallgrass Prairie, scattered elsewhere. *SN*

Arabis pycnocarpa
BRASSICACEAE

hairy rock cress, slender rock cress

Prairies, savannas, open woodlands, streambanks, gravel bars, shores. Spring, summer.

An upright, stiff-pubescent (at least below) biennial or short-lived perennial to 2½ ft. tall. Basal leaves to 3 in. long × 1 in. wide, entire or with a few low teeth, covered in stiff pustular-based hairs, usually shiny and dark green with yellowish midvein; stem leaves similar, alternate, sessile or auriculate-clasping, ascending to erect, to 2¼ in. long × ¾ in. wide. Flowers regular, ¼ in. wide, 4-petaled, in compact to elongated racemes at top of plant. Fruit a long, skinny, erect silique to 3 in. long. Syn. *A. hirsuta*. **IA, IL, IN, MI, MN, MO, OH, WI. Scattered throughout.** *GL*

Berteroa incana
BRASSICACEAE

hoary alyssum, hoary false madwort

Fields, pastures, roadsides, railroads, waste areas. Spring, summer, fall.

A gray-green, densely hairy annual or short-lived perennial to 3 ft. tall. Leaves basal initially, then alternate, simple, sessile, entire, to 3 in. long × ½ in. wide. Flowers regular, to ¼ in. wide, with 4 very deeply lobed petals, several in dense, rounded clusters to 2 in. wide near top of plant. Fruit an elliptic, somewhat flattened silicle with skinny beak, to ¼ in. long, fruiting portion expanding into long, cylindrical raceme. **IA, IL, IN, MI, MN, MO, OH, WI. Introduced from Europe. Abundant in the Northern Lakes and adjacent portions of the Tallgrass Prairie, less frequent in the Ozark Highlands and Eastern Forests.** *DT*

Boechera canadensis
BRASSICACEAE

sicklepod

Savannas, open woods, dry forests, forested dunes. Spring, summer.

An unbranched biennial to 3 ft. tall. Basal leaves to 5 in. long × 1¼ in. wide, shallowly sparsely bluntly toothed; stem leaves alternate, to 4 in. long × 1¼ in. wide, smaller above, sessile, sharply toothed to nearly entire. Flowers regular, ¼ in. wide, with 4 petals, spreading to reflexed on stalks to ½ in. in elongated raceme to 16 in. long at top of plant. Fruit a long, skinny, flattened, sickle-shaped silique to 4 in. long × ¹⁄₁₂ in. wide on drooping stalk. Syns. *Arabis canadensis, Borodinia canadensis*. IA, IL, IN, MI, MN, MO, OH, WI. Scattered throughout, with fewer occurrences northward. *MJH*

Boechera dentata
BRASSICACEAE

toothed rock cress

Mesic upland forests, stream terraces, mossy slopes. Spring.

An erect biennial, mostly unbranched, often spreading, sparsely covered with forked hairs, 8–18 in. tall. Basal leaves petiolate, obovate to oblanceolate; stem leaves alternate, sessile and clasping, toothed, oblong to oblanceolate, simple hairs above, forked ones beneath, 1–3 in. long × ½–1½ in. wide. Flowers regular, about ⅛ in. wide and 4-petaled, in mostly terminal racemes that elongate in fruit. Fruit a silique, ¾–1½ in. long and widely spreading. IA, IL, IN, MI, MN, MO, OH, WI. Widespread but uncommon to absent in the Northern Lakes. *BS*

Boechera grahamii
BRASSICACEAE

spreading-pod rock cress

Prairies, open woods, forest clearings
and borders, rock outcrops, shores.
Spring, summer.

An unbranched biennial to 3 ft. tall. Basal
leaves coarsely hairy, entire to sparsely
toothed, often purplish beneath, to 5 in. long
× ¾ in. wide; stem leaves alternate, to 2 in.
long × 1 in. wide, smaller above, clasping with
pointed basal lobes, entire, glabrous, glaucous, mostly erect. Flowers regular, to ⅓ in.
wide, with 4 petals, spreading to arching, in
elongated raceme at top of plant. Fruit a long,
skinny, flattened silique to 3 in. long × ¹⁄₁₂ in.
wide on spreading to somewhat ascending or
drooping stalk. Syn. *Arabis divaricarpa*. **MI,
MN, WI. Scattered primarily in the Northern
Lakes.** *KC*

Boechera laevigata
BRASSICACEAE

smooth rock cress

Rock ledges, steep slopes, open woodlands,
savannas. Spring.

An erect biennial, mostly unbranched, glabrous and glaucous, 1–2½ ft. tall. Basal leaves
obovate with short petioles; stem leaves narrowly lanceolate to linear, mostly entire (especially on upper stem) or somewhat toothed
(mostly those on lower stem), sessile and
clasping, 1–6 in. long × ¼–1 in. wide. Flowers
regular, about ⅛ in. wide with 4 petals, in terminal racemes that elongate in fruit. Fruit a
curved silique, 3–4 in. long and widely spreading. Syn. *Arabis laevigata*. **IA, IL, IN, MI, MN,
MO, OH, WI. Widespread and common but
mostly absent from the upper Northern
Lakes and western Tallgrass Prairie.** *SN*

Boechera stricta

BRASSICACEAE

Drummond's rock cress

Forests, dunes, rocky openings. Spring.

An unbranched biennial to 2½ ft. tall.
Basal leaves entire to sparsely toothed,
often purplish, to 3 in. long × ⅜ in. wide,
short-petiolate; stem leaves alternate, to 3 in.
long × ¾ in. wide, smaller above, clasping with
pointed basal lobes, entire or with few teeth,
glabrous, glaucous, mostly erect. Flowers regular, to ½ in. wide, with 4 petals, erect in elongated raceme at top of plant. Fruit a skinny
silique to 4 in. long on erect stalk. Syn. *Arabis
drummondii*. IA, IL, IN, MI, MN, OH, WI.
Scattered in the Northern Lakes, northern
portion of the Tallgrass Prairie, and eastern
portion of the Eastern Forests. *PD*

Capsella bursa-pastoris

BRASSICACEAE

shepherd's-purse

Agricultural fields, roadsides, pastures,
disturbed sites, waste areas. Spring.

An erect annual, mostly unbranched, hairy
toward the base, 4–24 in. tall. Basal leaves
petiolate, oblanceolate, hairy, and typically
pinnately lobed. Stem leaves comparatively
few, alternate, linear to lanceolate, sessile and
clasping, 1–4 in. long × ⅛–1 in. wide. Flowers
regular, ⅛ in. wide, 4-petaled, mostly occurring in terminal racemes that elongate when
in fruit. Fruit is a flat, triangular silicle about
⅓ in. long and wide. Shepherd's-purse is easily recognized by the distinctive "purse-like"
shape of its fruit. *Capsella* is Greek for "small
box." IA, IL, IN, MI, MN, MO, OH, WI.
Introduced from Europe. Widespread and
abundant. *KC*

Cardamine bulbosa
BRASSICACEAE

spring cress
―――――――――

Floodplain forests, springs, fens,
vernal pools, swamps. Spring.

An erect perennial, unbranched, mostly
glabrous, 1–2 ft. tall. Leaves alternate, lower
and basal leaves long-petiolate, rounded to
ovate, typically withered by flowering time,
stem leaves becoming sessile upward, oblong
to lanceolate, entire or wavy-margined or
with a few teeth, 1–3 in. long × ½–1½ in. wide.
Flowers regular, about ½ in. wide, 4-petaled,
in mostly terminal racemes that elongate
when in fruit, the latter an ascending, linear
silique up to 1¼ in. long. IA, IL, IN, MI, MN,
MO, OH, WI. Common and widely scattered,
becoming local or absent in the far northern
Midwest. *CB*

Cardamine concatenata
BRASSICACEAE

cutleaf toothwort
―――――――――

Rich, moist forests, ravines,
stream terraces. Spring.

An erect perennial, unbranched, mostly glabrous, especially above stem leaves, 4–12 in.
tall. Leaves basal as well as on the stem, petiolate, palmately 3-parted, lobes linear to lanceolate, sharply toothed, the lateral lobes often
further divided, commonly 3 in whorl, 1–5
in. long and about as wide (or wider). Flowers regular, about ½ in. wide, 4-petaled, in
terminal racemes. Fruit an ascending silique,
¾–1½ in. long. Syn. *Dentaria laciniata*. Fineleaf
toothwort (*C. dissecta*) occurs locally in OH and
IN; it has extremely narrow leaf segments and
usually just 2 leaves per flowering stem. IA, IL,
IN, MI, MN, MO, OH, WI. Common in most
of the Midwest. *SN*

Cardamine diphylla
BRASSICACEAE

crinkleroot

Rich, moist forests, ravines,
stream terraces, seeps. Spring.

An erect perennial, unbranched, mostly glabrous, 6–15 in. tall. Leaves positioned basally
and upon stem, petiolate, palmately 3-parted,
lobes broadly ovate-elliptic, sharply toothed,
stem leaves commonly 2 and opposite, 1½–4 in.
long and about as wide. Flowers regular, about
½ in. wide, 4-petaled, in terminal racemes. Fruit
a silique, spreading to ascending, ¾–1½ in. long.
Syn. *Dentaria diphylla*. Crinkleroot's large stem
leaves distinguish it from slender toothwort (*C.
angustata*), a species found in southern IN and
OH. The latter's stem leaves are somewhat like
those of cutleaf toothwort (*C. concatenata*). IL,
IN, MI, MO, OH, WI. **Mostly in the eastern
Northern Lakes and Eastern Forests.** *AG*

Cardamine pensylvanica
BRASSICACEAE

bitter cress

Streams, seeps, wetlands.
Spring, summer.

An erect annual or biennial, mostly
unbranched, mostly glabrous, 4–24 in. tall.
Leaves basal and alternate, petioles glabrous, deeply pinnately divided with 3–13
obovate-oblanceolate segments flared at
base, the terminal one broader. Main stem
leaves 1½–3 in. long × ¾–1¼ in. wide. Flowers regular, about ¼ in. wide, 4-petaled, in
terminal racemes. Fruit a silique. Similar are
small-flowered bitter cress (*C. parviflora*) and
hairy bitter cress (*C. hirsuta*). The terminal leaf
segment of the former is a little wider than the
lateral ones. Petioles of *C. hirsuta* have spreading hairs. IA, IL, IN, MI, MN, MO, OH, WI.
Common throughout the Midwest. *MAH*

Cardamine pratensis
BRASSICACEAE

cuckoo flower

Fens, bogs, swamps, wet meadows, lakeshores. Spring.

A weak perennial to 2½ ft. tall. Leaves to 3 in. long, basal and alternate, pinnately compound with up to 13 leaflets, basal and lower leaves long-petiolate, petioles shorter above, leaflets rounded, more linear above, terminal leaflet largest and usually lobed. Flowers regular, ½ in. wide, with 4 petals, in terminal raceme. Fruit an ascending silique to 1½ in. long. North American plants are var. *palustris* (syn. *C. dentata*), Old World plants are var. *pratensis*, which are larger with pink flowers, growing in drier conditions. The two are combined by some botanists and treated as non-native. IL, IN, MI, MN, OH, WI. Nearly exclusively in the Northern Lakes. *LC*

Draba brachycarpa
BRASSICACEAE

short-fruited whitlow grass

Prairies, fields, glades, roadsides, railroads. Spring.

A hairy annual to 4 in. tall with purplish stem. Leaves alternate, with few teeth on margins, petiolate below, sessile above, to 1 in. long × ¼ in. wide. Flowers regular, to ⅛ in. wide, with 4 spreading petals, in dense clusters at top of plant that elongate as fruit form. Fruit a stalked, spreading silique to ¼ in. long. Wedgeleaf whitlow grass (*D. cuneifolia*), with leaves more basally disposed, has sessile leaves with wedge-shaped bases, resulting in nearly diamond-shaped leaf blades; it is also covered in stellate pubescence. IL, IN, MO, OH. Widespread in the Ozark Highlands, adjacent Tallgrass Prairie, and adjacent Eastern Forests, with scattered occurrences elsewhere. *AG*

Draba verna
BRASSICACEAE

vernal whitlow grass

Lawns, prairies, roadsides, railroads.
Winter, spring.

An erect annual, unbranched, stem mostly
glabrous, 2–6 in. tall. Leaves a basal rosette,
oblanceolate to obovate, entire, with branched
hairs, ⅓–1¼ in. long × ¹⁄₁₆–⅓ in. wide. Flowers
regular, ⅛ in. wide, with 4 deeply cleft petals, in
a few-flowered terminal cluster that elongates
with age. Fruit a silicle. The native whitlow
grass (*D. reptans*) has entire or barely notched
petals. Common names notwithstanding, these
are not grasses. Whitlow is an infection of tissue
near fingernails or animal hooves, and *Draba*
species were believed to have curative properties. IA, IL, IN, MI, MN, MO, OH, WI. Introduced from Eurasia. A common weed but
absent from most of IA, MN, and WI. *MAH*

Leavenworthia uniflora
BRASSICACEAE

Michaux's gladecress

Flat, pavement-like limestone
and dolomite glades. Spring.

An erect annual, glabrous, 1–4 in. tall. Leaves
in basal rosette, pinnately divided into up
to 12 lateral lobes and a larger terminal one,
toothed, to 3 in. long × ¼ in. wide. Flowers
regular, about ¼ in. wide, with 4 oblanceolate
petals. Normally 1 flower atop each stalk
arising from the rosette. Fruit a flattened
silique, ¾–1½ in. long. Michaux's gladecress
is a winter annual: plants germinate in fall
and bloom early the following spring. IN,
MO, OH. Restricted in the Midwest to
southern IN, southern OH, and the Ozark
Highlands. *SN*

Lepidium campestre
BRASSICACEAE

field pepperweed, field cress

Fields (including cropland), roadsides, railroads, waste areas. Spring, summer.

A gray-hairy biennial to 2 ft. tall. Basal leaves in rosette, to 4½ in. long × 1 in. wide, broadest above, wavy-margined; stem leaves alternate, clasping with 2 lobes at base, toothed or entire, ascending, to 3 in. long × ¾ in. wide, smaller above. Flowers regular, ⅛ in. wide, with 4 petals, on spreading stalks in many-flowered terminal raceme. Fruit a notched, oval silicle to ¼ in. long. Edible with peppery taste. Hoary cress (*L. draba*) has larger flowers and unnotched fruit. IA, IL, IN, MI, MN, MO, OH, WI. **Introduced from Eurasia. Common and widespread except in IA, MN, and WI, where scattered.** *SN*

Lepidium virginicum
BRASSICACEAE

Virginia peppergrass, common peppergrass

Prairies, fields, open woods, vacant lots, gardens, roadsides, waste areas. Spring, summer, fall.

A branching annual to 1½ ft. tall. Basal leaves in rosette, to 3 in. long × 1 in. wide, with many narrow pinnate lobes; stem leaves alternate, sessile, simple, coarsely toothed (especially in upper half) to entire, to 3½ in. long × ¾ in. wide, smaller above. Flowers regular, 1/12 in. wide, with 4 petals, on ascending stalks in many-flowered terminal raceme. Fruit a shallowly notched, round silicle to ⅛ in. wide. Small peppergrass (*L. densiflorum*) has petals absent or much reduced and fruit broadest above middle. IA, IL, IN, MI, MN, MO, OH, WI. **Common and widespread, becoming scattered northward.** *SN*

Nasturtium officinale
BRASSICACEAE

watercress

Rivulets and springs in fens, swamps, and seeps; rivers, streams. Spring, summer, fall.

A glabrous, creeping, colony-forming perennial to 15 in. tall, floating or rooting at nodes in mud. Leaves to 4 in. long × 1½ in. wide, petiolate, alternate, pinnately compound with 3–11 leaflets, terminal leaflet often largest and shallowly lobed. Flowers regular, to ⅜ in. wide, with 4 petals, in racemes at ends of branches. Fruit an ascending, abruptly beaked silique to 1¼ in. long; seeds in 2 rows. Cultivated and sold commercially. Edible. One-row watercress (*N. microphyllum*) has narrower fruit (to ¹⁄₁₄ in.) and seeds in 1 row. **IA, IL, IN, MI, MN, MO, OH, WI. Introduced from Eurasia. Scattered, widespread in some regions.** *PR*

Planodes virginicum
BRASSICACEAE

Virginia rock cress

Fields, pastures, cropland, yards, roadsides, waste areas. Spring.

An annual to 8 in. tall. Basal leaves to 6 in. long, petiolate, in reddish-purple rosette, pinnately lobed or compound with 20+ blunt lobes/leaflets, terminal lobe/leaflet often with 3 lobes. Stem leaves few along erect to ascending green to brownish-red stem, alternate, similar to basal, with up to 29 lobes/leaflets, to 1 ⅗ in. long × ⅖ in. wide, short-petiolate. Flowers regular, ¼ in. wide, with 4 petals, in racemes at top of plant. Fruit an ascending silique to 1 in. long. **IA, IL, IN, MO, OH. Widespread in southern portion of the Midwest, scattered northward.** *KB*

Rorippa aquatica
BRASSICACEAE

lake cress

Ponds, oxbows, swamps, marshes. Spring, summer.

A perennial with stems submersed or emergent, sprawling, branched within inflorescence, glabrous, 1–2½ ft. tall. Leaves alternate, submersed leaves highly pinnately dissected, emergent leaves lanceolate, sessile, entire or with small teeth on margins, 1–2 in. long × ¼–¾ in. wide. Flowers regular, with 4 obovate petals, in elongated racemes. Fruit an inflated, obovoid silicle about ¼ in. long. Detached leaves and stem pieces have the unusual ability of rooting and forming new plants. IA, IL, IN, MI, MN, MO, OH, WI. Widespread but not common. *BJ*

Thlaspi arvense
BRASSICACEAE

field pennycress

Fields, pastures, cropland, roadsides, railroads, waste areas. Spring, summer.

A glabrous annual to 2½ ft. tall. Basal leaves in rosette, to 3¼ in. long × 1 in. wide, spatulate, long-petiolate; stem leaves alternate, becoming clasping above with 2 pointed basal lobes, broadly toothed, to 4 in. long × 1 in. wide. Flowers regular, ⅛ in. wide, with 4 petals, on spreading to ascending stalks in many-flowered racemes to 8 in. in upper part of plant. Fruit a shallowly notched, nearly round, winged silicle to ½ in. long. Perfoliate pennycress (*Microthlaspi perfoliatum*) has more-shallowly notched fruit to ¼ in. and leaves with rounded basal lobes. IA, IL, IN, MI, MN, MO, OH, WI. Introduced from Eurasia. Scattered to widespread. *SN*

Turritis glabra

BRASSICACEAE

tower mustard

Prairies, glades, fields, savannas, barrens, roadsides. Spring, summer.

A glaucous, unbranched, upright biennial to 5 ft. tall. Basal leaves to 3½ in. long × ¾ in. wide, sparsely and broadly toothed to lobed; stem leaves alternate, sessile and auriculate-clasping stem, entire, glabrous, to 3 in. long × 1 in. wide, sometimes purplish. Flowers regular, ⅙ in. wide, with 4 petals, ascending on stalks to ½ in. in terminal, elongated raceme to 2 ft. long. Fruit a stiffly erect silique to 2½ in. long, densely arranged. Syn. *Arabis glabra*. IA, IL, IN, MI, MN, MO, OH, WI. Frequent in the Northern Lakes, scattered in the Eastern Forests, more scattered to rare westward. *KC*

Campanula aparinoides

CAMPANULACEAE

marsh bellflower

Marshes, sedge meadows, fens, swamps, pond and lake margins. Summer.

A rough-textured, sprawling perennial to 3 ft. with slender, branched stems. Leaves alternate, linear, irregularly toothed, to 2½ in. long × ¼ in. wide, with tiny hooked hairs on margins and underside midvein, containing milky sap. Flowers regular, to ½ in. wide with 5 spreading to recurved white to very pale blue corolla lobes, solitary on long stalks, sometimes nodding. Plant texture similar to cleavers (*Galium aparine*), which has whorled leaves. Plants with larger flowers on longer stalks are var. *grandiflora*. IA, IL, IN, MI, MN, MO, OH, WI. Widespread in the Northern Lakes and adjacent portions of the Tallgrass Prairie and Eastern Forests; rare in the Ozark Highlands. *PR*

Lobelia dortmanna
CAMPANULACEAE

water lobelia

Lakes, especially with low pH.
Summer, fall.

A colony-forming aquatic perennial to 40 in.
tall, with stem emerging from water to 12 in.
Leaves numerous in submerged basal rosette
or growing on recently exposed soil, appearing as a pair of hollow tubes in cross section,
blunt-tipped, simple, entire, to 2 in. long × ¼
in. wide, containing white sap. Flowers irregular, with 2 small horizontally spreading upper
lobes and 3 larger spreading lower lobes at
apex of long tube, pale blue to white, to ¾ in.
long, few on short stalks in terminal raceme,
occasionally blooming and producing fruit
underwater. **MI, MN, WI. Restricted to northern portion of the Northern Lakes.** *PD*

Lonicera japonica
CAPRIFOLIACEAE

Japanese honeysuckle

Various forest types, forest edges, thickets,
disturbed sites. Spring, summer.

A sprawling and climbing perennial vine,
hairy (at least when young), to 20+ ft.
long. Leaves opposite, short-petiolate,
oblong-elliptic, sometimes pinnately lobed,
evergreen in mild winters, 1½–4 in. long ×
1–2 in. wide. Flowers irregular, about 1½ in.
long, white fading to cream or yellow with
age, tubular, 2-lipped, with 4 narrow lobes on
upper lip, the lower lip narrow but unlobed,
in pairs in leaf axils, very fragrant. Fruit a
black berry. This species is highly invasive.
**IL, IN, MI, MO, OH, WI. Introduced from
Asia. Widespread but very limited in the
north.** *MAH*

Symphoricarpos orbiculatus
CAPRIFOLIACEAE

coralberry, buckbrush

Open woodlands, mesic to dry forests, forest edges, thickets. Summer.

A finely branched shrub, with runners from base, young twigs hairy, 2–4+ ft. tall. Leaves opposite, short-petiolate, elliptic to ovate and rounded at tip, entire, short-hairy, 1–2 in. long × ¾–1½ in. wide. Flowers regular, ⅛–¼ in. wide, whitish to pinkish, tubular with 5 lobes, in tight axillary clusters of 10+ flowers. Fruit a pinkish-purple or coral-colored drupe. Coralberry fruit is generally avoided by wildlife. Deer are said to browse the twigs (hence buckbrush). IA, IL, IN, MI, MN, MO, OH, WI. Primarily in the lower Midwest. *PR*

Arenaria serpyllifolia
CARYOPHYLLACEAE

thyme-leaved sandwort

Glades, old fields, railroads, roadsides, pavement cracks. Spring, summer.

A branched annual to 10 in. tall (usually shorter) with finely hairy stems, often sprawling. Leaves ovate with pointed tip, opposite, entire, sessile, to ⅓ in. long × ⅙ in. wide. Flowers regular, ⅕ in. wide, with 5 petals about the same length as 5 sepals or shorter, on short stalks in small clusters at top of plant and in leaf axils. Can be confused with chickweeds in *Cerastium* and *Stellaria*, but plants in these genera have clearly lobed petals. IA, IL, IN, MI, MN, MO, OH, WI. Introduced from Eurasia. Throughout, with fewer occurrences in IA, MN, and WI. *SN*

Cerastium arvense
CARYOPHYLLACEAE

field chickweed

Gravel bluffs, riverbanks, shorelines, upland forests. Spring, summer.

A perennial branching from near base, stem hairy, often glandular, with a taproot, 3–8 in. tall. Leaves opposite, entire, sessile, lanceolate to linear, glabrous to hairy and sometimes glandular, often with axillary leaves, up to 2 in. long × ½ in. wide. Flowers regular, about ½ in. wide with 5 styles and 5 deeply 2-lobed petals. Inflorescence of spreading clusters. Fruit a cylindrical capsule tipped with 10 short teeth. Subspecies *strictum* is native; subsp. *arvense* is introduced from Eurasia. The latter is strongly rhizomatous and slightly taller. *Cerastium velutinum* has leaves about twice as wide as *C. arvense*. IA, IL, IN, MI, MN, MO, OH, WI. **Widely scattered and local. *PR***

Cerastium fontanum
CARYOPHYLLACEAE

common mouse-ear chickweed

Lawns, fields, vacant lots, roadsides, waste areas. Spring, summer, fall.

Matted, ascending, hairy perennial to 1 ft. tall. Leaves opposite, entire, sessile, with strongly impressed midvein, to 1 in. long × ½ in. wide. Flowers regular, ¼ in. wide with 5 deeply lobed petals and 10 stamens, in clusters at top of plant. Similar to other white-flowered, 5-petaled, small weedy plants in the pink family. Other mouse-ear chickweeds (*Cerastium*) are annual and/or have 5 stamens. Chickweeds (*Stellaria*) have stems and leaves glabrous or stems with hairs in lines only. IA, IL, IN, MI, MN, MO, OH, WI. **Introduced from Eurasia. Throughout, with fewer occurrences in western IA and western MN. *SN***

Cerastium nutans
CARYOPHYLLACEAE

nodding chickweed

Fallow fields, cliff bases, creek banks, forest edges, waste areas. Spring, summer.

An erect to sprawling annual, often branching near base, pubescent with glandular hairs, 3–15 in. tall. Leaves opposite, entire, sessile, oblong-lanceolate, ½–2 in. long × ⅛–½ in. wide. Flowers regular, about ¼ in. wide, usually with 10 stamens, 5 styles, 5 rounded and deeply 2-lobed petals, and glandular-hairy bracts. Pedicel twice as long or longer than capsule. Inflorescence in open clusters. Short-stalked mouse-ear chickweed (*C. brachypodum*) is similar, but its capsules and pedicels are of similar length. IA, IL, IN, MI, MN, MO, OH, WI. Common in most of the Midwest. *ST*

Gypsophila paniculata
CARYOPHYLLACEAE

panicled baby's-breath

Sandy or gravelly fields, dunes, shores, roadsides, waste areas. Summer.

A bushy, glaucous, tumbleweed-like perennial to 3 ft. tall. Leaves opposite, entire, sessile, with pointed tip, to 4 in. long × ½ in. wide, smaller up stem, reduced to scale-like bracts in inflorescence. Flowers regular, ¼ in. wide, bell-shaped with 5 spreading petals surrounding spreading-ascending stamens (see inset), short-stalked, 100+ in branching, dome-shaped terminal clusters. Low baby's-breath (*G. muralis*) is a pink-flowered annual to 16 in. tall with linear leaves. IA, IL, IN, MI, MN, OH, WI. Introduced from Europe. Scattered, most frequent in northern portion of the Northern Lakes, rare in the Tallgrass Prairie. *PD*

Holosteum umbellatum
CARYOPHYLLACEAE

jagged chickweed

Roadsides, pavement cracks, lawns, disturbed sites. Spring.

An erect annual, branched or not, mostly glabrous, 2–15 in. tall. Leaves opposite, sessile or short-petiolate, oblanceolate to ovate, entire, somewhat succulent, ½–1½ in. long and up to ½ in. wide. Flowers regular, about ¼ in. wide, with 5 jagged-tipped petals, umbellate on long stalks. Fruits cylindrical with 6 recurved teeth at tip. This species looks like some *Cerastium* chickweeds, but the latter's flowers are not umbellate and most possess deeply notched petals and capsules tipped with 10 teeth. IA, IL, IN, MI, MO, OH, WI. Introduced from Europe. Mostly in the lower Midwest. *ST*

Minuartia michauxii
CARYOPHYLLACEAE

rock sandwort, stiff sandwort

Prairies, glades, savannas, dunes, limestone ledges and pavement. Spring, summer.

A clustered, somewhat mat-forming perennial to 1 ft. tall. Leaves opposite, entire, simple, stiff, flat, sessile, to ¾ in. long, less than ⅛ in. wide, with needle-like tips, more abundant lower on stem, where clusters of secondary leaves occur in primary leaf axils. Flowers regular, to ⅓ in. wide, with 5 ascending petals longer than sepals, stalked and few to numerous in open, branched, leafless inflorescence at top of plant. Syns. *Arenaria stricta, Sabulina michauxii*. IA, IL, IN, MI, MO, OH, WI. In widely scattered, separated regions of the Midwest. *SN*

Minuartia patula
CARYOPHYLLACEAE
Pitcher's stitchwort, slender sandwort

Prairies, savannas, glades, limestone outcrops, dunes. Spring, summer.

A multi-stemmed perennial to 8 in. tall. Leaves opposite, entire, simple, stiff, roundish in cross section, sessile, to ⅗ in. long, less than ⅛ in. wide, evenly spaced along wiry stem, lacking clusters of secondary leaves from primary leaf axils. Flowers regular, to ⅕ in. wide, with 5 ascending petals longer than sepals, stalked and few to numerous in open, branched, leafless inflorescence at top of plant. Fruit a capsule. IL, IN, MO, OH. Widespread in the Ozark Highlands, rare elsewhere. *CB*

Moehringia lateriflora
CARYOPHYLLACEAE
wood sandwort, grove sandwort, bluntleaf sandwort

Moist forests, woodland edges, prairies. Spring, summer.

A colony-forming, tangled, creeping perennial to 8 in. tall. Leaves opposite, entire, sessile to short-petiolate, oval, to 1½ in. long × ½ in. wide, with conspicuous midvein. Flowers regular, to ⅓ in. wide, with 5 spreading petals longer than sepals, stalked in branching clusters from upper leaf axils and at top of plant. Fruit a roundish capsule to ⅓ in. long. Syn. *Arenaria lateriflora*. IA, IL, IN, MI, MN, MO, OH, WI. Widespread in the transition zone along the boundary of the Northern Lakes and the Tallgrass Prairie; uncommon to rare in the Eastern Forests. *SN*

Saponaria officinalis
CARYOPHYLLACEAE

bouncing bet, soapwort

Old fields, roadsides, sand savannas, railroads. Spring, summer.

An erect perennial with some branching above, glabrous, colonial, 1–3 ft. tall. Leaves opposite, entire, sessile, elliptic to oblance-olate, with 3 prominent veins, 1–4 in. long × ½–2 in. wide. Flowers about 1 in. wide, with 5 white to pinkish, spreading to reflexed pet-als, slightly notched at tip, in dense terminal clusters atop stems. This common species was used by early pioneers to make liquid soap for washing clothes and other items (hence soapwort). IA, IL, IN, MI, MN, MO, OH, WI. Introduced from Eurasia. Widespread throughout. *MAH*

Silene antirrhina
CARYOPHYLLACEAE

sleepy catchfly, sleepy silene

Glades, roadsides, railroads, fields, disturbed sites. Spring, summer.

An erect annual, stems single or branched above, glabrous or sparsely hairy, internodes typically with sticky reddish patches, 6–24 in. tall. Leaves opposite, entire, sessile, narrowly oblanceolate, elliptic to linear, 1–3 in. long × ⅛–½ in. wide. Flowers regular, up to ¼ in. wide, with 5 2-lobed, white to pinkish-purple petals (or petals lacking in forma *apetala*), in open panicles. Fruit a ribbed capsule with 6 terminal teeth. IA, IL, IN, MI, MN, MO, OH, WI. Widespread throughout. *PD*

Silene latifolia
CARYOPHYLLACEAE

white campion, bladder campion

Old fields, forest edges, open disturbed sites. Spring, fall.

An erect to ascending annual or short-lived perennial, branched above, hairy and glandular, 1–3 ft. tall. Leaves opposite, entire, sessile, lanceolate to elliptic, 1–5 in. long × ¼–1¼ in. wide. Flowers regular, about 1 inch wide, dioecious, females with 5 styles, with 5 2-lobed petals and an inflated calyx bearing reddish veins (10 veins in males and 20 in females), these hairy but smooth between them. Inflorescence of terminal branching panicles. Flowers are fragrant. White campion (*S. noctiflora*) is similar but has perfect flowers with 3 styles. IA, IL, IN, MI, MN, MO, OH, WI. Introduced from Eurasia. Widespread. *SN*

Silene nivea
CARYOPHYLLACEAE

snowy campion

Moist woods, floodplains, riverbanks. Spring, summer.

A weak-stemmed perennial to 3 ft. tall. Leaves opposite, to 4 in. long × 1 ⅓ in. wide, entire, pointed at tip, rounded at base, sessile or nearly so. Flowers regular, to ¾ in. wide, tubular with 5 spreading shallowly notched corolla lobes, with 10 stamens, 3-parted style, and appendages at base of corolla lobes short-exserted from tube, subtended by tubular calyx nearly as long as corolla tube with 5 short triangular lobes at top, calyx dimpled at base; 1 to few on stalks to 2 in. from upper leaf axils. IA, IL, IN, MI, MN, MO, OH, WI. Uncommon, most frequent in eastern portion of the Tallgrass Prairie, scattered elsewhere. *SN*

Silene stellata

CARYOPHYLLACEAE

starry campion

Dry upland forests, woodlands, savannas, prairies. Summer.

An erect perennial, unbranched, short-hairy, 1–3 ft. tall. Leaves in whorls of 4 or occasionally opposite, sessile, ovate to lanceolate, entire, 1–4 in. long × ¼–1½ in. wide. Flowers regular, about 1 in. wide, petals 5, each with multiple deeply cleft lobes, calyx tube bell-shaped, in a terminal panicle. The deep lobing of the petals gives flowers a fringed or "starry" look, hence both the common name and specific epithet (see inset). IA, IL, IN, MI, MN, MO, OH, WI. **Widespread but absent in most of the Northern Lakes.** *SN*

Silene vulgaris

CARYOPHYLLACEAE

bladder campion

Railroads, roadsides, dunes. Spring, summer, fall.

An erect perennial, branching, mostly glabrous and glaucous, 1–2½ ft. tall. Leaves opposite, sessile, entire, oblong to oblanceolate, 1–3 in. long × ¼–1½ in. wide. Flowers regular, about ¾ in. wide, petals 5, deeply cleft into 2 lobes widest at the tips, calyx tube becoming greatly inflated with 20 main veins and connecting lateral ones, glabrous, tipped with 6 triangular teeth, dimpled at base, in a spreading, terminal panicle. The similar Balkan campion (*S. csereii*) has smaller flowers and a barely inflated calyx tapered at base, with most veins parallel. IA, IL, IN, MI, MN, MO, OH, WI. **Introduced from Eurasia. Occurs mostly in the Northern Lakes.** *MJH*

Stellaria longifolia
CARYOPHYLLACEAE

long-leaved stitchwort, long-leaved
starwort, long-leaved chickweed

Wet prairies, marshes, swamps, shores,
bogs. Spring, summer.

A glabrous, branched, weak, sprawling perennial to 20 in. Leaves opposite, entire, linear,
broadest at middle, sessile, to 2 in. long × ⅛
in. wide. Flowers regular, ⅓ in. wide, with 5
petals, each 2-lobed to base, in branched and
stalked spreading clusters at top of plant and
from leaf axils. In grass-leaved chickweed (*S.
graminea*), introduced, leaves are broadest
just above base and inflorescence is distinctly
terminal, with more flowers. Northern stitchwort (*S. borealis*) has ciliate leaf margins. IA,
IL, IN, MI, MN, MO, OH, WI. Widespread in
the Northern Lakes, scattered in the adjacent
Tallgrass Prairie and the Eastern Forests. *KC*

Stellaria media
CARYOPHYLLACEAE

common chickweed

Mesic forests, stream terraces, pastures,
lawns, crop fields. Spring, summer, fall.

A usually reclining annual, branching, stems
often hairy in lines, to 1½ ft. long. Leaves
opposite, entire, short-petiolate to sessile,
ovate, mostly glabrous and somewhat fleshy,
¼–1½ in. long × ⅛–¾ in. wide. Flowers regular, about ¼ in. wide, with 3 styles and 5 petals
each deeply cleft into 2 oblong lobes, sepals 5,
exceeding the length of the petals, occurring
in terminal clusters or singly in leaf axils.
Water chickweed (*Myosoton aquaticum*) is similar but has 5 styles and sticky-textured leaves.
IA, IL, IN, MI, MN, MO, OH, WI. Introduced
from Eurasia. Common throughout. *SN*

Stellaria pubera
CARYOPHYLLACEAE

star chickweed, great chickweed

Rich mesic forests, ravines, stream terraces.
Spring.

A weakly erect perennial, stems hairy, branching, 4–12 in. tall. Leaves opposite, sessile to
short-petiolate, entire, lanceolate to elliptic,
glabrous to sparsely hairy, ½–4 in. long ×
¼–1½ in. wide. Flowers regular, about ½
in. wide, with 5 deeply 2-lobed petals and 5
pointed sepals mostly shorter than petals, in
open clusters at branch tips. Star chickweed is
a native member of our spring flora with large
showy flowers. Tennessee chickweed (*S. corei*)
occurs in a few counties mostly near the Ohio
River; its sepals are gradually slender-tipped
and as long or longer than the petals. IL, IN,
OH. Mostly in IN and OH. *MAH*

Polanisia dodecandra
CLEOMACEAE

clammyweed, red-whiskered clammyweed

Sandbars, gravel bars, glades, railroads.
Summer, fall.

An erect annual, branched, glandular-hairy,
sticky, 1–2 ft. tall. Leaves alternate, petiolate,
compound, leaflets 3, these elliptic to oblanceolate, entire, 1–2 in. long × ½–1 in. wide. Flowers irregular, about ½ in. long, petals 4, erect,
one pair longer than other pair, white to purple, narrowed at base, notched at tip, stamens
long-exserted, in a somewhat dense raceme.
Petals of the more common type (pictured)
are shorter than those in subsp. *trachysperma*.
Spiderflower (*Cleome hassleriana*), a popular
ornamental, is a tropical American relative.
IA, IL, IN, MI, MN, MO, OH, WI. Widely
scattered throughout. *ST*

Convallaria majalis
CONVALLARIACEAE
lily-of-the-valley

Homesites, cemeteries; occasionally in moist forests. Spring, summer.

A colony-forming perennial to 10 in. tall. Leaves 2 or 3, alternate (but appearing nearly opposite at top of plant), thick-textured, glabrous and sometimes glaucous, entire, to 10 in. long × 5 in. wide, upright, with strong midvein, parallel veins less deeply impressed, on sheathing petioles to 3 in. long. Flowers regular, white (rarely faintly pink), urn-shaped with 6 short, pointed, flaring lobes, to ⅓ in. long, drooping in erect to arching racemes, with up to 16 flowers per inflorescence. Fruit a red berry. IA, IL, IN, MI, MN, MO, OH, WI. **Introduced from Europe. Throughout.** *SN*

Maianthemum canadense
CONVALLARIACEAE
wild lily-of-the-valley,
Canada mayflower

Forests, swamps, bogs. Spring, summer.

A colony-forming perennial to 8 in. tall. Leaves simple, to 3 in. long × 2 in. wide, 2 or 3, sessile, clasping, alternate; sterile plants consist of single, petiolate, cordate leaf. Flowers regular, to ¼ in. wide, with 4 spreading to recurved tepals, 4 stamens, and single pistil protruding, paired in erect racemes at top of plant. Fruit a speckled red berry to ¼ in. wide. Stems and leaf undersides short-hairy in var. *interius*, hairless in var. *canadense*. **IA, IL, IN, MI, MN, OH, WI. Widespread and common in the Northern Lakes and adjacent portions of the Tallgrass Prairie, scattered in the Eastern Forests.** *PS*

Maianthemum racemosum
CONVALLARIACEAE

Solomon's plume, false Solomon's seal

Mesic to dry-mesic upland forests, stream terraces. Spring.

A mostly arching to erect perennial, unbranched, somewhat zigzag, glabrous to finely hairy, 1–2½ ft. tall. Leaves alternate, somewhat 2-ranked, sessile, mostly elliptical to ovate, entire, 3–6 in. long × 2–3 in. wide. Flowers regular, ⅛–¼ in. wide, with 6 spreading tepals and 6 stamens with conspicuous filaments. Flowers 50–100+ in a branching, terminal panicle. Fruit a round red berry. It has the general appearance of smooth Solomon's seal (*Polygonatum biflorum*), but flowers of the latter dangle beneath the leaves. Syn. *Smilacina racemosa*. IA, IL, IN, MI, MN, MO, OH, WI. Common throughout, possibly in every county. *EH*

Maianthemum stellatum
CONVALLARIACEAE

starry false Solomon's seal

Moist prairies, sand dunes, moist forests, swamps, floodplains, seeps. Spring, summer.

An unbranched perennial to 2½ ft. tall. Leaves sessile and clasping or very short-petiolate, simple, alternate, arching, to 6 in. long × 2 in. wide. Flowers regular, to ⅓ in. wide, with 6 spreading tepals, 6 stamens and pistil protruding, in terminal raceme. Fruit a striped ¼-in. berry that matures to a bright red. Syn. *Smilacina stellata*. IA, IL, IN, MI, MN, MO, OH, WI. Widespread in northern portion of the Northern Lakes and adjacent portions of the Tallgrass Prairie, scattered in the Eastern Forests and southern portions of the Tallgrass Prairie, rare in the Ozark Highlands. *MAH*

Maianthemum trifolium
CONVALLARIACEAE

false mayflower, three-leaf false Solomon's seal

Swamps, bogs. Spring.

A glabrous, unbranched perennial to 6 in. tall. Leaves 2–4 (usually 3), sheathing, simple, entire, alternate, oval, to 4½ in. long × 1½ in. wide, with abrupt rolled point at tip. Flowers regular, to ⅓ in. wide, with 6 spreading tepals, 6 stamens and single pistil protruding, up to 15 in raceme at top of plant. Fruit a spotted ¼-in. berry that matures to a bright red. Syn. *Smilacina trifolia*. **MI, MN, OH, WI. Almost entirely restricted to the Northern Lakes.** *MJH*

Calystegia sepium
CONVOLVULACEAE

hedge bindweed

Fields, thickets, fencerows, woodland edges, streambanks, marshes, ditches, roadsides. Spring, summer, fall.

A twining perennial vine to 10 ft. long, lacking tendrils. Leaves alternate, simple, entire, glabrous, to 5 in. long × 3 in. wide, sagittate, often with flared, squared-off bases, on 2½-in. petioles. Flowers regular, funnel-shaped, 3 in. long, with 5 white to pink (often bi-colored) fused corolla lobes and yellow throat, arising singly from leaf axils, subtended by 2 conspicuous green bracts, each with winged ridge down middle. Giant bindweed (*C. silvatica*) has bracts more overlapping and inflated at bases (see inset) and broader leaf sinus. **IA, IL, IN, MI, MN, MO, OH, WI. Widespread.** *MAH/SN*

Calystegia spithamaea
CONVOLVULACEAE

low false bindweed

Barrens, glades, dry upland woodlands. Spring, summer.

An erect to somewhat reclining perennial, generally not twining, hairy, 1–1½ ft. tall. Leaves alternate, short-petiolate, entire, oblong with a slightly cordate base, about 2–3 in. long × 1½ in. wide. Flowers regular, about 2 in. wide, funnel-shaped with 5 shallow lobes and 2 large overlapping bracts that more or less cover the calyx. Flowers occur singly in leaf axils. The name *Calystegia* ("calyx covering") alludes to the large bracts that conceal the calyx. IA, IL, IN, MI, MN, MO, OH, WI. **Scattered throughout, rare in the western counties.** *SN*

Convolvulus arvensis
CONVOLVULACEAE

field bindweed

Fields, lawns, gardens, roadsides, railroads, waste areas. Spring, summer, fall.

A perennial herbaceous vine to 7 ft. long, twining. Leaves alternate, simple, entire, glabrous, to 2 in. long × 1 in. wide, sagittate, short-petiolate. Flowers regular, funnel-shaped, to 1½ in. wide, with 5 white to pale pink fused corolla lobes and yellow throat, 1–3 on long stalks from leaf axils, each flower with 2 narrow bracts well below flower and not concealing calyx. *Calystegia* species have 2 large, egg-shaped bracts concealing calyx. *Fallopia* species lack showy flowers and have ocreae where petioles attach to stem. IA, IL, IN, MI, MN, MO, OH, WI. **Introduced from Eurasia. Widespread, less frequent in northern MN and northern WI.** *MJH*

Cuscuta glomerata
CONVOLVULACEAE

rope dodder

Prairies, sedge meadows, fens, marshes, swamps, roadsides. Summer, fall.

An annual herbaceous vine to 4 ft. long with orange stems that twine around and "root" into host plants but lack tendrils. Leaves virtually absent. Flowers regular, to ⅛ in. wide, with 5 spreading to reflexed lobes, surrounded by 8–15 tiny bracts, sessile, very numerous and arranged in extremely dense clusters that create rope-like appearance. Fruit a globose capsule, wider than tall. Hosts are most often in Asteraceae, especially sunflowers (*Helianthus*) and goldenrods (*Solidago*). IA, IL, IN, MI, MN, MO, OH, WI. Scattered in the Tallgrass Prairie and Ozark Highlands, with occasional occurrences in southern portion of the Northern Lakes and in the Eastern Forests. *CR*

Cuscuta gronovii
CONVOLVULACEAE

common dodder

Bottomland forests, swamps, gravel bars, wet prairies. Summer, fall.

An annual parasitic vine, branching and twining, yellowish-orange, glabrous, 3–6+ ft. long. Leaves minute, alternate, sessile and scale-like, same color as stems. Flowers regular, about ⅛ in. wide, bell-shaped with 5 spreading and slightly rounded lobes, in dense clusters on lateral branches. Dodders possess only minute amounts of chlorophyll, usually restricted to their developing capsules. Most of the several species of *Cuscuta* look like yellow or orange strands of cooked spaghetti. Many host plants belong to the aster family, but others can be parasitized, including poison ivy. IA, IL, IN, MI, MN, MO, OH, WI. Common in most of the Midwest. *SN*

Ipomoea lacunosa
CONVOLVULACEAE

small white morning-glory

Streambanks, swamp and pond borders, wet fields. Summer, fall.

An annual twining vine, branching, sparsely hairy, growing to 6 ft. long or more. Leaves alternate, petiolate, ovate, 3-lobed or unlobed, with a cordate base, 1–5 in. long and about as wide. Flowers regular, up to 1 in. wide, white (rarely pinkish/purple), shallowly 5-lobed and funnel-shaped, occurring singly or in small clusters from leaf axils. Of somewhat similar range and size is scarlet morning-glory (*I. coccinea*); its reddish-orange flowers are trumpet-shaped. **IA, IL, IN, MO, OH. Occurs mostly in the lower third of the Midwest.** *MAH*

Ipomoea pandurata
CONVOLVULACEAE

wild sweet-potato, man-of-the-earth, manroot

Fields, thickets, forests, roadsides, railroads. Spring, summer, fall.

A perennial herbaceous vine to 30 ft., climbing on vegetation or growing along the ground but lacking tendrils. Leaves alternate, glabrous, simple, entire, cordate, to 6 in. long × 4 in. wide, long-petiolate, often with purplish coloration to leaf blade or at least veins or margins. Flowers regular, funnel-shaped, to 3 in. wide and long, with 5 white fused corolla lobes and pink to purplish-red throat, stalked in few-flowered clusters from leaf axils. The edible tuber can resemble a human figure (hence the common names). **IA, IL, IN, MI, MO, OH. Widespread in the Ozark Highlands, southern Tallgrass Prairie, and Eastern Forests, rare in the Northern Lakes.** *SN*

Cornus canadensis
CORNACEAE

bunchberry, dwarf cornel

Moist forests, swamps, bogs.
Spring, summer.

A colony-forming perennial from woody
rhizome, to 8 in. tall. Leaves in 2 or 3 opposite
pairs, appearing in whorl of 4 (vegetative
plants) to 6 (flowering plants), sometimes with
additional pair below, to 3 in. long × 2 in. wide,
entire, short-petiolate. Flowers regular, green
to white, to ⅛ in. wide, with 4 or 5 reflexed
petals, in solitary, dense cluster to ½ in. wide
at top of plant, made conspicuous by 4 white
petal-like bracts, each to ½ in. long and wide.
Fruit a red, fleshy drupe, in dense cluster. IA,
IL, IN, MI, MN, OH, WI. Widespread in the
Northern Lakes, rare in adjacent portion of
the Tallgrass Prairie. *BS*

Sedum ternatum
CRASSULACEAE

mountain stonecrop

Mesic forests, cliffs, rocky slopes,
stream terraces. Spring, summer.

A succulent perennial, branched, glabrous, low-growing and mat-forming,
semi-evergreen, fertile shoots up to 5 in. tall.
Leaves opposite or commonly in whorls of 3
or 4, flat and rounded or spatulate on vegetative stems, elliptic on fertile shoots, ½–¾ in.
long × ¼–½ in. wide. Flowers regular, about
½ in. wide, with 4 elliptic-lanceolate petals,
arranged along the upper side of laterally
spreading branches atop the fertile stalk.
IA, IL, IN, MI, MO, OH. Fairly widespread
and common, mostly in the Eastern Forests. *MAH*

Echinocystis lobata
CUCURBITACEAE

wild cucumber, wild balsam apple

Floodplain forests, riverbanks. Summer, fall.

An annual climbing vine with tendrils, monoecious, glabrous, reaching 10 ft. or more. Leaves alternate, petiolate, broadly ovate with 5 triangular lobes, 3–4 in. long and about as wide. Flowers regular, about ½ in. wide and somewhat cup-shaped with 6 strap-shaped petals, females few in small axillary clusters, males on elongate racemes. Fruit smooth with scattered thin (but relatively soft) prickles. Bur cucumber (*Sicyos angulatus*) is similar but has hairy stems and leaves, leaves less deeply lobed and smaller, and densely prickled fruits tightly clustered together. IA, IL, IN, MI, MN, MO, OH, WI. **Scattered throughout, less common in the far south.** *SN*

Drosera rotundifolia
DROSERACEAE

round-leaved sundew

Bogs, fens, cedar swamps, wet sand prairies, mossy crevices. Summer.

A carnivorous perennial to 10 in. tall. Leaves basal, green, covered in sticky-tipped red glandular hairs used to trap and digest insects, to ½ in., round or wider than long, on petioles to 2 in. Flowers regular, ⅓ in. wide, 5-petaled, in an unfurling raceme at top of naked stem. Usually found growing in sphagnum moss. Spatulate-leaved sundew (*D. intermedia*; see inset) has leaves more ascending (horizontally spreading in round-leaved sundew), longer than wide, and broadest above middle. IA, IL, IN, MI, MN, OH, WI. **Throughout the Northern Lakes, with few occurrences in the Tallgrass Prairie and Eastern Forests.** *SN*

Andromeda glaucophylla
ERICACEAE

bog rosemary

Bogs, fens, conifer swamps, wet rock crevices. Spring, summer.

A shrub to 2 ft. tall. Leaves alternate, linear with margins rolled under, to 2 in. long × ⅓ in. wide, pointed upward and growing from upper side of branches, dark blue-green above with impressed veins, undersurface dense white-hairy, purplish in winter. Flowers regular, white to pink, broadly urn-shaped, to ¼ in. long, on dangling stalks in clusters of up to 9 at ends of branches. Syns. *A. polifolia*, *A. polifolia* var. *latifolia*. Leaves more linear than those of vegetatively similar *Kalmia polifolia*, with which it grows; latter has leaves with glossy top surface. IL, IN, MI, MN, OH, WI. Nearly restricted to the Northern Lakes. *SN*

Arctostaphylos uva-ursi
ERICACEAE

bearberry, kinnikinick

Dunes, gravel ridges, bare limestone, sandstone glades, woodlands, prairies. Spring, summer.

A creeping shrub to 8 in. tall. Leaves simple, alternate, to 1 in. long × ½ in. wide, short-petiolate, leathery, glossy and evergreen, turning purple in fall, crowded toward ends of branches. Flowers regular, white to pink, bell-shaped with fused (except at very tip) corolla, to ⅜ in. long × ¼ in. wide, in clusters of up to 15 at ends of branches. Fruit a spherical red drupe to ½ in. wide. IA, IL, IN, MI, MN, OH, WI. Throughout much of northern portion of the Northern Lakes, and in counties bordering the Great Lakes; uncommon in bordering Tallgrass Prairie. *SN*

Chamaedaphne calyculata

ERICACEAE

leatherleaf

Bogs, poor fens, swamps, rock crevices bordering Lake Superior. Spring.

A low, colony-forming shrub to 3½ ft. tall. Leaves simple, alternate, leathery, to 1½ in. long × ½ in. wide, short-petiolate, with margins rolled under, upper surface green, lower surface silvery with tiny rufous scales. Flowers regular, drooping, bell-shaped with fused (except at very tip) corolla, to ⅓ in. long × ¼ in. wide, in many-flowered arching racemes to 4 in. long at ends of branches. Fruit a green capsule. The rusty leaf undersides are distinctive. **IL, IN, MI, MN, OH, WI. Well distributed in the Northern Lakes, with few occurrences in the bordering Tallgrass Prairie and Eastern Forests.** *SN*

Epigaea repens

ERICACEAE

trailing arbutus

Sandy savannas, acid woods, conifer swamps, bogs. Spring.

A rusty-hairy, creeping shrub to 3 in. tall. Leaves alternate, evergreen, leathery, deeply reticulate-veined and wrinkly, ovate with cordate base, blunt tip, and wavy margins, to 4 in. long × 2 in. wide. Flowers regular, white to pink, tubular with 5 spreading lobes, to ½ in. wide, several in tight clusters in upper leaf axils. Leaves often have purplish tinge, and flowers have a very pleasant fragrance when fresh. As legend has it, trailing arbutus was one of the first plants seen by the Pilgrims after their first winter in America (hence an alternative common name, mayflower). **IL, IN, MI, MN, OH, WI. Widespread in the Northern Lakes and eastern portion of the Eastern Forests.** *SN*

Gaultheria hispidula
ERICACEAE

creeping snowberry, moxie-plum

Moist forests, mossy rocks, conifer swamps, bogs, thickets. Spring.

A creeping shrub with bristly-hairy mat-forming stems, to 1½ ft. long × 4 in. high (see inset). Leaves simple, alternate, leathery, evergreen, to ⅓ in. long × ¼ in. wide with margins rolled under, on short petioles. Flowers regular, hidden and inconspicuous, partially enclosed in 2 green bracts, hanging below stem individually on short stalks from leaf axils, bell-shaped with 4 pointed lobes, to ⅛ in. long. Fruit an oblong white berry with scattered brown hairs. Crushed leaves and fruit have a wintergreen scent and flavor. MI, MN, OH, WI. Widespread through much of the Northern Lakes, especially northward. PD

Gaultheria procumbens
ERICACEAE

wintergreen, eastern teaberry

Upland woods, savannas, forested and shrubby bogs, in sandy or acidic soils. Summer.

An upright shrub to 8 in. tall. Leaves typically 3–5, simple, alternate, leathery, dark green, glossy, evergreen, to 2 in. long × 1 in. wide, short-petiolate, with sparsely sharp-toothed margins. Flowers regular, 1 to few, dangling from upper leaf axils, urn-shaped, to ⅓ in. long, with 5 short reflexed lobes at tip. Fruit a round red berry. Crushed leaves and fruit have a wintergreen scent and flavor. IL, IN, MI, MN, OH, WI. Widespread in the Northern Lakes and eastern portion of the Eastern Forests, uncommon in remainder of the Eastern Forests and eastern portion of the Tallgrass Prairie. PR

Hypopitys monotropa
ERICACEAE

pine sap

Mesic to dry upland forests. Summer.

An erect perennial, fleshy, hairy, and variously colored, from white to tan to red, 3–8 in. tall. Stems naked except for clasping non-green, alternate, lanceolate scales. Flowers regular, ¼–½ in. long, tubular with 4 or 5 petals in a terminal inflorescence of 4–10 drooping flowers atop a stem of similar color, becoming erect as seed capsules develop and mature. Like Indian pipe, this non-photosynthetic species is a mycoheterotroph; in many treatments the two are placed in *Monotropa*. IA, IL, IN, MI, MN, MO, OH, WI. Widespread but nowhere common, apparently absent from much of western IA and southwestern MN. *SN*

Moneses uniflora
ERICACEAE

one-flowered pyrola, single delight

Moist coniferous forests, swamps, bogs. Spring, summer, fall.

An evergreen perennial to 6 in. tall (often shorter). Leaves restricted to base of plant, opposite to whorled in groups of 2 or 3, shallowly toothed, to ¾ in. long and nearly as wide, waxy, on petioles to as long as leaf blades. Flower regular, nodding, to ¾ in. wide with 4 or 5 spreading petals and 10 stamens around a conspicuous exserted green pistil, solitary and delightful at top of stem. Fruit a round capsule, held erect. MI, MN, OH, WI. Restricted to the Northern Lakes, widespread north, scattered to absent south. *SN*

Monotropa uniflora
ERICACEAE

Indian pipe, ghost flower

Mesic and dry-mesic forests, flatwoods, ravines. Summer, fall.

An erect perennial, unbranched, glabrous, 3–6 in. tall. Leaves bract-like scales, alternate, white, ¼–⅔ in. long × ⅛–¼ in. wide. Flowers regular, about ¾ in. long, with 3–6 petals, usually white but some pinkish, bell- or urn-shaped, nodding initially then becoming vertical with age, occurring singly atop stem. Indian pipe is a mycoheterotroph, lacking chlorophyll and getting nourishment from other plants via subterranean fungi. Indian pipe is generally evident only when it blooms; otherwise it lives contentedly and entirely underground. IA, IL, IN, MI, MN, MO, OH, WI. Widespread over most of the region. SN

Orthilia secunda
ERICACEAE

one-sided pyrola, sidebells wintergreen

Moist forests, bogs, in moss. Summer.

An evergreen perennial to 9 in. tall. Leaves restricted to base of plant, appearing whorled, sharply toothed, to 2 in. long × 1¼ in. wide, shiny, short-petiolate. Flowers regular, to ¼ in. long, urn-shaped with exserted stamens and style, dangling like bells from solitary, arching stem, each subtended by small green scale-like bract, several to numerous in terminal raceme. Fruit a depressed capsule to ¼ in. long, dangling from erect stem but not necessarily appearing 1-sided. IA, IL, IN, MI, MN, OH, WI. Restricted to the Northern Lakes and adjacent portions of the Tallgrass Prairie, more frequent northward. MJH

Pyrola americana
ERICACEAE

round-leaved pyrola, American
wintergreen

Forests, cedar swamps, fens. Summer.

A glabrous evergreen perennial to 1½ ft. tall.
Leaves basal, round, entire, to 1½ in. long
and wide, shiny, dark green, thick-textured,
veiny, on petioles about as long as leaves.
Flowers regular, downward-facing, to ¾ in.
wide, with 5 spreading petals surrounding
exserted stamens and curved style, subtended
by scale-like bract, on spreading stalks, 6–20
in raceme atop solitary naked stem. Fruit a
dangling, segmented capsule. IL, IN, MI, MN,
OH, WI. **Scattered in eastern portion of the
Eastern Forests and the Northern Lakes and
adjacent portions of the Tallgrass Prairie.** *SN*

Pyrola elliptica
ERICACEAE

shinleaf

Moist to dry acid forests and woodland
slopes. Summer.

An erect evergreen perennial, unbranched,
glabrous, 6–12 in. tall. Leaves in a basal rosette,
petiolate, blade elliptic to oblong, longer than
petiole, thin with relatively dull surface, 1–3
in. long × ½–1½ in. wide. Flowers regular,
⅓–½ in. wide, each with 5 nodding, oval,
rather waxy petals and a protruding, downcurved style that is upcurved at tip, several in
an elongated raceme. Flowers fragrant. Leaf
blades of green-flowered pyrola (*P. chlorantha*)
are rarely more than 1–1½ in. long and mostly
shorter than the petioles; it is restricted to the
far northern counties. IA, IL, IN, MI, MN,
OH, WI. **Most occurrences in the Northern
Lakes.** *PS*

Rhododendron groenlandicum
ERICACEAE

Labrador-tea

Bogs, rich conifer woods and swamps, lakeshores, interdunal swales, rock crevices. Spring.

A shrub to 3 ft. tall. Leaves evergreen, leathery, alternate, to 2¼ in. long × ⅔ in. wide, dark green above with wrinkled texture and margins rolled under, white-woolly below (becoming rusty-hairy by second year). Flowers regular, to ⅓ in. wide, with 5 spreading petals, the 5–10 stamens exserted and surrounding pistil, on long stalks from cone-like bud, forming tight, dome-shaped clusters at ends of branches in upper part of plant; flower stalks drooping with age and in fruit. Fruit an elliptic capsule with persistent style. Syn. *Ledum groenlandicum*. **MI, MN, OH, WI. Widespread in the Northern Lakes.** *SN*

Vaccinium angustifolium
ERICACEAE

lowbush blueberry

Acidic sand savannas, rocky upland forests, moist sand flats. Spring, summer.

A low-growing shrub, multi-branched, mostly glabrous, forming dense colonies, 4–24 in. tall. Leaves alternate, sessile, narrowly elliptic to lanceolate bearing fine teeth on margins, ½–1½ in. long × ¼–¾ in. wide. Flowers regular, about ¼ in. long, white to pinkish, bell-shaped with 5 recurved lobes, nodding in racemes at branch tips. Fruit a blue berry. This species is often sought for its tasty fruits. It is similar in appearance to Canada blueberry (*V. myrtilloides*), but the latter's twigs and leaves (undersurfaces) are densely hairy. **IA, IL, IN, MI, MN, OH, WI. Occurs primarily in the Northern Lakes.** *SN*

Vaccinium cespitosum
ERICACEAE

dwarf bilberry

Forest openings on rocky or sandy ground. Spring.

A low, mat-forming shrub to 8 in. tall. Leaves simple, alternate, thin-textured, to 1 in. long × ½ in. wide, broadest above middle and tapering to sessile base, minutely toothed, gland-dotted beneath. Flowers regular, pink to white, nodding, urn-shaped with 5 short spreading lobes at tip, to ¼ in. long and nearly as wide, solitary on short stalks from lower leaf axils. Fruit an edible, round, dark blue berry with waxy coating, to ⅓ in. wide, lacking the persistent sepals of blueberries. **MI, MN, WI. Scattered to rare in western portion of the Northern Lakes.** *PD*

Vaccinium macrocarpon
ERICACEAE

large cranberry

Swamps, bogs, fens, interdunal swales. Summer.

A mat-forming shrub, 6 in. tall. Leaves alternate, evergreen, to ⅔ in. long, blunt or notched at tip, sessile to short-petiolate, with margins rolled under, dark green above, whitened beneath. Flowers regular, white to pink, nodding, with 4 narrow, recurved petals and exserted stamens and pistil, ⅜ in. wide, on pendulous stalks from near middle of stem; bracts 2, green. Fruit a round red berry, ⅔ in. wide (see inset). Small cranberry (*V. oxycoccos*) has more pointed leaves, flowers toward top of plant, and reddish bracts. **IL, IN, MI, MN, OH, WI. Frequent in the Northern Lakes, scattered to rare in eastern portion of the Eastern Forests.** *CR/SN*

Vaccinium pallidum
ERICACEAE

hillside blueberry

Acidic dry forests and sand savannas. Spring, summer.

A mostly low-growing shrub, highly branched, mostly glabrous, normally 2 ft. tall or less, but also taller, colonial. Leaves alternate, sessile or short-petiolate, ovate to broadly elliptic, margins usually entire, pale green or glaucous beneath, 1–2 in. long × ½–1 in. wide. Flowers regular, about ¼ in. long, white to pinkish, bell-shaped with 5 recurved lobes, nodding in racemes at branch tips. Fruit a blue berry. Compare with lowbush blueberry (*V. angustifolium*). IL, IN, MI, MO, OH, WI. Mostly in the Ozark Highlands and Eastern Forests. *MJH*

Croton glandulosus
EUPHORBIACEAE

sand croton, tooth-leaved croton

Prairies, glades, old fields, open woods, roadsides, railroads. Summer, fall.

A stellate-hairy, monoecious annual to 1½ ft. tall, with grayish to brownish stems. Leaves alternate, coarsely toothed, short-petiolate, to 2½ in. long × 1 in. wide, with 2 saucer-shaped glands where blade attaches to petiole. Flowers regular; male flowers on stalks elevated above sessile female flowers; male to ⅛ in. wide with 4 or 5 white petals and exserted stamens; female green, lacking petals, to ⅛ in. wide with 3 styles. IA, IL, IN, MI, MN, MO, OH, WI. Primarily in the Ozark Highlands and southern portion of the Tallgrass Prairie, with scattered occurrences in the Northern Lakes (especially around southern tip of Lake Michigan) and Eastern Forests. *MAH*

Croton michauxii
EUPHORBIACEAE

common rushfoil

Sandstone glades, exposed sand flats, exposed clay. Summer.

An erect annual, monoecious, branched, hairy, 4–16 in. tall. Leaves alternate, short-petiolate, lanceolate-linear, entire, upper surface covered with stellate hairs, lower surface with brown scales as well as stellate hairs, ½–1½ in. long × ⅛–⅓ in. wide. Flowers minute, male with 5 sepals and petals, female with sepals but lacking petals, in short axillary and terminal spikes. Fruit an achene without prickles (var. *ellipticus*; pictured). The less-common var. *michauxii* has narrower leaves and prickle-tipped fruits. Syn. *Crotonopsis elliptica*. IA, IL, IN, MO, OH. **Principally in the Ozark Highlands and southern third of IL.** *ST*

Croton monanthogynus
EUPHORBIACEAE

prairie tea

Glades, blufftops, gravel roadsides. Summer.

An erect annual, monoecious, branched above, with stellate hairs, 4–12 in. tall. Leaves alternate, petiolate, ovate-oblong, entire, 1–2 in. long × ½–1 in. wide. Flowers tiny, males 5–10 with 5 sepals and petals, female commonly single with sepals and 2 deeply branched styles but no petals, in short axillary and terminal spikes. Fruit a capsule, typically 1-seeded. Woolly croton (*C. capitatus*) has lanceolate leaves that are longer (up to 3 in.) and female flowers with 3 branched styles; its fruit is 3-seeded. IA, IL, IN, MI, MO, OH, WI. **Mostly in Ozark Highlands and southern portions of the Eastern Forests.** *MAH*

Euphorbia corollata
EUPHORBIACEAE

flowering spurge

Prairies, old fields, glades, savannas, woodland openings, roadsides, railroads. Summer.

A monoecious perennial to 3 ft. tall. Leaves alternate but whorled immediately below inflorescence, entire, to 2½ in. long × ½ in. wide, sessile or nearly so, containing milky sap. Flowers to ¼ in. wide, lacking true petals but with usually 5 spreading white petal-like appendages, each with a green gland at base, branched in large, showy, terminal clusters. Fruit a stalked 3-parted capsule. Vegetative plants look similar to bastard toadflax (*Comandra umbellata*), but the latter lacks milky sap in the foliage. IA, IL, IN, MI, MN, MO, OH, WI. Widespread except in western MN and northern MI. *MAH*

Euphorbia maculata
EUPHORBIACEAE

spotted spurge, prostrate spurge

Prairies, fields, roadsides, pavement cracks, waste areas. Spring, summer, fall.

A monoecious or dioecious annual, to 1 in. tall, forming patches to 18 in. wide. Stems pink, hairy. Leaves opposite, scarcely toothed, to ¾ in. long, often with central reddish spot, containing milky sap. Flowers to ⅛ in. wide, with 4 spreading white to pinkish petal-like lobed appendages with green to red basal glands, few in axillary clusters. Fruit a stalked, hairy, 3-parted capsule. Thyme-leaved spurge (*E. serpyllifolia*) and eyebane (*E. nutans*) are glabrous; seaside sandmat (*E. polygonifolia*), bordering the Great Lakes, has entire, parallel-margined leaves. IA, IL, IN, MI, MN, MO, OH, WI. Widespread and common, less so in northern MI and northwestern MN. *MAH*

Euphorbia marginata
EUPHORBIACEAE

snow-on-the-mountain

Prairies, fields, roadsides, railroads,
waste areas. Summer, fall.

A branched, monoecious annual to 3 ft. tall.
Leaves alternate but opposite or whorled
on flowering stems, entire, sessile or
short-petiolate, to 3½ in. long × 1 in. wide, containing milky sap, light green, sometimes with
white margins, white margins wider on bracts
and leaves of flowering stems. Flowers to ⅜ in.
wide, with usually 5 spreading white petal-like
appendages, each with green gland at base, in
flat clusters at tips of branches in upper part
of plant. Fruit a stalked 3-parted capsule. The
conspicuous white-margined pale green bracts
provide most of the visual interest. IA, IL, IN,
MI, MN, MO, OH, WI. Native in western portion of Midwest. Scattered throughout. *MJH*

Baptisia alba
FABACEAE

white wild indigo, white false indigo

Prairies, glades, thickets, savannas.
Spring, summer.

A bushy, gray-green perennial to 5 ft. tall.
Leaves alternate, short-petiolate, with 2 small,
pointed, leaf-like stipules at base of petioles,
compound with 3 sessile leaflets to 2 in. long ×
¾ in. wide, glabrous, entire. Flowers irregular
with typical pea family shape, to 1 in. long,
numerous in erect terminal racemes to 18 in.
long. Fruit a black, oblong, inflated legume.
Midwestern plants are var. *macrophylla*. Syns.
B. leucantha, *B. lactea*. IA, IL, IN, MI, MN, MO,
OH, WI. Common in central and southern
Tallgrass Prairie and the Ozark Highlands,
with fewer occurrences in the Northern Lakes
and Eastern Forests. *SN*

Baptisia bracteata
FABACEAE

cream wild indigo

Prairies, dry woodlands, glades.
Spring, summer.

A perennial with stems erect or spreading, branched and hairy, 1½–2½ ft. tall.
Leaves alternate, divided into 3 leaflets,
short-petiolate, leaflets oblanceolate, 1–3½ in.
long × ⅓–1 in. wide, hairy, stipules 2, leafy and
persistent. Flowers irregular with typical pea
family shape, whitish to cream-colored, about
1 in. long, with pedicels bearing persistent
leafy bracts at base. Inflorescence a densely
flowered horizontal or reclining raceme. Fruit
a legume. Syns. *B. leucophaea*, *B. bracteata*
var. *leucophaea*. IA, IL, IN, MI, MN, MO, WI.
Occurs principally in the Tallgrass Prairie
and Ozark Highlands. *SN*

Dalea candida
FABACEAE

white prairie clover

Dry to moist prairies, savannas, glades,
woodland openings. Summer.

A perennial to 3 ft. tall. Leaves pinnately compound, alternate, glabrous, to 7 in. long, with
up to 9 elliptic leaflets. Flowers irregular with
typical pea family shape, numerous, to ¼ in.
wide, in cylindrical spikes to 3 in. long × ¾
in. wide at top of plant, each spike blooming
from bottom up. Syn. *Petalostemum candidum*.
Foxtail prairie clover (*D. leporina*) has more
numerous (15–35) leaflets that are smaller
and more tightly arranged on leaf; it is much
less common, occasional only in the Tallgrass
Prairie. IA, IL, IN, MN, MO, WI. Widespread
in the Ozark Highlands and Tallgrass Prairie,
uncommon in western portion of the Northern Lakes and Eastern Forests. *SN*

Desmanthus illinoensis

FABACEAE

Illinois bundleflower, prairie mimosa

Prairies, glades, floodplain meadows, roadsides, railroads. Summer.

A perennial to 4 ft. tall. Leaves twice pinnately compound, alternate, to 8 in. long with numerous leaflets to ⅛ in. long. Flowers regular, to ⅛ in. wide, with 5 inconspicuous petals and 5 long-exserted stamens with yellow anthers and conspicuous white filaments, in dense spherical clusters to ½ in., on stalks to 3 in. from upper leaf axils. Fruit a wavy reddish-brown legume, held persistently in dense, nearly round clusters. The leaves of this plant fold up at night and in bright sun. IA, IL, IN, MI, MN, MO, OH, WI. Widespread in the Ozark Highlands and southern portion of the Tallgrass Prairie, scattered elsewhere. *MJH*

Glycyrrhiza lepidota

FABACEAE

American licorice

Prairies, fields, roadsides, railroads, waste areas. Summer.

An upright perennial to 3½ ft. tall. Leaves alternate, short-petiolate, with 11–19 pointed, glabrous, entire, sessile leaflets to 1½ in. long × ½ in. wide. Flowers irregular with typical pea family shape, white to pale yellow, to ½ in. long, numerous in stalked, dense spikes usually to 3 in. long from leaf axils. Fruit a legume to ½ in. long, covered in hooked bristles, becoming coppery brown, reminiscent of those of unrelated cocklebur (*Xanthium strumarium*). Glycyrrhizin in the roots gives them a licorice odor and flavor. IA, IL, IN, MN, MO, WI. Primarily in northwestern portion of the Tallgrass Prairie, rare elsewhere. *PR*

Hylodesmum pauciflorum
FABACEAE

few-flowered tick-trefoil

Rich mesic upland and lowland forests.
Summer.

An erect or ascending to reclining perennial,
unbranched, hairy, 8–18 in. tall. Leaves alternate, petiolate, trifoliate, lateral leaflets ovate,
terminal leaflet diamond-shaped and 2–3½
in. long × 1½–2½ in. wide. Flowers irregular
with typical pea family shape, about ⅓ in.
wide, in mostly axillary racemes of fewer than
10 flowers extending above the plant. Fruit
a flattened, jointed, hairy legume with 1–3
segments. Syn. *Desmodium pauciflorum*. IL, IN,
MO, OH. Mostly confined to the Eastern Forests and Ozark Highlands. *MAH*

Lathyrus ochroleucus
FABACEAE

pale vetchling, cream pea

Forests and forest openings, savannas,
riverbanks, rocky ridges. Spring, summer.

An upright perennial herbaceous vine to 3 ft.
tall. Leaves alternate, petiolate, with 2 leafy
stipules to 1 in. long at base, pinnately compound with 3–5 pairs of entire leaflets to 2 in.
long × 1¼ in. wide, terminating in branched
tendril. Flowers irregular with typical pea
family shape, to ¾ in. wide, in racemes of up to
12 from leaf axils. Fruit a flattened legume to
3 in. long. IA, IL, IN, MI, MN, OH, WI. Widespread in the Northern Lakes and adjacent
portions of the Tallgrass Prairie, absent from
the Ozark Highlands, rare elsewhere. *CB*

Lespedeza capitata
FABACEAE

round-headed bush-clover

Prairies, barrens, glades, savannas. Summer.

An erect perennial, unbranched except for inflorescence, with appressed silvery hairs, 2–5 ft. tall. Leaves alternate, short-petiolate, with 3 narrowly elliptic to oblong leaflets, the leaflets 1–2 in. long × ¼–½ in. wide. Flowers irregular with typical pea family shape, about ⅓ in. long, with purple streaking at base of banner petal. Inflorescence of compact, short-stalked and rounded heads nested in leaf axils. Fruit a legume. Hairy bush-clover (*L. hirta*) has nearly round leaflets and flower heads on stalks extending well beyond the main leaves. IA, IL, IN, MI, MN, MO, OH, WI. Widespread mostly throughout the Tallgrass Prairie and Ozark Highlands. *SN*

Lespedeza cuneata
FABACEAE

sericea lespedeza

Fields, pastures, roadsides, utility line rights-of-way. Summer, fall.

An erect perennial, branched and shrub-like, stem ridged, with appressed silvery hairs, 2–5 ft. tall. Leaves alternate, short-petiolate, with 3 leaflets, these narrowly oblong, slightly wider toward a somewhat straight-edged tip, ¾–1 in. long × ¼–⅓ in. wide. Flowers irregular with typical pea family shape, about ⅓ in. long, banner petal with purple streaking at base. Inflorescences of 1–4 flowers on short stalks from leaf axils. Fruit a legume. Sericea lespedeza is a highly invasive weed. IA, IL, IN, MI, MN, MO, OH, WI. Introduced from Asia. Occurs primarily in the lower Midwest. *MJH*

Lespedeza leptostachya
FABACEAE

prairie bush-clover

Prairies. Summer, fall.

A slender, silver velvety-hairy perennial to 3 ft. tall. Leaves compound with 3 short-stalked leaflets to 1¾ in. long × ⅓ in. wide, alternate, on short petioles. Flowers irregular with typical pea family shape, numerous, white to pale pink with reddish throat, to ⅓ in. wide, loosely arranged in spikes to 3 in. long on stalks from upper leaf axils. Fruit a small, densely hairy, oval, pointed legume. Round-headed bush-clover (*L. capitata*) is much more common and widespread, with broader leaflets and shorter and more round spikes on shorter stalks. IA, IL, MN, WI. A very rare Midwest endemic in the Tallgrass Prairie and Northern Lakes. *PD*

Melilotus albus
FABACEAE

white sweet clover

Fields, pastures, roadsides, waste areas. Summer.

An erect biennial, branched, glabrous, 3–5 ft. tall. Leaves alternate, petiolate, compound, divided into 3 oval to oblanceolate leaflets with finely saw-toothed margins, ⅓–1½ in. long × ¼–⅔ in. wide. Flowers irregular with typical pea family shape, ⅛–¼ in. long, occurring on erect, narrow racemes. Fruit is a wrinkled black legume. Except for flower color and slight differences in flower size this species is almost identical to yellow sweet clover (*M. officinalis*). Both are highly invasive. IA, IL, IN, MI, MN, MO, OH, WI. Introduced from Eurasia. Abundant, probably in every county. *SN*

Trifolium reflexum
FABACEAE

buffalo clover

Rocky wooded slopes, acidic glades. Spring, summer.

An erect to ascending annual or biennial, branched, glabrous to hairy, 6–12 in. tall. Leaves alternate, petiolate with green leafy stipules, compound with 3 broadly elliptic to oblong-obovate leaflets, these finely toothed, ¾–1½ in. long × ¼–½ in. wide. Flowers irregular with typical pea family shape, tubular, white to deep pink, calyx lobes 2–4 times length of tube, numerous in heads 1–1½ in. wide from upper leaf axils. Fruit a legume, hidden within withered flower, on strongly reflexed pedicels. The rare running buffalo clover (*T. stoloniferum*) is similar but stoloniferous. IA, IL, IN, MO, OH. **Mostly in Ozark Highlands. Uncommon.** *SN*

Trifolium repens
FABACEAE

white clover

Lawns, fields, roadsides. Spring, summer, fall.

A perennial with creeping stolons, usually glabrous, 3–10 in. tall. Leaves alternate, petiolate with membranous stipules clasping the stolon, compound with 3 broadly elliptic to rounded leaflets, finely toothed, and some with a whitish chevron, ½–1 in. long and about as wide. Flowers irregular with typical pea family shape, calyx lobes about equal in length to tube, numerous in heads ½–¾ in. wide on leafless flower stalks arising from the stolons. Fruit a legume hidden in withered flower. The native running buffalo clover (*T. stoloniferum*), also stoloniferous, possesses a pair of leaves on the flowering stalk. IA, IL, IN, MI, MN, MO, OH, WI. **Introduced from Eurasia. Likely in every county.** *SN*

Vicia caroliniana
FABACEAE

wood vetch, pale vetch

Dry woodlands, rocky slopes, bluffs, clearings, on acid soils. Spring.

A perennial climbing vine, branched, sparsely to moderately hairy, 1–4 ft. long. Leaves alternate, petiolate with stipules, pinnately compound with a terminal tendril and 5–14 pairs of elliptic to oblong-lanceolate, finely hairy leaflets ⅓–¾ in. long × ¼ in. wide. Flowers irregular with typical pea family shape, ¼–½ in. long, tubular, keel tip may have a tinge of blue, in long 1-sided racemes from leaf axils. Fruit a flattened legume. IL, IN, MI, MO, OH, WI. **Occurs mostly in the southern Northern Lakes plus the Ozark Highlands and unglaciated areas of IN and OH.** *CR*

Bartonia virginica
GENTIANACEAE

yellow screwstem

Moist, often mossy sand and peaty hummocks. Summer.

An erect annual with twisted stems, glabrous, yellowish-green, 3–8 in. tall. Leaves mostly opposite or nearly so, scale-like, scarcely measurable. Flowers regular, about ⅛ in. long, with 4 oblong, whitish petals pressed against the ovary within. Inflorescence a terminal panicle with opposite branching. This species is easily overlooked due to its small size and lack of sizable leaves. The similar twining screwstem (*B. paniculata*) differs in having mostly alternate leaves and stem and branches that are sometimes tinted with purple; it is much rarer in our region. IL, IN, MI, MN, MO, OH, WI. **Found primarily in the Northern Lakes.** *MAH*

Gentiana alba
GENTIANACEAE

white gentian, cream gentian

Prairies, savannas, glades, barrens.
Summer, fall.

An erect perennial, unbranched, glabrous, 1–2 ft. tall. Leaves opposite, sessile,
ovate-lanceolate, entire, 2–4 in. long × 1–2 in.
wide. Flowers regular, 1½–2 in. long, tubular
with 5 lobes closed at tip to slightly spreading,
yellowish-white with green veins, clustered
tightly atop stem and occasionally in upper
leaf axils. White gentian is a favorite food of
herbivores, most likely deer, as evidenced by
commonly seen nipped stems. Syn. *G. flavida*.
IA, IL, IN, MI, MN, MO, OH, WI. Mostly
in the Tallgrass Prairie and Ozark Highlands. *SN*

Obolaria virginica
GENTIANACEAE

pennywort

Rich mesic forests. Spring.

An erect perennial, typically unbranched,
glabrous, and somewhat fleshy, 1–6 in. tall.
Leaves opposite, sessile, obovate with blunt
tips, purplish-green, ½–¾ in. long and about
as wide. Flowers regular, about ½ in. long,
funnel-shaped with 4 white to purplish
oblong-obovate lobes, 1–3 in the leaf axils.
This species, the only member of its genus
in the world, is not easily seen. Not only is it
small, its color blends into the forest floor, and
although it does possess some chlorophyll, it
is thought to receive most of its nourishment
from mycorrhizal fungi. IL, IN, MO, OH.
Confined primarily to the southern and far
eastern areas of the Eastern Forests. *PR*

Geranium carolinianum
GERANIACEAE

Carolina cranesbill

Fields, pastures, roadsides, waste areas.
Spring, summer.

An erect to sprawling annual, branched, hairy,
mostly non-glandular, 6–18 in. tall. Leaves
opposite, petiolate, rounded, palmately
divided into 5+ cleft oblong to obovate lobes,
1–3 in. long and wide. Flowers regular, ¼–⅓
in. wide, white to pale pink, with 5 petals that
are obovate with shallow notched tips, on
pedicels equal to or shorter than calyx, in an
umbel-like inflorescence at branch tips. Fruit
narrow with a long "beak" (hence cranesbill).
Although native, this geranium is quite weedy
and more commonly encountered in disturbed than natural areas. IA, IL, IN, MI, MN,
MO, OH, WI. Widespread and common, less
so in the Northern Lakes. *MJH*

Ornithogalum umbellatum
HYACINTHACEAE

star-of-Bethlehem

Mesic upland forests, floodplain forests,
lawns, roadsides. Spring.

An erect perennial from poisonous bulbs,
glabrous, 5–12 in. tall. Leaves in a basal rosette,
linear, somewhat fleshy, with a white central
stripe, 4–12 in. tall. Flowers regular, ¾–1 in.
wide, with 6 spreading, lanceolate to oblong
tepals possessing a broad green stripe on the
outer surface, in a flat-topped, umbel-like
raceme. Flowers open for only a few daylight
hours. This attractive plant has escaped cultivation and is spreading into natural areas.
IA, IL, IN, MI, MN, MO, OH, WI. Introduced
from Europe. Widespread but absent from
most of the upper Midwest. *MJH*

Sisyrinchium albidum
IRIDACEAE

white blue-eyed grass, common blue-eyed grass

Prairies, glades, fields, savannas, open forests. Spring, summer.

A gray-green perennial to 1 ft. tall. Leaves basal in fan-like arrangement, entire, sword-shaped, erect, to 10 in. long × ⅛ in. wide. Flowers regular, to ½ in. wide, with 6 spreading white to pale blue tepals with yellow throat and abrupt tips, on stalks to ½ in. long, up to 6 in umbel from 2 bracts to 1 in. long that are partially contained within sessile leaf-like bract to 3 in. long at top of naked stem, outer bract barely fused at base. IA, IL, IN, MI, MO, OH, WI. Widespread in eastern portion of the Tall-grass Prairie and southern portion of the Northern Lakes, scattered elsewhere. *DT*

Triglochin maritima
JUNCAGINACEAE

common bog arrowgrass, seaside arrowgrass

Sedge meadows, fens, bogs, shores, beaches, pannes. Spring, summer.

A slender, inconspicuous perennial to 2½ ft. tall. Leaves mostly basal, entire, linear, sessile, to 16 in. long, filiform, erect or nearly so. Flowers regular, to ⅛ in. long, with 6 tepals surrounding 6 feathery stigmas (see inset), numerous in terminal raceme to 16 in. long on naked stalk. Fruit 6-parted, oblong. Slender bog arrowgrass (*T. palustris*) is more slender and has narrowly cylindrical flowers with 3 feathery stigmas and 3-parted, nearly parallel-margined fruit. Both species easily overlooked. IA, IL, IN, MI, MN, OH, WI. Uncommon, primarily in proximity to the Great Lakes, frequent throughout much of MN. *MJH*

Blephilia hirsuta
LAMIACEAE

hairy wood mint, hairy pagoda-plant

Rich mesic forests, stream terraces, floodplains. Spring, summer.

An erect perennial, branched above, with spreading hairs, 2–3 ft. tall. Leaves opposite, petiolate, ovate-lanceolate with small teeth, variously hairy, 3–4 in. long × 1 in. wide. Flowers irregular, ¼–½ in. long, tubular, upper lip 2-lobed and lower lip 3-lobed, clustered in 1-in.-wide heads spaced in 2–4 tiers atop stem (like a pagoda). Corolla and calyx tubes long-hairy, floral bracts linear-lanceolate. The leaves of this species have a strong but pleasant minty scent when bruised. IA, IL, IN, MI, MN, MO, OH, WI. Scattered throughout, but rare to mostly absent in far northern and western counties. *MAH*

Leonurus cardiaca
LAMIACEAE

motherwort

Fields, thickets, disturbed woods, floodplains, around buildings, waste areas. Spring, summer, fall.

A square-stemmed perennial to 5 ft. tall. Leaves opposite, to 4 in. long × 3 in. wide, variable, deeply 5-lobed and coarsely toothed below, smaller and unlobed with pair of coarse teeth above, deeply veined with wrinkled appearance, long-petiolate. Flowers irregular, white to pale pink with purplish spots inside, long-hairy, ⅓ in. long, tubular and flaring with 1 upper lobe and 3 lower lobes (2 lateral much reduced), subtended by calyx with 5 sharp-pointed spiny lobes (making inflorescence bristly), sessile in whorls of up to 12 in leaf axils. IA, IL, IN, MI, MN, MO, OH, WI. Introduced from Eurasia. Scattered. *PD*

Lycopus americanus
LAMIACEAE

American water horehound, American bugleweed

Lowland forests, shallow swamps, wet prairies, marshes. Summer.

An erect to ascending perennial, unbranched to branched, glabrous to hairy, stems 4-angled, 1–2½ ft. tall. Leaves opposite, short-petiolate, lanceolate, at least the lower ones deeply and irregularly pinnately lobed, 1–4 in. long × ½–1½ in. wide. Flowers nearly regular, ⅛–¼ in. long, short-tubular with 4 lobes and 5 narrowly triangular, long-pointed sepals, several in compact axillary whorls. Similar species include stalked water horehound (*L. rubellus*) and rough water horehound (*L. asper*). They differ from *L. americanus* in having 5-lobed flowers, and sessile leaves, respectively. IA, IL, IN, MI, MN, MO, OH, WI. **Widespread, perhaps in every county.** *SN*

Lycopus virginicus
LAMIACEAE

Virginia water horehound

Lowland forests, shallow swamps, slough and stream borders, ditches. Summer.

An erect to ascending perennial, branched, glabrous to hairy, stems 4-angled, 1–2½ ft. tall. Leaves opposite, short-petiolate, lance-ovate to elliptic, coarsely but somewhat regularly toothed, 2–4 in. long × ½–1½ in. wide. Flowers nearly regular, ⅛–¼ in. long, short-tubular with 4 lobes and 5 rather blunt, broadly triangular sepals, several in compact axillary whorls. Northern bugleweed (*L. uniflorus*) shares the trait of having broadly triangular sepals, but flowers are 5-lobed. It typically has narrower leaves as well. IA, IL, IN, MI, MN, MO, OH, WI. **Widespread and common, less so northward.** *MAH*

Marrubium vulgare
LAMIACEAE

common horehound

Fields, yards, old homesites, waste areas.
Summer, fall.

A hairy, gray-green perennial to 1½ ft. tall.
Leaves opposite, simple, to 2 in. long × 1 ⅗ in.
wide, toothed and wavy-margined, deeply
veined, short-petiolate. Flowers irregular, ⅖ in.
long, tubular and flaring with upper lip shallowly lobed and lower lip with 2 narrow, slightly
spreading lateral lobes and broader, shallowly
lobed central lobe, subtended by calyx with
10 sharp-pointed spiny lobes, sessile in dense
whorls in leaf axils. Hemp-nettle (*Galeopsis tetra-
hit*) has fewer, larger, white to lavender, 4-lobed
flowers with hood-like upper lobe in axillary
clusters and less deeply veined leaves. IA, IL,
IN, MI, MN, MO, OH, WI. **Introduced from
Eurasia. Scattered, less frequent to absent
northward.** *ME*

Mentha arvensis
LAMIACEAE

wild mint, field mint

Marshes, wet prairies, fens, swamps,
wet shores, floodplains, thickets, ditches,
disturbed wet areas. Summer, fall.

A square-stemmed upright to sprawling
perennial to 1½ ft. tall. Leaves opposite,
simple, sharply toothed, strongly veined,
short-petiolate, to 2½ in. long × 1 in. wide,
potently pleasantly fragrant (minty) when
bruised. Flowers irregular, to ⅛ in. long, white
to pale pink or lavender (sometimes white with
lavender stripes or spots), tubular and flaring
with 2-lobed upper lip and 3-lobed lower lip,
densely packed in globular clusters in leaf axils
throughout plant. North American plants are
var. *canadensis* (syn. *M. canadensis*). IA, IL, IN,
MI, MN, MO, OH, WI. **Widespread and common, except in MO and southern IL, where
scattered.** *SN*

Mentha spicata
LAMIACEAE

spearmint

Moist fields, shores, thickets, ditches, roadsides, disturbed sites. Summer, fall.

A square-stemmed upright perennial to 2 ft. tall. Leaves opposite, simple, toothed, deeply veined and appearing wrinkled, sessile or nearly so, to 2½ in. long × 1½ in. wide, potently pleasantly fragrant (spearminty) when bruised. Flowers irregular, to ⅛ in. long, white to pale pink, tubular, 5-lobed, densely packed in terminal spike to 6 in. long × ⅖ in. wide, often with 2 smaller lateral spikes, spikes made up of tight verticels. Peppermint (M. ×piperita) has petiolate leaves and shorter, wider spikes, and bruised leaves smell like peppermint. IA, IL, IN, MI, MN, MO, OH, WI. Introduced from Eurasia. Scattered, more widespread in eastern portion of the Midwest. *AH*

Monarda punctata
LAMIACEAE

dotted horsemint, spotted beebalm

Sand prairies, fields, savannas, dunes, roadsides, railroads. Summer, fall.

A square-stemmed perennial to 3 ft. tall. Leaves opposite, toothed (to entire above), short-petiolate, deeply veined, to 3 in. long × ¾ in. wide, fragrant (like oregano) when bruised; upper often with whitish or pinkish, at least at base. Flowers irregular, surprisingly inconspicuous, creamy yellow with purple spots, to 1 in. long, tubular with unlobed arching upper lip and 3-lobed lower lip, sessile in whorls in upper part of plant, subtended by whorled white to pinkish bracts smaller than leaves, often entire. IA, IL, IN, MI, MN, MO, OH, WI. Common and widespread in portions of the Tallgrass Prairie and Northern Lakes, widely scattered and uncommon elsewhere. *SN*

Nepeta cataria
LAMIACEAE

catnip, catmint

Fields, pastures, open woodlands, shores, thickets, fencerows, gardens, parking lots, roadsides, railroads, disturbed sites. Summer, fall.

A gray-green, hairy, square-stemmed perennial to 4 ft. tall. Leaves opposite, simple, bluntly toothed, deeply veined and appearing wrinkled, short-petiolate into shallowly cordate base, to 4 in. long × 2 in. wide, potently fragrant (distinctively sweetly minty; cats love it!) when bruised. Flowers irregular, to ⅓ in. long, tubular with 2 spreading lips, upper lip shallowly 2-lobed, lower lip 3-lobed with middle lobe largest, in tight, densely flowered whorls at ends of branches and top of plant. IA, IL, IN, MI, MN, MO, OH, WI. Introduced from Eurasia. Widespread and common, less frequent southward. *MJH*

Pycnanthemum pilosum
LAMIACEAE

hairy mountainmint

Prairies, glades, fields, savannas, forest openings, thickets. Summer.

A hairy gray-green perennial to 4 ft. tall. Leaves opposite, entire, sessile to short-petiolate, to 3 in. long × ¾ in. wide, fragrant (minty) when bruised. Flowers irregular, white with purple spots, ¼ in. wide, tubular with unlobed upper lip and 3-lobed lower lip, stalkless in tight 1-in. heads, forming terminal dome to flat-topped cluster. Whorled mountainmint (*P. verticillatum*) is very similar but has leaf undersides with hairs only on veins (hairy throughout in hairy mountainmint). IA, IL, IN, MI, MO, OH. Widespread in the Ozark Highlands and southern portion of the Tallgrass Prairie, scattered in the Eastern Forests and southern portion of the Northern Lakes. *SN*

Pycnanthemum pycnanthemoides
LAMIACEAE

southern mountainmint

Dry upland woodlands, clearings, forest edges. Summer.

An erect perennial, square-stemmed, branched above, hairy, 3–5 ft. tall. Leaves opposite, petiolate, lance-ovate to oblong, toothed, the upper ones conspicuously whitened with dense hairs, strongly mint-scented when bruised, 1½–4 in. long × ½–1½ in. wide. Flowers irregular, about ½ in. long, tubular, 2-lipped, the lower 3-lobed with purple spots, teeth of the calyx narrowed to a point and more than half as long as the tube, crowded in a compact head with whitish leaf-like bracts. Fruit with wrinkled or pitted nutlets hairy at the tip. IL, IN, OH. **Locally common in southern counties and northeastern OH.** *MAH*

Pycnanthemum virginianum
LAMIACEAE

common mountainmint, Virginia mountainmint

Prairies, fens, sedge meadows, woodland openings, swamps, thickets. Summer.

A square-stemmed perennial to 3 ft. tall with hair on angles of stem. Leaves opposite, entire, sessile, to 2½ in. long × ½ in. wide, fragrant (minty) when bruised. Flowers irregular, white with purple spots, to ⅛ in. wide, tubular with shallowly notched upper lip and 3-lobed lower lip, stalkless in tight heads to ¾ in. across, heads forming terminal flat-topped cluster. Slender mountainmint (*P. tenuifolium*) has glabrous stem angles and leaves to ¼ in. wide that are nearly odorless when bruised. The two form hybrids in plantings. IA, IL, IN, MI, MN, MO, OH, WI. **Scattered to widespread.** *SN*

Synandra hispidula
LAMIACEAE

synandra, Guyandotte beauty

Rich mesic forests, ravines, stream terraces. Spring.

An erect biennial, unbranched, hairy, 8–18 in. tall. Leaves opposite, basal leaves and lower stem leaves long-petiolate, upper stem leaves sessile, ovate with cordate base, toothed, 1–3 in. long × ¾–2¾ in. wide. Flowers irregular, about 1½ in. long and almost as wide, tubular, 2-lipped, upper forming a hood, the lower 3-lobed, the middle lobe lowermost and purple-striped, in 4–8 pairs on a terminal spike. This is the only *Synandra* species in the world. IL, IN, OH. Restricted to the Eastern Forests, locally common where found. *MAH*

Erythronium albidum
LILIACEAE

white trout lily

Mesic forests, flatwoods, slopes. Spring.

An erect to reclining perennial, glabrous, strongly colonial by stolons, 4–7 in. tall. Leaves basal, single for non-flowering plants, 2 for flowering ones, elliptic, entire, sometimes brown-splotched, 4–6 in. long × 1–2 in. wide. Flowers regular, to 1½ in. long and wide, trumpet-shaped, with 6 strongly reflexed tepals, occurring singly atop flowering stem. Prairie trout lily (*E. mesochoreum*; see inset), mostly of grasslands in our southwestern counties, is similar but not colonial, with leaves normally unspotted and folded lengthwise when in bloom. IA, IL, IN, MI, MN, MO, OH, WI. Widespread and common except in farthest north counties. *SN/MAH*

Floerkea proserpinacoides
LIMNANTHACEAE

false mermaid

Moist forested lowlands and flatwoods. Spring.

An annual, weakly erect to sprawling, branched, glabrous, 4–10 in. tall. Leaves alternate, petiolate, pinnately divided into 3–7 linear-elliptic entire segments, 1–3 in. long × ½–1½ in. wide. Flowers regular, about ⅛ in. wide, with 3 minute, linear-oblanceolate petals and 3 much broader lance-ovate spreading sepals, occurring singly from leaf axils. False mermaid completely disappears by early summer. It is the only member of the genus in the world and known only in North America. IA, IL, IN, MI, MN, MO, OH, WI. **Widespread but uncommon or absent in far southern counties and areas west of the Mississippi River.** *KB*

Rotala ramosior
LYTHRACEAE

toothcup

Open swamps, prairie swales, mudflats, shorelines. Summer.

An erect annual, branched, glabrous, stems are commonly reddish with 4 rounded angles, 5–12 in. tall. Leaves opposite, sessile to short-petiolate, linear to narrowly elliptic, entire, ½–1½ in. long × ⅛–½ in. wide. Flowers regular, about ¼ in. long, tubular, with 4 tiny and ephemeral petals, occurring singly in upper leaf axils. Two somewhat similar species of redstem, *Ammannia coccinea* and *A. robusta*, differ in having leaves with ear-like lobes at base that clasp the stem and typically 3 or more pinkish flowers per axillary cluster. IA, IL, IN, MI, MN, MO, OH, WI. **Patchy distribution, less common in the north.** *SN*

Hibiscus trionum

MALVACEAE

flower-of-an-hour

Disturbed places, gardens, cultivated fields. Summer.

An erect to sprawling annual, moderately branched, hairy, to 1½ ft. tall. Leaves alternate, petiolate, deeply 3-lobed, the segments oblong with rounded teeth, mostly glabrous above, hairy below and along margins, 1–3 in. long × ½–2 in. wide. Flowers regular, about 2 in. wide, 5-petaled with a deep purple throat, calyx purple-veined and inflated, occurring singly from leaf axils. Its common name is telling, as its flowers are fully open for only a brief time. **IA, IL, IN, MI, MN, MO, OH, WI. Introduced from Europe. Widespread and mostly common except for the far northern counties.** *SN*

Malva neglecta

MALVACEAE

common mallow, cheeses

Agricultural fields, lawns, roadsides, railroads, dumps, waste areas. Spring, summer, fall.

A hairy, branching, creeping biennial to 12 in. tall and 3 ft. long. Leaves simple, 5- to 9-lobed with pleated appearance, toothed, alternate, to 3 in. long and wide, cordate at base, long-petiolate. Flowers regular, to ¾ in. wide, with 5 shallowly lobed white to pale violet or pale violet-striped petals, with stigma-tipped column covered with stamens in center, in clusters of 1–3 on short axillary stalks. One common name comes from supposed resemblance of the fruit to a round wheel of cheese, cut into wedges. **IA, IL, IN, MI, MN, MO, OH, WI. Introduced from Eurasia. Scattered to widespread, less frequent in IA and MN.** *SN*

Napaea dioica
MALVACEAE

glade mallow

Riverbanks and terraces, ditches, fencerows, roadsides, railroads. Summer.

An erect, dioecious perennial, glabrous or sparsely hairy, 4–8 ft. tall, colonial. Leaves basal and alternate on upright stems, petiolate, rounded in outline, very deeply 5- to 9-lobed and sharply toothed, 10–18 in. long and about as wide. Flowers regular, ¼–½ in. wide, 5-petaled, numerous on widely spreading terminal panicles, lightly fragrant. This special plant is a Midwest endemic, occurring naturally nowhere else on earth. IA, IL, IN, MN, OH, WI. Very local but often common where found. *MAH*

Aletris farinosa
MELANTHIACEAE

white colic root

Sand prairies, savannas, forest openings. Summer.

A perennial to 3½ ft. tall. Leaves basal and alternate; basal leaves lanceolate, to 8 in. long × 1 in. wide; stem leaves few, linear, to ¾ in. long. Flowers regular, granular-textured (see inset), tubular with 6 lobes, to ½ in. long × ¼ in. wide, numerous in dense spike-like raceme at top of plant. Rosettes of yellow-green, parallel-veined leaves are unmistakable but can be easily overlooked early in growing season. IL, IN, MI, OH, WI. Mostly in the Northern Lakes, in proximity to the Great Lakes with a coastal plain disjunct distribution; a few occurrences in the Tallgrass Prairie and Eastern Forests. *SN*

Anticlea elegans
MELANTHIACEAE

white camas, mountain death camas

Prairies, wet meadows, fens, shores, dolomite bluffs. Summer.

A glabrous (usually glaucous) perennial to 3 ft. tall. Leaves mostly basal, entire, to 1 ft. long × ½ in. wide; stem leaves few, shorter, sheathing stem. Flowers regular, ¾ in. wide, with 6 spreading tepals with yellowish-green heart-shaped gland near middle of each; stalked in elongated many-flowered cluster to 10 in. long. Western plants with flowers in racemes (var. *elegans*) shorter than eastern plants with flowers in panicles (var. *glaucus*). Syn. *Zigadenus elegans*. **IA, IL, IN, MI, MN, MO, OH, WI. Mostly in northern portion of the Tallgrass Prairie and eastern portion of the Northern Lakes; also scattered in the Eastern Forests and rare in the Ozark Highlands.** *SN*

Melanthium virginicum
MELANTHIACEAE

Virginia bunchflower

Moist prairies, fens, meadows. Summer.

An erect perennial, unbranched except in inflorescence, stems short-hairy, 2–5 ft. tall. Leaves mostly basal with smaller ones above, linear, sheathing, entire and glabrous, 1–2½ ft. long × ¼–1 in. wide. Flowers regular, about 1 in. wide, with 6 oblong tepals, each bearing a pair of yellowish glands. Inflorescence a branched panicle. The somewhat similar featherbells (*Stenanthium gramineum*) has glabrous stems and ½-in.-wide flowers with narrowly lanceolate tepals; it occurs mostly in lower Midwest woodlands. **IA, IL, IN, MO, OH. Mostly in the lower Tallgrass Prairie, primarily IA and MO.** *CB*

Triantha glutinosa
MELANTHIACEAE

sticky false asphodel

Fens, wet marly shores, pannes, rock crevices. Summer.

A perennial, 1½ ft. tall. Leaves mostly basal, simple, entire, to 7 in. long × ¼ in. wide, ascending, sword-shaped; stem leaves alternate, few, smaller, sheathing. Flowers regular, to ⅓ in. wide, with 6 spreading-ascending tepals surrounding 6 stamens and 3-parted pistil, in sparse to dense terminal raceme on solitary, unbranched stem; upper stem covered in short, sticky, gland-tipped spreading hairs. Fruit a red, conical capsule. Syn. *Tofieldia glutinosa*. IL, IN, MI, MN, OH, WI. Scattered in the Northern Lakes, mostly in proximity to the Great Lakes, widely scattered in eastern portion of the Eastern Forests and in western MN. *SN*

Menispermum canadense
MENISPERMACEAE

moonseed

Mesic upland and lowland forests, streambanks. Spring, summer.

A twining, climbing perennial vine, can be woody in the south, herbaceous farther north, glabrous to hairy, 5–25+ ft. long. Leaves alternate, petiole attached near to but not on margin of leaf base, palmately veined with 5 lobes, or unlobed, 2–6 in. long × 2½–7 in. wide. Flowers regular, about ¼ in. wide, normally dioecious, petals 6–9, sepals typically 6 and longer than petals, in drooping axillary panicles. Fruit a blue-black drupe, poisonous, with crescent-shaped seeds. Flowers and fruit infrequently produced in the far north. IA, IL, IN, MI, MN, MO, OH, WI. Common in most of the Midwest. *ST*

Menyanthes trifoliata
MENYANTHACEAE

buckbean, bogbean

Fens, bogs, swamps, swales. Spring, summer.

A colony-forming perennial with flowering stems to 1 ft. tall. Leaves appearing basal, alternate along horizontal rhizome, often not immediately next to flowering stem, compound with 3 stalkless or very short-stalked gray-green entire to scalloped-margined leaflets each to 3 ⅛ in. long × 2 in. wide, glabrous, shorter than flowering stem. Flowers regular, to ½ in. wide, tubular with 5 spreading lobes covered in scraggly white hairs, up to 20 in solitary raceme, flower stalks subtended by small simple, entire bracts. Leaves may suggest Fabaceae (hence the common names). IA, IL, IN, MI, MN, MO, OH, WI. **Widespread in the Northern Lakes and adjacent portions of the Tallgrass Prairie, scattered elsewhere.** *PS*

Mollugo verticillata
MOLLUGINACEAE

carpetweed

Cultivated fields, shores, roadsides, waste areas. Spring, summer, fall.

A creeping to slightly ascending annual, glabrous, branched and mat-forming, 3–6+ in. long. Leaves whorled, 3–8 per node, sessile or short-petiolate, linear to oblanceolate, entire, ¼–1½ in. long × ⅛–⅓ in. wide. Flowers regular, ⅛–¼ in. wide, with 5 petal-like sepals, inner surface white, outer green (see inset), on stalks originating in leaf axils. Although often stated to be an introduction from tropical America, archaeological evidence from Tennessee shows that carpetweed was present in pre-Columbian North America. IA, IL, IN, MI, MN, MO, OH, WI. **Mostly common throughout except for the far northwest.** *CR*

Claytonia virginica
MONTIACEAE

spring beauty

Mesic and dry-mesic upland forests and high stream terraces. Spring.

An erect to reclining perennial, unbranched, glabrous, 2–6 in. tall. Leaves basal or opposite in a single pair, linear to narrowly elliptic, petiole indistinct, somewhat fleshy, 1–4 in. long × ¼–½ in. wide. Flowers regular, about ½–¾ in. wide, with 5 white to pink petals with or without dark pink stripes, in a loose terminal raceme. In our range Carolina spring beauty (*C. caroliniana*) occurs mostly in northern MI. Compared to *C. virginica*, its leaves are broader and possess a distinct petiole. IA, IL, IN, MI, MN, MO, OH, WI. Common and widespread except in western parts of IA and MN. *PR*

Trientalis borealis
MYRSINACEAE

starflower

Moist or swampy forests, hummocks in bogs. Spring, summer.

A glabrous perennial to 8 in. tall. Leaves simple, entire, 5–9 of unequal size whorled at top of stem, to 4 in. long × 1½ in. wide, short-petiolate, few alternate scale-like leaves sometimes on stem. Flowers regular, to ¾ in. wide, with 5–9 (often 7) spreading, minutely toothed petals; 1–3 on ascending stalks to 2½ in. from whorl of leaves. Recently reclassified as *Lysimachia borealis*. IL, IN, MI, MN, OH, WI. Widespread in the Northern Lakes and adjacent portions of the Tallgrass Prairie, with few scattered occurrences in the Eastern Forests. *SN*

Nelumbo lutea
NELUMBONACEAE

American lotus

Lakes, ponds, river backwaters, sloughs.
Summer.

An aquatic perennial with leaves and flowers
from rhizomes rooted in mud, unbranched,
glabrous, 3–6+ ft. tall, colonial. Leaves grow
from submerged rhizomes, rounded, peltate on
a long petiole, looking like a floppy umbrella
at or above water surface, growing to a diameter of about 2 ft. Flowers regular, up to 8 in.
wide, with 15–20 creamy white to yellowish oval
tepals surrounding a receptacle that enlarges
into a pitted seedpod housing large pea-sized
nuts. Sacred lotus (*N. nucifera*), an introduced
species from Asia with pinkish to white tepals,
has naturalized in a few locations. IA, IL, IN,
MI, MN, MO, OH, WI. Populations scattered,
often in sites bordering large river systems. *SN*

Mirabilis albida
NYCTAGINACEAE

white four-o'clock, hairy four-o'clock, pale umbrellawort

Prairies, glades, fields, roadsides, railroads.
Summer, fall.

A sticky-hairy perennial to 3 ft. tall. Leaves
opposite, spreading to upright, mostly in lower
half of plant, to 4 in. long × 1½ in. wide, with
entire, wavy margins, sessile or short-petiolate.
Flowers regular, to ½ in. wide, with white to
pink corolla-like calyx, trumpet-shaped with
5 notched lobes, subtended by papery bract
with 5 spreading triangular lobes, stalked in
clusters at tips of branches in upper half of
plant, opening in late afternoon and closing by
mid-morning. IA, IL, IN, MI, MN, MO, OH,
WI. Frequent in western half of the Midwest,
where native; scattered in eastern half of the
Midwest, where introduced. *KC*

Nymphaea odorata

NYMPHAEACEAE

American white waterlily, fragrant waterlily

Bogs, marshes, lakes, ponds, rivers. Summer, fall.

A colony-forming aquatic perennial to over 6 ft. tall from thick rhizome, with leaves floating on water surface. Leaves horizontal, round with slit to middle on one side, to 12 in. long and wide, simple, entire, shiny, with petiole attached at tip of slit. Flowers regular, to 7½ in. wide, with numerous white (rarely pink) spreading to ascending petals surrounding numerous stamens and short disk-like stigma, solitary on long stalks, floating. Two subtly different subspecies are sometimes recognized, differing in color of stem and leaf undersides. IA, IL, IN, MI, MN, MO, OH, WI. Widespread and common in the Northern Lakes, scattered elsewhere. *SN*

Circaea canadensis

ONAGRACEAE

enchanter's nightshade

Mesic and dry-mesic upland forests, stream terraces, thickets. Summer.

An erect perennial, stem unbranched and hairy, 10–24 in. tall. Leaves opposite, petiolate, ovate, somewhat cordate base, mostly glabrous and few teeth, 2–6 in. long × 1–2½ in. wide. Flowers regular, about ⅛ in. wide, with 2 deeply notched petals (giving appearance of 4 petals), 2 reflexed sepals, and a grooved ovary. Fruit a capsule with grooves and hooked hairs, borne on a widely spreading raceme. The hooked hairs of the fruits easily attach to clothing. Small enchanter's nightshade (*C. alpina*), a delicate species mostly in the northern states, lacks grooves on its fruit. IA, IL, IN, MI, MN, MO, OH, WI. Widespread and common throughout the Midwest. *MJH*

Epilobium coloratum
ONAGRACEAE

cinnamon willowherb

Wet meadows, marshes, bogs, fens,
seeps, swamps, streambanks, ditches.
Summer, fall.

A branched perennial to 4 ft. tall. Leaves
opposite or some alternate, often infused with
red, to 4 in. long × ½ in. wide, with irregularly
sharp-toothed margins. Flowers regular, to
⅓ in. wide, with 4 white to pale pink notched
petals at top of long, skinny ovary, numerous
from upper leaf axils. Fruit a linear, erect capsule opening from tip to expose small seeds
topped with long silky cinnamon-colored hairs.
Fringed willowherb (*E. ciliatum*), more northern in our region, has white hairs on seeds and
leaves more ovate with regular teeth. IA, IL,
IN, MI, MN, MO, OH, WI. Widespread, with
fewer occurrences in western Midwest. *MJH*

Epilobium leptophyllum
ONAGRACEAE

bog willowherb, fen willowherb

Bogs, fens, marshes, sedge meadows,
swamps. Summer, fall.

A grayish perennial to 3½ ft. tall. Leaves
mostly opposite, to 2¾ in. long × ½ in. wide,
with margins rolled under. Flowers regular, to
⅓ in. wide, with 4 white to pale pink notched
petals atop skinny ovary, several in upper part
of plant. Fruit a linear, erect pod opening from
tip to expose small seeds topped with long
silky hairs. Downy willowherb (*E. strictum*) has
spreading rather than incurved stem hairs.
Marsh willowherb (*E. palustre*) has hairs on
leaf midveins (throughout in *E. leptophyllum*).
IA, IL, IN, MI, MN, MO, OH, WI. Widespread through most of the Northern Lakes,
less frequent in the Tallgrass Prairie and
Eastern Forests. *NP*

Cypripedium candidum
ORCHIDACEAE

small white lady's slipper

Fens, wet prairies, sedge meadows, glades.
Spring.

A short-hairy perennial to 1½ ft. tall. Leaves
alternate, sheathing, few, to 6 in. long × 2½ in.
wide, parallel veins conspicuous. Flowers irregular, with 3 green to brownish-yellow sepals usually spotted and/or striped with reddish-brown,
the upper hood-like, the lower 2 fused, 2 lateral
petals similar to sepals but spreading and spi-
rally twisted, lip petal white, slipper-like, to 1 in.
long, often with faint reddish-purple veins and
speckling in slipper opening, with red-spotted
yellow staminode at back of opening; flowers
terminal, 1 (rarely 2). IA, IL, IN, MI, MN, MO,
OH, WI. Rare, primarily in northern portion
of the Tallgrass Prairie and southern portion
of the Northern Lakes. *PG*

Goodyera pubescens
ORCHIDACEAE

downy rattlesnake plantain

Mesic to dry-mesic mostly acidic upland
forests. Summer.

An erect perennial, unbranched, hairy, 8–15
in. tall. Leaves in a basal rosette, petiolate,
elliptic-ovate, evergreen, with a network of
white veins, 1–3 in. long × ½–1½ in. wide (see
inset). Flowers irregular, about ¼ in. wide, the
lip petal pouch-shaped, cleft above forming
a short spout, surrounded by 3 sepals and 2
similar petals, 20–40 in a spike-like terminal
raceme. This species is easily distinguished
by its ovate leaves patterned with a large white
median vein and branching side ones. Leaves
are evident year-round. IA, IL, IN, MI, MN,
MO, OH, WI. Widespread, locally common
but absent from most of the western Midwest. *DT/SN*

Goodyera repens
ORCHIDACEAE

creeping rattlesnake plantain, lesser rattlesnake plantain

Moist to dry coniferous and mixed forests, swamps, bogs. Summer, fall.

A diminutive single-stemmed evergreen perennial to 7 in. tall. Leaves entire, primarily basal, to 1¼ in. long × ¾ in. wide, dark green with silvery-white reticulate pattern (see inset); stem leaves few, entire, alternate, scale-like. Flowers irregular, somewhat tubular, to ⅛ in. long, with lip petal pouch-shaped with short spout-like tip, surrounded by 3 sepals and 2 similar petals, short-stalked, up to 25 in loose raceme with flowers mostly on one side of stalk. Tesselated rattlesnake plantain (*G. tesselata*) is taller; its larger more gray-green leaves have a more elaborate silvery-white pattern. **MI, MN, WI. Almost entirely restricted to the Northern Lakes.** *PD*

Platanthera clavellata
ORCHIDACEAE

club spur orchid

Acid bogs, seep springs, vernal pool margins, flatwoods, upland woods. Summer.

An erect perennial, unbranched, glabrous, 5–12 in. tall. Leaves 2 or 3, alternate, the lowest considerably larger and oblanceolate, 3–6 in. long × ½–1½ in. wide. Flowers irregular, about ½ in. long, lip petal lowermost, oblong, wedge-shaped, shallowly 3-toothed at apex, with a club-shaped spur at its base, remaining petals and sepals ovate. Inflorescence a terminal cluster to 2 in. long bearing 7–15 flowers, the flowers tilted at an angle from main stem. **IA, IL, IN, MI, MN, MO, OH, WI. Occurs principally in the Northern Lakes.** *PS*

Platanthera dilatata
ORCHIDACEAE

tall white bog orchid, bog-candle

Fens, bogs, springy meadows, openings in
conifer swamps. Summer.

A glabrous perennial to 2½ ft. tall. Leaves few,
alternate, entire, sheathing stem, to 8 in. long
× ¾ in. wide, smaller up stem and becoming
bracts in inflorescence. Flowers irregular, to ½
in. wide, with green spur, with upper sepal and
lateral petals forming hood and lateral sepals
and lip petal spreading, with clove-like spicy
fragrance, short-stalked, up to 60 in loose to
dense slender terminal raceme. Syn. *Piperia
dilatata*. **IL, IN, MI, MN, WI. Scattered, nearly
restricted to the Northern Lakes, rare in the
Tallgrass Prairie.** *PR*

Platanthera leucophaea
ORCHIDACEAE

eastern prairie fringed orchid

Tallgrass prairies, sedge marshes, fens.
Summer.

An erect perennial, unbranched, glabrous, 1–3
ft. tall. Leaves alternate, oblong-lanceolate
to elliptic, sheathing, 4–7 in. long × 1–3 in.
wide. Flowers irregular, about 1 in. long, with 3
petals and 3 ovate sepals, lip petal lowermost,
broadly ovate with 3 deeply fringed lobes and
a basal spur 1–1¼ in. long, other petals broadly
wedge-shaped with finely toothed margins.
Inflorescence 15–30 flowers in a terminal
raceme. This and the very similar western
prairie fringed orchid (*P. praeclara*) of our
western counties are both threatened species.
**IA, IL, IN, MI, MO, OH, WI. Most occur-
rences in Tallgrass Prairie and southeastern
Northern Lakes.** *LC*

Platanthera orbiculata
ORCHIDACEAE

round-leaved orchid

Rich forests, conifer swamps, bogs.
Summer.

An easily overlooked glabrous perennial to
2 ft. tall. Leaves 2, basal, entire, lying flat on
ground or nearly so, broadly oval, to 6 in. long
and wide (see inset). Flowers irregular, to 1 in.
long, long-spurred with spur thickest at tip,
with upper sepal hood-like and lateral petals,
lateral sepals, and lip petal spreading radially,
lip petal longer and narrower than others,
stalked and each subtended by a bract, up to
35 in loose terminal raceme. IL, IN, MI, MN,
OH, WI. **Scattered, primarily in the Northern
Lakes and eastern portion of the Eastern
Forests, rare in the Tallgrass Prairie.** PS

Spiranthes cernua
ORCHIDACEAE

nodding ladies'-tresses

Wet prairies, glades, fields, savannas, rocky
ledges, pannes, fens, bogs. Summer, fall.

A perennial to 14 in. tall. Basal leaves to 8 in.
long × ⅓ in. wide, usually withering by flowering; stem leaves alternate, scale-like, not
overlapping just below inflorescence. Flowers
irregular, spreading or nodding, to ½ in. long,
in tight spiraling terminal spike. Case's ladies'-
tresses (*S. casei*) has loose spike with flowers to
¼ in. long. Yellow ladies'-tresses (*S. ochroleuca*)
has cream-colored flowers. Recent work has
described *S. arcisepala* and *S. incurva*, the former with downward-pointing lateral sepals,
the latter with cream-colored lip centrally. IA,
IL, IN, MI, MN, MO, OH, WI. **Scattered to
widespread.** NP

Spiranthes lacera
ORCHIDACEAE

slender ladies'-tresses

Prairies, fields, savannas, open forests, conifer thickets, disturbed sites. Summer, fall.

A perennial to 20 in. tall. Basal leaves to 2 in. long × ¾ in. wide; stem leaves few, alternate, small, scale-like, sheathing. Flowers irregular, white with green central portion of lip, spreading or nodding, to ¼ in. long, with spreading lateral sepals, in a single spiraling column along terminal spike. The more northern var. *lacera* has basal leaves present at flowering, more loosely arranged spikes, and flowers earlier; the more southern var. *gracilis* has basal leaves absent at flowering, more densely arranged spikes, and flowers later. IA, IL, IN, MI, MN, MO, OH, WI. Scattered throughout. *PG*

Spiranthes lucida
ORCHIDACEAE

shining ladies'-tresses

Fens, seeps, shores, riverbanks, moist calcareous soils, including excavations in sandy or gravelly soil, disturbed sites, especially deer trails. Spring, summer.

A perennial to 14 in. tall. Leaves entire, 3 or 4 in basal rosette, to 4¾ in. long × ⅗ in. wide, stem leaves absent or few, alternate, small, scale-like, sheathing. Flowers irregular, white with majority of lip petal bright yellow, spreading or nodding, to ¼ in. long, crowded in rather tight spiraling terminal spike. Fruit an erect capsule. IA, IL, IN, MI, MO, OH, WI. Scattered in the Ozark Highlands, the Eastern Forests, and the Northern Lakes, rare in the Tallgrass Prairie. *CR*

Spiranthes magnicamporum
ORCHIDACEAE

Great Plains ladies'-tresses

Calcareous mesic to dry prairies, glades, sand dunes. Summer, fall.

An erect perennial, unbranched, stem glandular-hairy, 10–18 in. tall. Leaves basal, linear-lanceolate to oblanceolate, glabrous, disappearing by flowering time, 3–6 in. long × ¼–½ in. wide, upper stem bracts overlapping. Flowers irregular, to ½ in. long, lip petal oblong-ovate with a dull yellowish center, wavy-margined, upper petals 2, linear-lanceolate, sepals 3, petal-like, the lateral ones widely spreading and arching upward. Inflorescence of 15–40 flowers in a tightly spiraled terminal spike. Flowers are deliciously fragrant with a scent of vanilla or coumarin. IA, IL, IN, MI, MN, MO, OH, WI. Occurs mostly in the Tallgrass Prairie and Ozark Highlands. *PG*

Spiranthes ovalis
ORCHIDACEAE

oval ladies'-tresses

Stream terraces, upland forests, thickets. Summer, fall.

An erect perennial, unbranched, glandular-hairy on upper stem into inflorescence, 8–18 in. tall. Leaves mostly basal and present during flowering, oblong to oblanceolate, glabrous, 2–6 in. long × ¼–½ in. wide. Flowers irregular, to ¼ in. long, lip petal ovate, wavy-margined and somewhat inrolled toward tip, upper 2 petals lanceolate, sepals 3, the lateral ones slightly spreading. Inflorescence of 20–40 flowers in 3 tight rows spiraling on a terminal spike. It is the most shade tolerant of our ladies'-tresses. Ours is var. *erostellata*. IA, IL, IN, MI, MO, OH, WI. Widely scattered but mostly absent from the Northern Lakes. *PG*

Spiranthes romanzoffiana
ORCHIDACEAE

hooded ladies'-tresses

Wet meadows, fens, conifer thickets, shores, interdunal swales, sandy excavations. Summer, fall.

A perennial to 20 in. tall. Leaves entire, basal to 9 in. long × ⅓ in. wide, stem leaves alternate, small, scale-like, sheathing. Flowers irregular, white with cream throat, spreading to ascending, to ½ in. long, tubular and appearing 2-lipped, with hood formed by lateral petals, upper sepal, and ascending lateral sepals, and with lower lip fiddle-shaped, abruptly bent downward, and ruffled at tip, in rather tight spiraling terminal spike. Almost finishing flowering when nodding ladies'-tresses (*S. cernua*) is beginning to flower. IA, IL, IN, MI, MN, OH, WI. **Scattered in the Northern Lakes, rare in the Tallgrass Prairie and Eastern Forests.** *PG*

Spiranthes tuberosa
ORCHIDACEAE

little ladies'-tresses

Barrens, dry acidic woodlands, sandstone glades, old fields. Summer, fall.

An erect perennial, unbranched, glabrous, 6–12 in. tall. Leaves a basal rosette, ovate, absent by flowering, about 1 in. long × ½ in. wide. Flowers irregular, to ⅛ in. long, lip petal ovate, broad and wavy-jagged at tip, upper petals 2, linear-oblong, sepals 3, linear-lanceolate, lateral ones slightly spreading and nodding. Inflorescence a terminal spike of 10–20 flowers in a single, graceful spiral. This species is the smallest of our ladies'-tresses. Its tiny, pure white flowers are especially attractive when observed with a hand lens. IL, IN, MI, MO, OH. **Mostly in the unglaciated hills of lower Midwest, plus southern MI.** *EH*

Spiranthes vernalis
ORCHIDACEAE

spring ladies'-tresses

Grassy barrens, prairies, old fields, pastures. Summer.

An erect perennial, unbranched, with pointed hairs on upper stem and within inflorescence, 1½–2½ ft. tall. Leaves primarily basal, linear-lanceolate, mostly glabrous, 2–10 in. long × ¼–⅓ in. wide. Flowers irregular, to ½ in. long, lip petal ovate, wavy at apex with a creamy yellow center, upper petals 2, oblong-lanceolate, sepals 3, lanceolate, the lateral ones slightly spreading and inrolled. Inflorescence a terminal spike of 30–40 flowers either in a single, graceful spiral, or tightly congested. The pointed hairs of this species are unique within our region's ladies'-tresses. IA, IL, IN, MO, OH. **Mostly in the lower Midwest.** *BL*

Triphora trianthophora
ORCHIDACEAE

three birds orchid

Rich mesic forests. Summer.

An erect perennial, glabrous, stem purplish-green, 3–8 in. tall. Leaves alternate, ovate and clasping, ¼–½ in. long and about as wide. Flowers irregular, about ⅔ in. wide, white to pale pinkish-purple, lip petal obovate, 3-lobed, with 3 green parallel ridges in middle, the 2 lateral petals and 3 spreading sepals elliptic-obovate. Flowers commonly 3 in a loose terminal raceme. The common name refers to the usual number of flowers per plant and their superficial resemblance to flying birds. A flower lasts only one day, and when conditions are right all plants of a population bloom on the same day. IA, IL, IN, MI, MO, OH, WI. **Widely scattered, mostly in lower Midwest.** *DT*

Conopholis americana
OROBANCHACEAE

bear corn, cancer-root

Mesic to dry-mesic upland forests. Spring.

An erect perennial, unbranched, glabrous, 2–8 in. tall. Scale-like leaves about ¾ in. long, ovate, fleshy and lacking chlorophyll. Flowers irregular, about ½ in. long, tubular, 2-lipped, upper lip hood-like, lower lip downwardly curved and 3-lobed, grouped in several cone-like spikes. It is a harmless parasite on the roots of various species of oaks. Superficially resembling a pine cone or an ear of corn, it is reported to be a food source for bears emerging from spring hibernation. **IA, IL, IN, MI, OH, WI. Occurs mostly in the eastern half of the Midwest.** *SN*

Melampyrum lineare
OROBANCHACEAE

American cow-wheat

Forests (often with sparse canopy), fens, bogs, rock cliffs. Summer.

An annual to 14 in. tall. Leaves variable, opposite, sessile or nearly so, to 2½ in. long × ⅓ in. wide, pointed at tip, entire to irregularly and sparsely toothed, those below flowers sometimes with 2–6 long, narrow teeth near base. Flowers irregular, to ½ in. long, cream-colored with yellow interior, tubular, constricted just below 2 lips, upper lip shallowly 2-lobed, lower lip shallowly 3-lobed, solitary and diverging horizontally in leaf axils in upper part of plant. Hemiparasitic. **IL, IN, MI, MN, OH, WI. Nearly restricted to the Northern Lakes (widespread north), infrequent in the Eastern Forests.** *SN*

Orobanche uniflora

OROBANCHACEAE

one-flowered broomrape

Mesic and dry-mesic upland forests, savannas, prairies. Spring, summer.

A perennial mostly consisting of a thickened underground stem from which emerge erect, tan-colored, unbranched and glandular-hairy flowering stalks, 2–6 in. tall. Leaves absent. Flowers irregular, ¾–1 in. long, tubular, arching, and somewhat nodding, glandular-hairy, with 5 flaring lobes, 1 flower per stalk. This and related species are root parasites, deriving nourishment from their host plants. Syn. *Aphyllon uniflorum*. Clustered broomrape (*O. fasciculata*), which occurs in our northern counties, bears multiple flowers per stem. IA, IL, IN, MI, MN, MO, OH, WI. Widely scattered but mostly uncommon, rare in the far northern counties. *BS*

Pedicularis lanceolata

OROBANCHACEAE

swamp lousewort, fen betony

Wet prairies, sedge meadows, fens, marshes, swamps, seeps, moist limestone ledges, shores. Summer, fall.

A perennial to 3 ft. tall. Leaves opposite, to 4 in. long × 1¼ in. wide, fern-like with numerous regularly spaced rounded, toothed lobes, short-petiolate. Flowers irregular, tubular with hooded upper lip and hidden 3-lobed lower lip, to ¾ in. long, subtended by bracts with hairy margins, in spikes to 4 in. long in upper part of plant, with pinwheel-like arrangement viewed from above. Hemiparasitic. IA, IL, IN, MI, MN, MO, OH, WI. Widespread in broad transition zone between the Northern Lakes and the Tallgrass Prairie, scattered in the Eastern Forests and periphery; also in parts of the Ozark Highlands. *KC*

Oxalis montana
OXALIDACEAE

northern woodsorrel, mountain woodsorrel, common woodsorrel

Moist forests (often with conifers), hummocks in cedar swamps.
Spring, summer.

A sparsely hairy, colony-forming perennial to 6 in. tall. Leaves basal, to ¾ in. long and wide, compound, with 3 often deflexed, stalkless, cordate leaflets attached at their points, on petioles to about 5 in. long. Flowers regular, to ¾ in. wide, with 5 shallowly notched white to pale pink petals striped with pinkish-purple and with yellow spot at base, solitary and often slightly nodding at tips of naked stems. Fruit a depressed capsule. Syn. *O. acetosella* subsp. *montana*. MI, MN, OH, WI. Widespread in northern portion of the Northern Lakes, rare in easternmost portion of Eastern Forests. *SN*

Adlumia fungosa
PAPAVERACEAE

climbing fumitory, Allegheny vine

Moist rocky forests, cliffs, shores.
Summer.

A slender, scrambling or climbing biennial vine to 10 ft. Leaves glabrous, thin-textured, twice compound into rounded, lobed, abundant leaflets to ¾ in. long. Flowers irregular, pink to white, to ¾ in. long, dangling, elongate heart-shaped with spreading corolla lobes. Characteristically shows up after disturbance (fire, clearing, soil disturbance) and persists in areas with little competition on shallow soils; seeds also persist in seedbank for long periods of time until adequate conditions are present. IA, IL, IN, MI, MN, OH, WI. Rare with spotty distribution throughout the Northern Lakes, with a few occurrences in the Eastern Forests and Tallgrass Prairie. *BS*

Dicentra canadensis
PAPAVERACEAE

squirrel corn

Moist deciduous upland forests,
ravines. Spring.

A perennial to 1 ft. tall. Leaves basal, 1 or 2 per
stem, fern-like, 4 times pinnately compound
and lobed to rounded tips, to 7 in. long and
wide, gray-green above and whitened beneath.
Flowers irregular, white to pale pink, to ¾ in.
long, dangling, elongate and heart-shaped, in
ascending raceme at top of plant. Fragrant.
Yellow-orange corms resembling corn kernels
develop at or just below ground surface and
distinguish vegetative individuals from similar Dutchman's breeches (*D. cucullaria*). IA, IL,
IN, MI, MN, MO, OH, WI. **Throughout, but
most abundant in the Eastern Forests and
Northern Lakes.** *SN*

Dicentra cucullaria
PAPAVERACEAE

Dutchman's breeches

Moist deciduous upland forests,
ravines. Spring.

A perennial to 1 ft. tall. Leaves basal, 1 or 2
per stem, fern-like, 3 or 4 times pinnately
compound and lobed to rounded tips, to
7 in. long and wide, gray-green above and
whitened beneath. Flowers irregular, white
to pale pink, to ¾ in. long, dangling, shaped
like upside-down "Dutchman's breeches," in
ascending raceme at top of plant. Pink clusters
of teardrop-shaped corms are at or just below
ground surface, distinguishing vegetative
individuals from similar squirrel corn (*D.
canadensis*). IA, IL, IN, MI, MN, MO, OH, WI.
Widespread throughout. *MAH*

Sanguinaria canadensis
PAPAVERACEAE

bloodroot

Rich mesic forests, stream terraces, north-facing slopes. Spring.

An erect perennial, unbranched, glabrous, from a rhizome with reddish sap, 4–10 in. tall. Leaf 1, basal, petiolate, rounded in outline, somewhat rubbery, generally with 5–7 deep lobes, wrapped around flowering stalk early in development, then expanding, 4–6 in. long and about as wide. Flowers regular, 1–2 in. wide, with 8–12 elliptic petals surrounding a center with numerous yellow stamens, solitary on a leafless stalk. The orange-red sap looks remarkably like blood. The flower's scent has been likened to everything from new automobile tires to chocolate. IA, IL, IN, MI, MN, MO, OH, WI. **Common in most counties.** *SN*

Parnassia glauca
PARNASSIACEAE

fen grass-of-Parnassus

Fens, marshes, seeps, interdunal wetlands, marl lakeshores. Summer, fall.

A perennial to 16 in. tall. Basal leaves in rosette, 2½ in. long × 2 in. wide, glabrous, thick, cordate, petiolate. Flowers regular, to 1¼ in. wide, with 5 spreading green-veined white petals and yellow-tipped 3-pronged staminodes shorter than stamens (see inset), solitary atop stems. Marsh grass-of-Parnassus (*P. palustris*), northern, has shorter petals and thinner leaves. Largeleaf grass-of-Parnassus (*P. grandifolia*), in the Ozark Highlands, has staminodes longer than stamens. IA, IL, IN, MI, MN, OH, WI. **Transition zone between Northern Lakes and Tallgrass Prairie, also eastern portion of the Northern Lakes, scattered in Eastern Forests.** *LC/SN*

Phryma leptostachya
PHRYMACEAE

American lopseed

Mesic upland forests, stream terraces,
ravines, thickets. Summer.

An erect perennial, mostly unbranched,
hairy, stems 4-angled, 1–2 ft. tall. Leaves
opposite, petiolate, ovate, toothed, blade 3–5
in. long × 1½–2½ in. wide. Flowers irregular,
about ¼ in. long, tubular, 2-lipped, upper lip
pinkish-purple, slightly notched in middle and
curved upward, lower lip larger, whitish and
3-lobed, occurring in pairs spaced more or less
evenly along axillary and terminal spike-like
racemes. Fruit reflexed, pressed tightly pointing
downward against main stem. Lopseed's common name alludes to the downward-pointing
fruits, like the ears of a lop-eared rabbit. IA, IL,
IN, MI, MN, MO, OH, WI. Occurs commonly
except in the far north. *MJH*

Phytolacca americana
PHYTOLACCACEAE

pokeweed, pokeberry, poke

Mesic forests, thickets, roadsides,
disturbed sites. Summer.

An erect perennial, branching, glabrous, green
or reddish-green with hollow stems, 4–6+ ft.
tall. Leaves alternate, petiolate, lance-oblong,
entire, 6–12 in. long × 3–6 in. wide. Flowers
regular, ⅛–¼ in. wide with 5 white to pinkish
petal-like sepals, numerous on long cylindrical racemes. Fruit a juicy, dark purple berry in
dangling clusters. Although the boiled young
leaves have been eaten as greens by some
people, many sources consider all parts of the
plant to be poisonous. IA, IL, IN, MI, MO,
OH, WI. Common in the region's southern
half, becoming infrequent to absent northward. *MAH*

Chelone glabra
PLANTAGINACEAE

white turtlehead

Swamps, fens, spring runs, marshes.
Summer, fall.

An erect perennial, unbranched, glabrous,
2–4 ft. tall. Leaves opposite, mostly sessile,
linear to lanceolate with serrated margins,
1–8 in. long × ¼–1 in. wide. Flowers irregular,
about 1½ in. long, 2-lipped, the upper one
helmet-like, the lower one shallowly 3-lobed
and bearing whitish hairs in the throat, clustered in dense terminal spikes. Flowers may
have various amounts of pink on the outer
"lips," causing possible confusion with pink
turtlehead (*C. obliqua*). Flowers are more fully
pink in the latter, with broader leaves that are
usually petiolate. IA, IL, IN, MI, MN, MO,
OH, WI. Most occurrences in the Northern
Lakes and Eastern Forests. *SN*

Gratiola neglecta
PLANTAGINACEAE

clammy hedge-hyssop

Mudflats, wet depressions, agricultural
fields. Spring, summer, fall.

A glandular-hairy annual to 10 in. tall. Leaves
opposite, sessile, to 1¾ in. long × ½ in. wide,
glabrous, sparsely toothed. Flowers irregular,
tubular, whitish with purplish-striped yellow
tube and throat, to ⅓ in. long, with 1- or 2-lobed
upper lip and 3-lobed lower lip, subtended
by 5 green sepals and 2 slightly larger green
bracts immediately below flower, solitary on
long stalks from leaf axils. Round-fruited
hedge-hyssop (*G. virginiana*) has thicker flower
stalks up to or barely longer than the sepal
length. Often confused with false pimpernel
(*Lindernia dubia*). IA, IL, IN, MI, MN, MO,
OH, WI. Widespread through most of the
Midwest, with fewer occurrences in IA, MI,
and MN. *MJH*

Lindernia dubia
PLANTAGINACEAE

false pimpernel

Mudflats, shallow pools, ditches.
Summer, fall.

An erect annual, branched, stems 4-angled,
mostly glabrous, 2–8 in. tall. Leaves opposite,
sessile or nearly so, elliptical to ovate, entire
or with few small teeth, ½–1 in. long × ¼–½
in. wide. Flowers irregular, ¼–⅓ in. long,
tubular, 2-lipped, upper one slightly 2-lobed,
lower with 3 spreading lobes, solitary in upper
leaf axils on bractless stalks about as long or
longer than the leaves beneath them. Some
botanists have called those with flower stalks
extending beyond the leaves either a variety
(var. *anagallidea*) or a distinct species, namely
slender false pimpernel (*L. anagallidea*). IA,
IL, IN, MI, MN, MO, OH, WI. Mostly common throughout. *MJH*

Penstemon digitalis
PLANTAGINACEAE

foxglove beardtongue

Open woodlands, prairies, meadows,
roadsides, clearings. Spring, summer.

An erect perennial, unbranched, mostly glabrous to sparsely hairy, 1½–3 ft. tall. Leaves
basal and petiolate, stem leaves opposite,
sessile and clasping, lanceolate to lance-ovate,
with a few scattered teeth, 3–7 in. long × 1–2 in.
wide. Flowers irregular, about 1 in. long, tubular, widest toward the opening, 2-lipped with
5 spreading lobes, 2 upper and 3 lower, sepals
lanceolate-ovate, less than ⅓ in. long, anthers
hairy on backside. Similar in appearance to
smooth beardtongue (*P. calycosus*). IA, IL, IN,
MI, MN, MO, OH, WI. Common but mostly
absent from the northwestern counties. *MJH*

Penstemon tubaeflorus
PLANTAGINACEAE

trumpet beardtongue

Prairies, glades, savannas, open woods, railroads. Spring, summer.

A glabrous perennial to 3 ft. tall. Leaves in basal rosette in first year, petiolate, to 5 in. long × 2 in. wide, entire. Stem leaves primarily in lower part of stem, opposite, sessile, similar to basal, to 5 in. long × 2 in. wide. Flowers irregular, tubular with flaring lobes, to 1 in. long, upper lip 2-lobed, lower lip 3-lobed, inside covered in glandular hairs, in whorls of 6–12 along narrow, terminal inflorescence to 1½ ft. long. Fruit a teardrop-shaped capsule to ⅓ in. IA, IL, IN, MO, OH, WI. Primarily in southwestern corner of Midwest in the Ozark Highlands and the Tallgrass Prairie, scattered to rare elsewhere. *ST*

Plantago aristata
PLANTAGINACEAE

large-bracted plantain

Glades, sand barrens, dry prairies, old fields, railroads, waste areas. Summer.

An erect annual, unbranched, hairy, 4–12 in. tall. Leaves basal, petiolate and somewhat clasping, linear, entire, upper surfaces glabrous, 3–5 in. long × ⅛–¼ in. wide. Flowers about ⅛ in. wide, short-tubular, corolla papery, 4-lobed, in a spike with narrow bracts extending well beyond the flowers. The long length of the floral bracts is distinctive, making this species unlike any other plantain in the region. IA, IL, IN, MI, MN, MO, OH, WI. Native from IN westward, most occurrences elsewhere likely introduced. *SN*

Plantago cordata
PLANTAGINACEAE

heart-leaved plantain

Gravel bars, shallow stream channels, seeps, woodland ponds. Spring.

An erect, glabrous perennial. Leaves in a basal rosette, elliptic-ovate, cordate, petiolate, entire, principal veins branching away from midrib well above leaf base, to 1½ ft. long × 10 in. wide. Flowers about ⅛ in. wide, corolla short-tubular and papery with 4 lobes, arranged upon an unbranched, hollow spike 6–12 in. long. Where water quality of its habitat declines, this impressive plantain likewise declines, thus accounting for its rarity. IA, IL, IN, MI, MO, OH, WI. Widely scattered and extremely rare except for MO. *EU*

Plantago elongata
PLANTAGINACEAE

prairie plantain, slender plantain

Prairies, glades, savannas, forest openings, streambanks, roadsides. Spring.

An annual to 5 in. tall. Leaves in basal rosette, simple, linear, sessile, entire, erect to ascending, to 2½ in. long, less than ⅛ in. wide. Flowers inconspicuous, regular with 4 spreading lobes or closed, less than ⅛ in. long, with 2 purple-tipped stamens, each subtended by green bract with translucent margins, bract as long as flower; in loose terminal spike to 2¾ in. long on leafless stem. As treated here, includes *P. pusilla*. Woolly plantain (*P. patagonica*) is hairier with irregular flowers in a dense spike. IL, IN, MN, MO. Primarily southern, in the Ozark Highlands, the Tallgrass Prairie, and the Eastern Forests; rare in extreme southwestern MN. *JT*

Plantago lanceolata
PLANTAGINACEAE

English plantain, buckhorn plantain

Lawns, fields, roadsides, waste areas. Summer.

A glabrous to sparsely hairy perennial, to 1½ ft. tall. Leaves in a basal rosette, short-petiolate, narrowly elliptic to lanceolate with 3–5 parallel veins, entire, 4–12 in. long × ½–2 in. wide. Flowers about ⅛ in. wide, corolla short-tubular and papery, 4-lobed, peduncle deeply ridged, erect, unbranched, densely arranged on a somewhat cone-shaped to cylindrical spike 1–3 in. long. This lawn weed is likely known to nearly everyone even if its name is unfamiliar. IA, IL, IN, MI, MN, MO, OH, WI. **Introduced from Eurasia. Abundant.** *SN*

Plantago rugelii
PLANTAGINACEAE

Rugel's plantain

Fields, gravel bars, waste areas, untreated lawns. Spring, summer.

A typically glabrous perennial, 6–10 in. tall. Leaves a basal rosette, petioles commonly purple at base, ovate to elliptic to cordate-ovate, entire, principal veins originate from leaf base and run parallel to midvein, 2–7 in. long × 1–4 in. wide. Flowers about ⅛ in. wide, corolla short-tubular and papery, 4-lobed, peduncle solid, erect and unbranched, arranged densely upon a narrowly cylindrical spike 4–6 in. long, the floral bracts with pointed tips. The non-native common plantain (*P. major*) typically has all green petioles and blunt-tipped floral bracts; it is mostly northern in our region. IA, IL, IN, MI, MN, MO, OH, WI. **Abundant throughout.** *MAH*

Plantago virginica
PLANTAGINACEAE

dwarf plantain, Virginia plantain

Prairies, fields, pastures, glades, roadsides, waste areas. Spring, summer.

A densely hairy annual to 8 in. tall. Leaves in basal rosette, simple, short-petiolate, entire to shallowly toothed, erect to spreading, to 6 in. long × 3 in. wide (often smaller). Flowers inconspicuous, male, female, or perfect, regular with 4 erect to spreading whitish to tan lobes, to ⅛ in. long, with 4 purplish-tipped stamens, each subtended by green bract with translucent margins, bracts less than ⅛ in., in dense terminal spike to 4 in. long on leafless stem. IA, IL, IN, MI, MN, MO, OH, WI. Widespread in southern portion of Midwest in the Ozark Highlands, the Tallgrass Prairie, and the Eastern Forests, scattered in the Northern Lakes. *PD*

Veronica peregrina
PLANTAGINACEAE

purslane speedwell

Cultivated fields, gardens, riverbanks, disturbed sites. Spring, summer.

An erect to ascending annual, often branched, somewhat fleshy, glabrous to glandular-hairy, 2–10 in. tall. Leaves opposite, mostly sessile, oblong or oblanceolate, entire or with few teeth, ¼–1¼ in. long × ¹⁄₁₆–¼ in. wide. Flowers irregular, ¹⁄₁₆–⅛ in. wide, tubular, with 4 oval lobes, in alternate leafy-bracted terminal racemes. Fruit a heart-shaped capsule. Stems and fruits of the widespread subsp. *peregrina* are glabrous; those of the more northern and western subsp. *xalapensis* are glandular-hairy. IA, IL, IN, MI, MN, MO, OH, WI. Abundant, possibly in every county. *MAH*

Veronica serpyllifolia
PLANTAGINACEAE

thyme-leaved speedwell

Lawns, forest openings, disturbed sites.
Spring.

A creeping to ascending branching perennial,
stems short-hairy, 4–10 in. long when creeping, less when upright. Leaves opposite (but
upper commonly alternate), nearly sessile,
broadly elliptic or ovate, entire or barely with
a few teeth, mainly glabrous, ⅓–1 in. long
× ¼–⅔ in. wide. Flowers irregular, ⅛–¼ in.
wide, tubular, with 4 rounded lobes, lower
one noticeably smaller than the others and
unstriped, the others and especially the upper
lobe with dark purple striping, on a spike-like
terminal raceme. Fruit a heart-shaped capsule.
IA, IL, IN, MI, MN, MO, OH, WI. Introduced
from Eurasia. Primarily found east of the
Mississippi River. Common. *MAH*

Veronicastrum virginicum
PLANTAGINACEAE

Culver's root

Prairies, savannas, woodlands, barrens,
fens. Summer.

A mostly glabrous perennial, 3–6 ft. tall.
Leaves typically in whorls of 4 or 5, petiolate, lanceolate to narrowly oblong, sharply
toothed, 2–5 in. long × ¼–1 in. wide. Flowers nearly regular, ¼–⅓ in. long, tubular
with 4 short, oblong-lanceolate lobes and
long-exserted stamens, numerous and densely
packed in erect, slender, showy whorls of terminal spikes 3–8 in. long. Some individuals
exhibit pinkish flowers. IA, IL, IN, MI, MN,
MO, OH, WI. Locally common, especially in
the Tallgrass Prairie. *SN*

Collomia linearis
POLEMONIACEAE

tiny trumpet, slender gilia

Prairies, pastures, thickets, shores, outcrops, roadsides, railroads, disturbed sites. Spring, summer.

A glandular-hairy annual to 2 ft. tall. Leaves alternate, entire, sessile, pointed at tip, with strongly impressed midvein, to 3½ in. long × ½ in. wide, often smaller in axillary clusters. Flowers regular, white, pink, or blue-violet, inconspicuous, to ¼ in. wide, tubular with 5 spreading lobes atop ½-in.-long tube, 5 stamens slightly exserted, subtended by 5-lobed calyx covered in stalked glands, sessile in dense clusters of up to 20 at top of plant and ends of branches, clusters subtended by glandular-hairy bracts. IA, IL, IN, MI, MN, MO, OH, WI. **Scattered in IA, MN, and WI, rare elsewhere.** *PR*

Polygala senega
POLYGALACEAE

Seneca snakeroot

Dry, sandy or rocky (usually limestone) woodlands, savannas, prairies. Spring, summer.

An erect perennial, mostly unbranched, hairy, 1–1½ ft. tall. Leaves alternate, short-petiolate, entire or margins finely toothed, lance-linear to elliptic, mostly glabrous, 1–3 in. long × ⅛–1¼ in. wide. Flowers irregular, about ⅛ in. long, with 3 small petals, the lower one keel-shaped, and 5 sepals, 2 of which are enlarged and petal-like (wings), crowded in elongated, narrow racemes 1–2 in. long. This species exhibits a wide range of leaf width; plants with the widest leaves are classified as var. *latifolia*. IA, IL, IN, MI, MN, MO, OH, WI. **Throughout, but patchy and only locally common.** *KC*

Polygala verticillata
POLYGALACEAE

whorled milkwort

Prairies, glades, savannas, sandy fields, shores, open woods. Spring, summer, fall.

An inconspicuous, glabrous annual to 10 in. tall. Leaves in whorls of 4 or 5 (often with some alternate below), entire, sessile, to 1 in. long × ⅜ in. wide. Flowers irregular, white to pale pink, to ⅛ in. wide, with 3 petals and 3 sepals forming tube and 2 sepals enlarged and petal-like (wings), lower petal with fringed tip at tube opening, in dense spike-like racemes to ¾ in. long in upper part of plant. Fruit drops soon after maturing, leaving portion of inflorescence with remains of flower stalks. Alternate milkwort (*P. ambigua*) has white to greenish-white flowers and upper leaves alternate. IA, IL, IN, MI, MN, MO, OH, WI. **Scattered throughout.** *EH*

Fallopia scandens
POLYGONACEAE

climbing false buckwheat, climbing black bindweed

Moist forests, floodplains, thickets, fencerows, gardens, roadsides. Summer, fall.

A perennial herbaceous vine to 20 ft., twining but lacking tendrils. Leaves alternate, simple, entire, to 4 in. long × 2 in. wide, cordate to sagittate, on petioles becoming shorter above, with ocreae (nodal sheaths). Flowers regular, white to green, to ⅙ in. long, with 5 tepals, the 3 outer winged, soon nodding, in axillary racemes to 8 in. Black bindweed (*F. convolvulus*), introduced, lacks wings on tepals and fruit. Fringed black bindweed (*F. cilinodis*), more northern, has branched inflorescences and downward-pointing hairs at nodes. IA, IL, IN, MI, MN, MO, OH, WI. **Widespread, with fewer occurrences in IA and northern MI.** *MAH*

Persicaria hydropiperoides
POLYGONACEAE

mild waterpepper

Swamps, bogs, lakes, ponds, rivers, ditches.
Summer, fall.

A colony-forming perennial to 3 ft. tall. Leaves
alternate, simple, entire, to 8 in. long × 1½ in.
wide, sessile or short-petiolate, with ocreae
(nodal sheaths) with summits fringed. Flowers
regular, white to very pale pink, to ⅛ in. long,
with 5 spreading-ascending smooth tepals,
numerous in delicate, erect, spike-like racemes
to 4 in. long in upper part of plant. Often in
areas with standing water (at least in spring),
producing floating leaves. Bristly smartweed
(*P. setacea*) has hairy leaves and spreading hairs
covering ocreae. IA, IL, IN, MI, MN, MO, OH,
WI. Widespread, with fewer occurrences in
northwestern portion of the Midwest. *ST*

Persicaria lapathifolia
POLYGONACEAE

pale smartweed, nodding smartweed

Shorelines bordering rivers and lakes,
marshes, disturbed sites. Summer, fall.

An erect or ascending annual, branched,
mostly glabrous, 1–3+ ft. tall. Leaves alternate, short-petiolate, lanceolate, 1½–5 in.
long × ¼–2 in. wide, ocreae (nodal sheaths)
glabrous and lacking bristly tips but usually
torn. Flowers regular, about ⅛ in. long, with
5 rather weakly spreading tepals, numerous,
arranged in several nodding or drooping slender spike-like racemes. This is a large smartweed with distinctive drooping racemes of
pale-colored flowers. It occurs worldwide, and
our populations may consist of both native
and introduced strains. IA, IL, IN, MI, MN,
MO, OH, WI. Common, possibly in every
county. *SN*

Persicaria punctata
POLYGONACEAE

dotted waterpepper

Marshes, fens, seeps, swamps, floodplains, lake and stream margins, ditches. Summer, fall.

A somewhat sprawling annual to 2½ ft. tall. Leaves alternate, simple, entire, to 6 in. long × 1 in. wide, short-petiolate, with ocreae (nodal sheaths) with summits fringed. Flowers regular, greenish-white, mostly closed, to ⅛ in. long with 5 pitted tepals, loose in slender erect to slightly nodding racemes to 5 in. long in upper part of plant. Fruit a shiny, smooth achene. Leaves have sharp, peppery taste (wait for it!). Waterpepper (*P. hydropiper*) differs in having some flowers concealed in ocreae, nodding racemes, and dull-textured achenes. IA, IL, IN, MI, MN, MO, OH, WI. Widespread, less frequent in northern MI. *SN*

Persicaria sagittata
POLYGONACEAE

arrow-leaved tearthumb

Marshes, seeps, wet meadows. Summer, fall.

A sprawling annual, glabrous or hairy with hooked prickles throughout, stems 4-angled, 1–6 ft. long. Leaves alternate, petiolate, narrowly sagittate, the base with pointed to rounded parallel "earlobes" straddling the main stem, 1–3½ in. long × ½–1¼ in. wide, ocreae (nodal sheaths) glabrous. Flowers ⅛–¼ in. long, with 5 rather weakly spreading tepals, white to pink, in small, rounded clusters arranged on long, terminal or axillary stalks. The plant's prickly stems can tear skin. Leaves of the also prickly halberd-leaved tearthumb (*P. arifolia*) have sharply pointed basal lobes flaring perpendicular to the main blade. IA, IL, IN, MI, MN, MO, OH, WI. Mostly common except for far western counties. *PR*

Persicaria virginiana
POLYGONACEAE

jumpseed, Virginia knotweed

Rich mesic forests, flatwoods, thickets.
Summer.

An erect to arching perennial, unbranched,
glabrous to hairy, 1–3 ft. tall. Leaves alternate,
petiolate, lanceolate to ovate-elliptical, entire,
some young leaves with a purplish splotch or
chevron on the upper surface, ocreae (nodal
sheaths) hairy, ¾–7½ in. long × 1–4 in. wide.
Flowers regular, about ⅛ in. long, with 4 often
weakly spreading tepals, white to reddish,
arranged widely spaced on a long, slender,
terminal spike-like raceme. The mature fruit
"jumps" off the stem when disturbed (hence
jumpseed). IA, IL, IN, MI, MN, MO, OH, WI.
**Common except in far northern and north-
western counties.** *MAH*

Polygonum articulatum
POLYGONACEAE

jointweed, coast jointweed

Sand prairies, savannas, plains, dunes,
banks, roadsides. Summer.

A glabrous, ascending annual to 20 in. tall.
Leaves alternate, simple, entire, ascending to
erect, to ¾ in. long and very narrow, somewhat
fleshy, sessile, with ocreae (nodal sheaths)
giving stem jointed appearance; often falling
early. Flowers regular, white to pinkish, to ⅛
in. wide with 5 spreading tepals, numerous
on spreading to arching stalks in racemes to
1½ in. long at ends of branches in upper part
of plant; each flower subtended by sheathing
scale (ocreole). Syn. *Polygonella articulata*. IA,
IL, IN, MI, MN, WI. Scattered in the Northern Lakes and eastern portion of the Tallgrass Prairie. *SN*

Polygonum aviculare
POLYGONACEAE

common knotweed

Roadsides, flowerbeds, pavement cracks, lawns, waste areas. Summer.

A highly variable annual plant, ascending to prostrate and mat-forming, branched, glabrous, 6–24+ in. long. Leaves alternate, short-petiolate, lance-ovate to elliptic, entire, bluish-green, 1–2½ in. long × ¼–¾ in. wide, summit of ocreae (nodal sheaths) entire or torn. Flowers regular, about ⅛ in. long, with 5 greenish-white to pinkish unkeeled tepals in axillary clusters. Found mostly in the upper Midwest, the similar leathery knot-weed (*P. achoreum*) has boat-shaped (keeled) outer tepals. IA, IL, IN, MI, MN, MO, OH, WI. Introduced from Europe or perhaps native in part. Abundant, probably in every county. *MAH*

Polygonum erectum
POLYGONACEAE

erect knotweed

Floodplains, cropland, roadsides, waste areas. Spring, summer, fall.

A glabrous, gray-green, erect to ascending, mostly unbranched annual to 30 in. tall. Leaves alternate, simple, entire, to 3 in. long × 1 in. wide along stems, these clearly larger than branch leaves, short-petiolate, with ocreae (nodal sheaths). Flowers regular, to ⅙ in. long, closed, with 5 tepals, in short-stalked clusters of 1–5 from leaf axils. Bushy knot-weed (*P. ramosissimum*) is more branched and usually has narrower leaves. Many other *Polygonum* species are superficially similar. IA, IL, IN, MI, MN, MO, OH, WI. Scattered throughout. *JM*

Polygonum tenue
POLYGONACEAE

slender knotweed

Barrens, dry prairies, sandstone glades, sand savannas. Summer.

An erect annual with mostly upright branching, stems angled, sometimes roughened, bluish-green, 4–18 in. tall. Leaves alternate, sessile to short-petiolate, linear to narrowly lanceolate, firm and somewhat folded, sharply pointed, erect with finely toothed margins, ½–1¼ in. long × ¹⁄₁₆–¼ in. wide, ocreae (nodal sheaths) green to brown. Flowers regular, about ⅛ in. long, with 5 greenish-white to pinkish tepals, usually single or small clusters in axils of upper bracteal leaves. Usually grows on acidic substrates. IA, IL, IN, MI, MN, MO, OH, WI. **Patchy distribution with most occurrences in the Ozark Highlands.** *CR*

Reynoutria japonica
POLYGONACEAE

Japanese knotweed

Forest edges, streambanks, ditches, roadsides, waste areas. Summer, fall.

A mostly glabrous, colony-forming, bushy perennial to 9 ft. tall with hollow stems. Leaves alternate, simple, entire, to 6 in. long × 4 in. wide, ovate, flat across base with pointed tip, short-petiolate, with ocreae (nodal sheaths). Flowers regular, trumpet-shaped, to ⅓ in. long, with 5 tepals, in several spike-like panicles to 4 in. long from each upper leaf axil. Giant knotweed (*R. sachalinensis*) has leaves over 6 in. long with cordate bases. Bohemian knotweed (*R. ×bohemica*) is an intermediate hybrid between the two. All are invasive. Sometimes placed in *Fallopia*. IA, IL, IN, MI, MN, MO, OH, WI. **Introduced from Japan. Scattered.** *PR*

Androsace occidentalis
PRIMULACEAE

western rockjasmine

Gravelly, sandy, or shallow-soil prairies, fields, waste areas. Spring.

An annual to 4 in. tall. Leaves in basal rosette to 1½ in. across, simple, sessile, to ¾ in. long × ½ in. wide, green to red with blunt teeth toward tips. A whorl of relatively broad bracts subtends flowering branches. Flowers regular, 5-parted, to ⅛ in. wide, with calyx forming cup around petals, solitary or branched on stalks to 3 in. long radiating from single point. Easily overlooked, diminutive. Does not persist in areas with competition from other plants (requires some disturbance, natural or otherwise). IA, IL, IN, MI, MN, MO, OH, WI. **Nearly restricted to the Tallgrass Prairie and Ozark Highlands.** *ST*

Dodecatheon meadia
PRIMULACEAE

eastern shooting star

Prairies, glades, savannas, woodlands, blufftops. Spring.

An erect, unbranched perennial, glabrous, 10–18 in. tall. Leaves in a basal rosette, oblong to oblanceolate, entire, gradually tapering to a petiole often tinged with red, 4–8 in. long × 1–3 in. wide. Flowers regular, ½–1 in. long, white to purplish, short-tubular, with 5 narrow lobes strongly reflexed to expose a pointed "beak" of yellow anthers. Inflorescence a cluster of downwardly pointed flowers atop stem. Jeweled shooting star (*D. amethystinum*) and French's shooting star (*D. frenchii*) differ from *D. meadia* by fruit and foliar characteristics, respectively. Sometimes placed in *Primula*. IA, IL, IN, MI, MN, MO, OH, WI. **Mostly in the central and southern Midwest.** *SN*

Samolus parviflorus
PRIMULACEAE

water pimpernel

Seeps, spring runs, sandbars, streambanks,
ephemeral pools. Spring, summer, fall.

An erect to ascending perennial, branched
above, glabrous, 4–12 in. tall. Leaves basal and
alternate, short-petiolate, elliptic to obovate,
entire, 1–3 in. long × ⅓–2 in. wide. Flowers
regular, about ⅛ in. wide, short-tubular with
5 oval-elliptic lobes slightly notched at tip,
occurring on open, widely branching terminal
and axillary racemes. Sometimes considered
a subspecies of the cosmopolitan brookweed
(*S. valerandi*), as *S. valerandi* subsp. *parviflorus*.
**IL, IN, MI, MO, OH, WI. Occurs mostly south
and east of a line from Detroit, MI, to Kansas
City, MO.** *SN*

Actaea pachypoda
RANUNCULACEAE

doll's eyes, white baneberry

Rich, moist upland forests, ravines,
flatwoods, mesic savannas. Spring.

An erect, branching perennial, normally
glabrous, 12–24 in. tall. Leaves 2 or 3 times
ternately compound, divided into sharply
toothed, oval to elliptic leaflets 2–4 in. long ×
1–2 in. wide. Flowers ⅛–¼ in. wide, with 4–8
ephemeral petals and numerous spreading
stamens, numerous in a cylindrical cluster 1–6
in. long that terminates a stalk extending well
above the foliage. Fruit a round white (rarely
red) berry, each with a dark "eye" and a stout
red pedicel (see inset). **IA, IL, IN, MI, MN, MO,
OH, WI. Occurs widely throughout the Midwest, less common to absent in far western
counties.** *MAH/SN*

Actaea racemosa
RANUNCULACEAE

black cohosh, black bugbane

Moist woods. Summer.

A perennial to over 7 ft. tall. Leaves basally disposed, glabrous, 2 or 3 times compound with sharply toothed leaflets 4 in. long × 3 in. wide. Flowers regular, to ½ in. long and wide, lacking true petals but with many showy stamens to ½ in. long, numerous in slender, dense, terminal, branched racemes to 2½ ft. long. Fruit a dry follicle. Syn. *Cimicifuga racemosa*. Used medicinally to treat a variety of ailments. Fruit and overall size of plant and inflorescence can be used to distinguish it from superficially similar *A. pachypoda* and *A. rubra*. **IL, IN, MI, MO, OH. Primarily in the Eastern Forests and Ozark Highlands; scattered in the Northern Lakes.** *MJH*

Actaea rubra
RANUNCULACEAE

red baneberry

Rich, moist or swampy forests, upland forests, mesic savannas. Spring.

An erect, branched perennial 12–24 in. tall. Leaves 2 or 3 times ternately compound, divided into sharply toothed, oval to elliptic leaflets 2 in. long and wide, often hairy along leaf veins. Flowers with 4–8 ephemeral petals, about ¼ in. wide, with numerous spreading stamens. Inflorescence of numerous flowers in a cluster 1–4 in. long at the tip of a stalk extending well above the foliage. Fruit a shiny red (rarely white) berry with a dark "eye" and a slender green pedicel. **IA, IL, IN, MI, MN, OH, WI. Principally in the Northern Lakes.** *SN*

Anemone canadensis
RANUNCULACEAE

Canada anemone, meadow anemone

Wet prairies, sedge meadows, open
floodplains, streambanks, shorelines.
Spring, summer.

A 2½-ft.-tall colonial perennial. Basal leaves
long-petiolate, 6 in. long and wide, deeply palmately 3- to 5-lobed and coarsely toothed; involucral bracts similar in shape, sessile. Flowers
regular, to 1½ in. wide, with 5 petal-like sepals
(true petals absent), 1–3 on long stalks sometimes with smaller sessile bracts. Larger flowers,
sessile involucral bracts, and lack of cottony
hairs in fruiting head distinguish it from similar thimbleweed (*A. cylindrica*) and tall thimbleweed (*A. virginiana*). Syn. *Anemonastrum
canadense*. IA, IL, IN, MI, MN, MO, OH, WI.
Widespread but mostly absent from the Ozark
Highlands and southern portion of the Eastern Forests. *MAH*

Anemone cylindrica
RANUNCULACEAE

thimbleweed

Dry prairies and barrens, woodlands,
roadsides, old fields. Spring, summer.

A perennial, erect and unbranched, hairy,
averaging 2 ft. tall. Leaves basal, petiolate,
deeply divided and toothed. Stem with 5–9
petiolate, leaf-like involucral bracts, 3 usually
large and the others small, each 3-parted with
toothed margins somewhat concave at base.
Peduncle usually bractless. Flowers regular,
½–¾ in. wide, with 5 petal-like sepals, fruiting
head cylindrical, 1–1½ in. tall × ½ in. wide.
Differs from tall thimbleweed (*A. virginiana*)
in having a longer fruiting head and more narrowly wedge-shaped leaf segments with teeth
restricted to upper margins. IA, IL, IN, MI,
MN, MO, OH, WI. Mostly in the Tallgrass
Prairie and Northern Lakes. *MJH*

Anemone quinquefolia
RANUNCULACEAE

wood anemone

Various forest types, upland and lowland, moist to relatively dry. Spring.

A low-growing perennial, erect, hairy, 4–8 in. tall, rhizomatous. Basal leaves solitary or absent, leaf-like involucral bracts on stem 3 in a single whorl, petiolate and divided into mostly 3 segments, these commonly lobed, toothed toward the tips, oval to elliptic, 1–1½ in. long × ½–¾ in. wide. Flowers regular, about 1 in. wide with 5 white petal-like sepals, on a slender stalk 1–2 in. long perched above involucral bracts. IA, IL, IN, MI, MN, MO, OH, WI. Most common in the upper Midwest. *SN*

Anemone virginiana
RANUNCULACEAE

tall thimbleweed

Dry forests, forest edges, roadsides, fields. Summer.

An erect perennial, unbranched, hairy, to 2–4 ft. tall. Leaves basal, petiolate and deeply lobed. Stems with usually 3 petiolate, leaf-like involucral bracts midway or so on stems, these mostly 3-parted with coarsely toothed segments convex at base. Peduncles usually with a pair of bracts. Flowers regular, about 1 in. wide, greenish-yellow to white, with 5 petal-like sepals. Fruiting head cone- or thimble-shaped, 1–1½ in. tall × ½ in. wide. See the similar thimbleweed (*A. cylindrica*) for comparison. IA, IL, IN, MI, MN, MO, OH, WI. Common throughout the Midwest. *MAH*

Clematis terniflora
RANUNCULACEAE

sweet autumn clematis,
sweet autumn virgin's-bower

Forest edges, thickets, riverbanks,
homesites. Summer, fall.

A glabrous, perennial vine to 20 ft. Leaves
opposite, long-petiolate, pinnately compound
into 3 or 5 ovate, entire, long-stalked leaflets
to 2½ in. long × 1½ in. wide. Flowers regular,
numerous in stalked clusters of 3–12 in leaf
axils, lacking true petals, with 4 showy, spreading, narrow sepals to 1 in. long and many
protruding stamens and pistils. Fruits form a
fluffy-headed cluster, styles feathery. Flowers
are late blooming and very fragrant. IL, IN,
MI, MN, MO, OH. Introduced from Asia.
Scattered (and increasing) throughout. *DT*

Clematis virginiana
RANUNCULACEAE

virgin's-bower

Thickets, woodlands, streambanks. Summer.

A climbing perennial vine, dioecious, the
stem somewhat branching, mostly glabrous to
sparsely hairy and angled, to 15+ ft. long. Leaves
opposite, petiolate, compound with 3 leaflets,
these ovate to lanceolate and commonly coarsely
toothed, sparsely hairy, about 1½–3½ in. long
× ½–3 in. wide. Flowers regular, about ¾ in.
wide, unisexual or rarely perfect, petals lacking,
with 4 showy sepals and numerous stamens or
pistils (rarely both), in dense axillary panicles,
styles of fruits conspicuously feather-like. The
non-native and invasive sweet autumn clematis
(*C. terniflora*) has entire leaflets and flowers typically perfect. IA, IL, IN, MI, MN, MO, OH, WI.
Common throughout. *SN*

Coptis trifolia
RANUNCULACEAE

threeleaf goldthread

Swamps, moist forests, bogs. Spring.

A colony-forming perennial to 6 in. tall. Leaves basal, evergreen, compound with 3 shiny leaflets each to 1 in. long, leaflets with small lobes at tip, reticulate-veined. Flowers regular, with 5–7 spreading sepals forming solitary ½-in.-wide flower on leafless stem, elevated above leaf; 4–7 true petals are tiny, bright yellow-orange, knob-tipped, somewhat inconspicuous; numerous white stamens and green curled pistils ascend above sepals. Foliage sometimes mistaken for swamp dewberry (*Rubus hispidus*), which is also evergreen but has thicker-textured leaflets broadest above middle. Common name references thin yellow roots. IN, MI, MN, OH, WI. **Widespread in the Northern Lakes and the WI portion of the Tallgrass Prairie.** *PS*

Enemion biternatum
RANUNCULACEAE

false rue anemone

Rich mesic forests, especially ravine bottoms and stream terraces. Spring.

A perennial spring ephemeral, erect, glabrous, colonial, 5–12 in. tall. Leaves basal as well as 1–4 alternate compound leaves on stem, these divided into 3 commonly 3-lobed leaflets about 1 in. long and about as wide, each with a tiny needle-like tip. Flowers regular, ½–¾ in. wide, petals lacking but typically with 5 petal-like sepals, occurring 1–3 in leaf axils. Often confused with rue anemone (*Thalictrum thalictroides*), but that species usually has more than 5 petal-like sepals. Syn. *Isopyrum biternatum*. IA, IL, IN, MI, MN, MO, OH, WI. **Relatively common throughout except in far northern and northwestern counties.** *SN*

Hepatica acutiloba
RANUNCULACEAE

sharp-lobed hepatica, liverleaf

Moist forests. Spring.

A perennial to 6 in. tall. Leaves basal, 3 in.
long and wide, 3-lobed more than halfway to
base with pointed tips, thick-textured, mottled with silvery-green and/or reddish-brown,
undersides often purplish. Flowers regular,
1 in. wide, solitary at ends of naked, hairy
stalks, petals absent but with 5–11 white, pink,
blue, or purple sepals. Syn. *Anemone acutiloba*.
Round-lobed hepatica (*H. americana*) has
blunt-tipped leaf lobes cut halfway to base
(see inset). The two are sometimes treated as
varieties of one species. IA, IL, IN, MI, MN,
MO, OH, WI. Widespread, less frequent in
extreme western and northern portions of
Midwest. *MAH/KB*

Hydrastis canadensis
RANUNCULACEAE

goldenseal

Forests. Spring.

A colonial perennial to 1 ft. tall. Basal leaf 1,
10 in. long and wide, stem leaves 2, alternate,
near top of stem, 8 in. long and wide, palmately 5- to 7-lobed, cleft and sharply toothed,
deeply veined. Flower regular, solitary,
short-stalked, ¾ in. wide, petals absent, with
numerous spreading-ascending white filaments tipped with yellow anthers surrounding
approximately 10 upright green pistils. Fruit
a ½-in.-wide cluster of bright red berries reminiscent of a raspberry (see inset). Roots commonly used medicinally. IA, IL, IN, MI, MN,
MO, OH, WI. Throughout the Ozark Highlands, Eastern Forests, southeastern portion
of the Northern Lakes, and eastern portion
of the Tallgrass Prairie. *MAH/SN*

Myosurus minimus
RANUNCULACEAE

mousetail

Cultivated fields, pastures, pond shores.
Spring.

An erect to ascending annual, glabrous, 1–5 in.
tall. Leaves all basal, linear to oblanceolate,
entire, 1–4 in. long × ⅛–¼ in. wide. Flowers
regular, about ¼ in. wide, with 5 small petals
and sepals, the latter bearing a spur pointing
downward and closely enveloping the pedicel.
Fruit an achene, numerous on an elongated
and "tail-like" receptacle located singly atop
the corolla. This little plant is often found in
agricultural fields before crops are planted.
The botanical name means "little mouse tail."
IA, IL, IN, MI, MN, MO, OH, WI. Occurs
mostly in the lower Midwest. *KB*

Ranunculus longirostris
RANUNCULACEAE

white water crowfoot, white water buttercup

Lakes, ponds, streams, ditches. Summer.

A glabrous, colony-forming, submerged
aquatic perennial to 3 ft. long, with naked
flower stalk emergent to 3 in. above water.
Leaves short-petiolate, alternate, to 1½ in. long
and wide, fan-shaped, very deeply palmately
lobed and divided several times into numerous filiform segments. Flowers regular, to ¾
in. wide, with 5 spreading white petals with
yellow spot at base and many stamens surrounding numerous pistils, solitary on long
stalks opposite upper leaves. Threadleaf crowfoot (*R. trichophyllus*) is less frequent but very
similar, with shorter styles. IA, IL, IN, MI,
MN, MO, OH, WI. Scattered to widespread,
more frequent in the Northern Lakes. *EH*

Thalictrum dasycarpum
RANUNCULACEAE

purple meadow-rue, tall meadow-rue

Wet prairies, marshes, swamps, fens, ditches. Spring, summer.

A dioecious perennial to 7 ft. tall. Leaves alternate, 2 ft. long and wide, long-petiolate, smaller and sessile above, 3 times pinnately compound, leaflets to 2 in. long × 1½ in. wide, with several lobes at tip. Flowers regular, to ⅓ in. wide, petals absent, with 4 or 5 sepals surrounding numerous pistils or stamens (anthers on filiform filaments), in showy, branched, terminal cluster to 2 ft. long. Late meadow-rue (*T. pubescens*) has stamens with thicker filaments. Waxy meadow-rue (*T. revolutum*) has glands on leaflet undersides. Veiny meadow-rue (*T. venulosum*) has toothed leaflet lobes. IA, IL, IN, MI, MN, MO, OH, WI. Widespread except in OH, where scattered eastward. *PR*

Thalictrum dioicum
RANUNCULACEAE

early meadow rue

Mesic forested slopes, cliffs, ravines, creek banks. Spring.

An erect, dioecious perennial, branched, glabrous or glandular, 1–2 ft. tall. Leaves alternate, long-petiolate, compound, 3 or 4 times divided to form a roughly triangular shape, leaflets rounded or kidney-shaped, 3-lobed with scalloped margins, 1–2 in. long and about as wide. Flowers about ¼ in. wide, petals absent, sepals 4, either greenish or purple, males with dangling yellow anthers, female with purplish stigmas, occurring on widely spreading branches from leaf axils and atop main stem. IA, IL, IN, MI, MN, MO, OH, WI. Locally common but rather sparse westward. *PD*

Thalictrum thalictroides

RANUNCULACEAE

rue anemone

Mesic and dry-mesic upland forests. Spring.

An erect perennial, unbranched, glabrous, 4–8 in. tall. Leaves basal, long-petiolate, 2 or 3 times divided into 3 leaflets, these nearly round and bluntly toothed or shallowly lobed, upper leaflike bracts opposite, each divided into 3 leaflets forming a whorl below the flowers, about 1 in. long and wide (or wider). Flowers regular, about ¾ in. wide, with 6–8 sepals (petals absent), numbering 3–6 atop the stem. The petal-like sepals of rue anemone are long-persistent and may be pink-tinged. Syn. *Anemonella thalictroides*. Compare with *Enemion*. **IA, IL, IN, MI, MN, MO, OH, WI. Relatively common, except for far northern and northwestern counties.** *SN*

Ceanothus americanus

RHAMNACEAE

New Jersey tea

Prairies, savannas, glades. Summer.

A low-growing shrub, highly branched, usually hairy, 1–3 ft. tall. Leaves alternate, ovate, short-petiolate, finely toothed, 1–4 in. long × ½–2½ in. wide. Flowers regular, ⅛–¼ in. wide, with 5 white petals and 5 white sepals arranged in dense panicles from leaf axils and branch tips. The common name comes from its use as a tea substitute during the Revolutionary War. A similar species, redroot (*C. herbaceus*), occurs primarily in our western counties; it has narrow, elliptic leaves and flower clusters mostly at branch tips. **IA, IL, IN, MI, MN, MO, OH, WI. Widely scattered but local throughout much of the Midwest.** *SN*

Aruncus dioicus

ROSACEAE

Goat's beard

Mesic, often rocky forested slopes and bluffs. Spring.

An erect perennial with several stems, glabrous or sparsely hairy, dioecious, 2–5 ft. tall. Leaves alternate, petiolate, 2 or 3 times compound, broadly ovate, leaflets lanceolate-ovate, with margins sharply serrated, 2–6 in. long × ¾–3 in. wide. Flowers regular, about ⅛ in. wide with 5 elliptic white petals, male bearing numerous stamens, female with 3 pistils. Inflorescence with 100+ flowers on widely branched, spike-like racemes. Goat's beard occurs almost exclusively on steep hillsides and streambanks. IA, IL, IN, MI, MO, OH, WI. Principally in the Ozark Highlands and unglaciated portions of the Eastern Forests. *MJH*

Drymocallis arguta

ROSACEAE

tall cinquefoil, prairie cinquefoil

Prairies, savannas, fields, roadsides. Summer.

A hairy perennial to 3 ft. tall. Leaves pinnately compound with up to 11 coarsely toothed leaflets to 3 in. long × 2 in. wide that are largest at end of leaf, mostly basal and petiolate, becoming smaller and short-petiolate above. Flowers regular, with 5 white to cream-colored petals, to ¾ in. wide, numerous in small clusters at top of plant. Syn. *Potentilla arguta*. IA, IL, IN, MI, MN, MO, OH, WI. Frequent in the Northern Lakes and Tallgrass Prairie, more commonly in the northwestern portion of the Midwest, with scattered occurrences in the Ozark Highlands and Eastern Forests. *MJH*

Fragaria virginiana
ROSACEAE

wild strawberry

Prairies, open woodlands, old fields, roadsides. Spring.

A perennial with a low-growing cluster of leaves and runners forming new plants, colonial, 3–6 in. tall. Leaves basal, petiolate, with 3 oval to obovate leaflets, glabrous to sparsely hairy, and sharply toothed, 1–3 in. long × ½–1½ in. wide. Flowers regular, ½–1 in. wide, with 5 rounded petals and 5 lanceolate sepals, numbering 3–7 in terminal branching clusters. Fruiting receptacle rounded and fleshy with achenes (seed-like) embedded in pits. Woodland strawberry (*F. vesca*), a mostly northern species with us, has the classic cone-shaped fruiting receptacle with achenes on the surface. IA, IL, IN, MI, MN, MO, OH, WI. **Common and widespread.** *SN*

Geum canadense
ROSACEAE

white avens

Mesic forests, thickets, disturbed sites. Spring, summer.

An erect to ascending perennial, branched, glabrous to downy, 1–2½ ft. tall. Basal leaves 4–10 in. long × 2–5 in. wide, petiolate, mostly divided into 3 or more obovate and serrated leaflets, stem leaves alternate, stipulate, upper ones sessile, simple to trifoliate, 1½–4 in. long, variable. Flowers regular, about ½ in. wide, with 5 oblong petals longer than sepals, on widely spreading leafy racemes. Fruits in a rounded bristly head. Cream avens (*G. virginianum*) is similar, but its cream-colored petals are shorter than sepals. IA, IL, IN, MI, MN, MO, OH, WI. **Widespread and common throughout.** *SN*

Geum laciniatum
ROSACEAE

rough avens

Moist fields, swamps, floodplains,
thickets, ditches. Spring, summer.

A glistening hairy perennial to 2½ ft. tall.
Leaves variable, pinnately compound with
coarsely toothed leaflets of variable size and
large leafy stipules at connection to stem, basal
leaves long-petiolate, to 6 in. long × 3 in. wide,
stem leaves alternate, short-petiolate to sessile,
often with narrower leaflets and lobes than
other avens, terminal leaflet largest. Flowers
regular with 5 petals shorter than sepals, to ½
in. wide, few in upper part of plant. Fruiting
heads spherical, to ¾ in. wide; beaked fruit
with hooked tips. IA, IL, IN, MI, MN, MO,
OH, WI. Widespread in the Northern Lakes,
northern portion of the Tallgrass Prairie, and
eastern portion of the Eastern Forests. SN

Gillenia stipulata
ROSACEAE

Indian physic, American ipecac

Dry to mesic upland forests. Spring,
summer.

An erect perennial, branched, glabrous to
sparsely hairy, 1–3 ft. tall. Leaves alternate,
nearly sessile, undersurface somewhat glandular, compound with 3 lanceolate and toothed
leaflets 2–3 in. long × ½–¾ in. wide, large
leaf-like stipules at base. Flowers nearly regular, about 1 in. wide, with 5 oblong-lanceolate
petals spreading somewhat unequally in length
and spacing, in a loose terminal panicle. Its
common names allude to its use as an emetic
and in treating digestive problems. Some place
it in *Porteranthus*. IL, IN, MI, MO, OH. Occurs
mostly in the lower Midwest. Occurrences in
MI may be introductions. SN

Rubus flagellaris
ROSACEAE

common dewberry

Dry upland woods, old fields, savannas, prairies, disturbed sites. Spring, summer.

A trailing perennial with a woody cane, branched, prickles hooked with broad bases, glabrous or hairy, up to 6+ ft. long, rooting at tip. Leaves alternate, petiolate with stip-ules, compound with mostly 3 rather sharply toothed leaflets, the latter ovate-elliptic, 1–4 in. long × ¾–3 in. wide. Flowers regular, about 1 in. wide, with 5 elliptic to obovate petals, occurring 1–5 at tips of vertical stalks. "Berry" an aggregate of numerous juicy drupelets, edible and delicious. IA, IL, IN, MI, MN, MO, OH, WI. Widespread and mostly common. *PR*

Rubus hispidus
ROSACEAE

swamp dewberry, bristly dewberry

Wet prairies, moist sandy woods, swamps, bogs, swales. Summer.

A trailing woody perennial with stems to 8 ft. long, upright flowering branches to 8 in. tall. Stems bristly-hairy, with weak narrow-based prickles. Leaves alternate, long-petiolate, with 3 shiny leaflets to 2½ in. long × 1¼ in. wide and broadest above middle, coarsely toothed (especially toward apex). Flowers regular, to ¾ in. wide, with 5 petals, in stalked few-flowered clusters from leaf axils. Fruit a dark purple aggregate berry. IA, IL, IN, MI, MN, OH, WI. Scattered to widespread in eastern portion of the Northern Lakes, scattered in eastern portion of the Tallgrass Prairie and in the Eastern Forests. *SN*

Rubus pubescens
ROSACEAE

dwarf raspberry

Moist coniferous forests, swamps, bogs, rock outcrops, cliffs. Spring, summer.

A hairy, trailing perennial to 8 ft. long, flowering branches to 6 in. tall. Lacking prickles. Leaves alternate, long-petiolate, with 3 coarsely toothed leaflets to 3 in. long × 2 in. wide with pointed tips. Flowers regular, ½ in. wide, with 5–7 curled petals, in terminal or axillary stalked clusters of 1–3. Fruit a red aggregate berry. Arctic raspberry (*R. arcticus*), restricted to northern MN and the Upper Peninsula of MI, has purplish-pink flowers and is less hairy (see inset). IA, IL, IN, MI, MN, OH, WI. Widespread in the Northern Lakes, scattered in adjacent portions of the Tallgrass Prairie and the Eastern Forests. *CR/BS*

Sanguisorba canadensis
ROSACEAE

American burnet, Canadian burnet

Wet prairies, fens, sedge meadows. Summer, fall.

A glabrous perennial to 5 ft. tall. Leaves alternate, long-petiolate, to 1½ ft. long, mostly along lower portion of stem, pinnately compound with up to 15 abundantly sharply toothed blunt-tipped short-stalked leaflets each to 2½ in. long × 1 in. wide. Flowers regular, to ¼ in. wide, with 4 petal-like sepals (true petals lacking) surrounding usually 4 long-exserted stamens and single pistil, numerous in several showy, solitary, bottlebrush-like densely flowered spikes to 8 in. long × 1 in. wide on naked stems at top of plant. IL, IN, MI, OH. Rare to scattered, more frequent eastward. *CR*

Sibbaldia tridentata

ROSACEAE

three-toothed cinquefoil, shrubby fivefingers

Savannas, dunes, rock outcrops, rocky shores. Summer.

A matted perennial to 1 ft. tall with reddish stems. Leaves dark green above, alternate but more basally disposed, petiolate below, becoming sessile above, palmately compound with 3 leaflets to 1½ in. long × ¾ in. wide, typically 3-toothed at apex but otherwise entire. Flowers regular with 5 spreading petals, to ¾ in. wide, with numerous exserted stamens and pistils, stalked in clusters of 3–25 at ends of branches at top of stem. Syn. *Sibbaldiopsis tridentata*. IA, IL, MI, MN, WI. Widespread in portions of the Northern Lakes, rare in nearby portions of the Tallgrass Prairie. *SN*

Spiraea alba

ROSACEAE

meadowsweet

Wet prairies, marshes, bogs, shores, sandy swales. Summer.

An erect but often low-growing shrub, sparsely branched, mostly glabrous, 1–5 ft. tall. Leaves alternate, short-petiolate, lance-elliptic to oblanceolate, sharply serrate, somewhat thick, 1–3 in. long × 1 in. wide. Flowers regular, ⅛–¼ in. wide, 5-petaled, numerous and showy, arranged in dense, finely hairy panicles. A few plants recognized as var. *latifolia* have been recorded for northeastern MI. Compared to the common type, its leaves are somewhat broader and the inflorescence glabrous; some consider it a distinct species. IA, IL, IN, MI, MN, MO, OH, WI. Most common in the Northern Lakes. *SN*

Galium aparine
RUBIACEAE

cleavers

Upland and lowland forests, thickets, gardens, weedy sites. Spring, summer.

A sprawling annual, climbing or mat-forming, branched, stems 4-angled with stiff downward-pointing hairs, 1–4 ft. long. Leaves whorled, usually 8 per whorl, narrowly oblanceolate, with stiff hairs on the margins and midvein beneath, 1–2 in. long × ⅛–¼ in. wide. Flowers regular, 1/16–⅛ in. wide, short-tubular with 4 pointed, spreading lobes, 1–3 per stalk from leaf axils. Fruit 2-lobed, rounded and bristly. All parts of this plant easily cleave to clothing and fur. It belongs to the same family as coffee and has been used as a substitute for it. IA, IL, IN, MI, MN, MO, OH, WI. Widespread and common. *MAH*

Galium asprellum
RUBIACEAE

rough bedstraw

Swamps, wet woods, fens, marshes. Spring, summer.

A sprawling, multi-branched perennial with 4-angled stems bearing downwardly pointed bristles on the angles, up to 6 ft. long. Main stem with leaves in whorls of 6, narrowly elliptic, widest above the middle, margins and midvein underneath bristly, tip sharply pointed, to ¾ in. long × ¼ in. wide. Flowers regular, about ⅛ in. wide, short-tubular with 4 pointed, spreading lobes, in axillary and terminal clusters. Fruit 2-lobed, rounded and glabrous. This plant readily clings to clothing and fur. Main stems of the introduced white bedstraw (*G. album*) are somewhat erect and glabrous. IA, IL, IN, MI, MN, MO, OH, WI. Most occurrences are in the Northern Lakes. *PC*

Galium boreale
RUBIACEAE

northern bedstraw

Prairies, sedge meadows, fens, conifer
swamps, upland forest openings. Summer.

An upright perennial with 4-angled, glabrous (rarely hairy) stems to 2 ft. tall. Leaves
spreading, in whorls of 4 with 3–5 prominent
parallel veins, entire, to 2 in. long × ¼ in. wide,
sessile, pointed but blunt-tipped. Flowers
regular, to ⅓ in. wide, with 4 spreading lobes,
numerous in showy panicle at top of plant.
Fruit bristly-hairy (rarely glabrous). The
long, narrow leaves in whorls of 4, coupled
with showy terminal panicle, distinguish this
bedstraw from similar species. IA, IL, IN, MI,
MN, MO, OH, WI. **Primarily in the Northern
Lakes and adjacent northern portion of the
Tallgrass Prairie, with few occurrences elsewhere.** *SN*

Galium circaezans
RUBIACEAE

forest bedstraw, wild licorice

Mesic to dry-mesic upland forests.
Spring, summer.

An erect or ascending perennial, simple or
branched, glabrous or hairy, stems 4-angled,
8–18 in. tall. Leaves in whorls of 4, ovate to
elliptic, entire, with 3 prominent veins, 1–2 in.
long × ⅓–1 in. wide. Flowers regular, 1/16–⅛ in.
wide, short-tubular with 4 pointed, spreading
lobes, sessile on terminal and upper axillary
spreading branches. Fruit 2-lobed, rounded
and bristly. Two varieties exist, the hairy var.
hypomalacum and the mostly glabrous var.
circaezans. Leaves of the similar hairy bedstraw
(*G. pilosum*) usually possess only 1 distinct
vein. IA, IL, IN, MI, MN, MO, OH, WI. **Common and widespread except in region's northern third.** *SN*

Galium concinnum
RUBIACEAE

shining bedstraw

Mesic and dry-mesic upland forests, well-drained lowland forests. Spring, summer.

An ascending or reclining perennial, often weakly branched, glabrous to hairy, 8–18 in. tall. Leaves in whorls of 4–6, sessile, linear-elliptic, entire, mostly glabrous and shiny, ½–1 in. long × ⅛ in. wide. Flowers regular, ⅟₁₆ in. wide, short-tubular with 4 pointed, spreading lobes, on wiry, widely spreading branches from stem tips and upper leaf axils. Fruit 2-lobed, rounded and smooth. In upland habitats a bedstraw with a whorl of 6 narrow shiny leaves and smooth fruits is most likely this species. IA, IL, IN, MI, MN, MO, OH, WI. Widespread and common except rare to absent in the far northern Midwest. *KB*

Galium obtusum
RUBIACEAE

wild madder, bluntleaf bedstraw

Wet prairies, moist forests, floodplains, swamps, fens, seeps, thickets, ditches. Spring, summer.

A weak, ascending to sprawling, branched perennial with 4-angled, glabrous stems to 1½ ft. tall. Leaves spreading, in whorls of 4 or 5 (rarely 6), entire, to 1 in. long × ¼ in. wide, sessile, blunt-tipped, with harsh hairiness along underside midvein. Flowers regular, to ⅛ in. wide, prevailingly with 4 (some with 3) spreading lobes, in small, few-flowered branched clusters from upper leaf axils. Fruit glabrous. The more northern bog bedstraw (*G. labradoricum*) has more erect stems with less branching and shorter leaves that are reflexed, at least at lower nodes. IA, IL, IN, MI, MN, MO, OH, WI. Scattered throughout. *MJH*

Galium tinctorium

RUBIACEAE

stiff marsh bedstraw

Wet prairies, marshes, fens, swamps, thickets, ditches. Spring, summer.

A weak, ascending to sprawling, branched perennial with 4-angled, rough stems to 2 ft. Leaves spreading, in whorls of 4 or 5 (rarely 6), entire, to 1 in. long × ¼ in. wide, sessile, blunt-tipped, with harsh hairiness along underside midvein. Flowers regular, to ¹⁄₁₆ in. wide, prevailingly with 3 (some with 4) spreading lobes, in stalked clusters of 3 from upper leaf axils. Fruit glabrous. Small bedstraw (*G. trifidum*) and short-stalked bedstraw (*G. brevipes*) are very similar, differing primarily in length and hairiness of flower stalks. IA, IL, IN, MI, MN, MO, OH, WI. Widespread, with greater affinity to the Northern Lakes. *KC*

Galium triflorum

RUBIACEAE

fragrant bedstraw

Mesic upland and lowland forests. Spring, summer.

An ascending to trailing perennial, simple to branching, stems 4-angled, glabrous except for downward-pointing bristles on angles, 1–2 ft. long. Leaves in whorls of 6, sessile, elliptic to oblanceolate with a rigid tip, with 1 main vein, glabrous but margins and veins beneath minutely hairy, ½–2 in. long × ½ in. wide. Flowers regular, ⅛ in. wide, short-tubular with 4 pointed, spreading lobes, on widely spreading branches. Fruit 2-lobed, rounded and bristly. The dried foliage of this bedstraw is said to be vanilla-scented (hence the common name), but the fragrance is not evident to everyone. IA, IL, IN, MI, MN, MO, OH, WI. Widespread and common throughout. *KC*

Houstonia longifolia
RUBIACEAE

long-leaved bluets

Prairies, glades, savannas, open woods, usually in rocky or sandy ground. Spring, summer.

A square-stemmed perennial to 10 in. tall. Leaves opposite, entire, with single vein, to 1 in. long × ¼ in. wide, sessile, with whitish papery triangular stipule at base. Basal leaves absent or if present glabrous. Flowers regular, tubular with 4 spreading lobes, purplish-pink to white, to ¼ in. wide, in few-flowered clusters at ends of branches in upper part of plant. Purple bluets (*H. purpurea*) has leaves with 3 veins. Canadian summer bluets (*H. canadensis*) has persistent basal rosettes of partially hairy leaves. IL, IN, MI, MN, MO, OH, WI. Primarily in the Ozark Highlands, Northern Lakes, and eastern portion of the Eastern Forests, scattered elsewhere. *MJH*

Houstonia nigricans
RUBIACEAE

glade bluets, diamond-flowers

Limestone glades, dry exposed bluffs. Summer.

An erect to sprawling perennial, branched, glabrous to hairy, 8–12 in. tall. Leaves opposite, sessile, linear, hairy, ¾–1¼ in. long × ¹⁄₁₆–⅛ in. wide, the lower ones often with axillary leaves present. Flowers regular, about ⅓ in. wide, funnel-shaped and hairy within, with 4 lanceolate lobes about ⅔ the length of the floral tube, in rather compact clusters borne at stem tips. Fruit a capsule, longer than wide. Syn. *Stenaria nigricans*. IA, IL, IN, MI, MO, OH. Occurs mostly in the Ozark Highlands. *MAH*

Mitchella repens

RUBIACEAE

partridgeberry

Acid soils of mesic to dry-mesic forests, rock ledges, steep slopes, hummocks. Summer.

A creeping, branching perennial that commonly forms mats, mostly glabrous, runners rooting at nodes, 4–12 in. long. Leaves opposite, evergreen, thick and glossy with a whitish midvein, short-petiolate, round to ovate, ⅓–¾ in. long and about as wide. Flowers regular, ⅓–½ in. long, trumpet-shaped, with 4 spreading to recurved lobes with long hairs on the inner surface. The flowers occur in pairs that together form a single scarlet, berry-like drupe, the scars of the 2 flowers evident (see inset). IA, IL, IN, MI, MN, MO, OH, WI. **Occurs mostly in the Eastern Forests and Northern Lakes.** *SN/MAH*

Spermacoce glabra

RUBIACEAE

smooth buttonweed

Floodplain forests, swamps, riverbanks, lakeshores. Summer.

An erect or ascending perennial, stems branched, glabrous or finely hairy, squarish, 8–24 in. tall. Leaves opposite, sessile or short-petiolate, with bristle-like stipules at base, blades narrowly elliptic to oblong-lanceolate with impressed veins on upper leaf surfaces, entire, 1–3 in. long × ¼–1 in. wide. Flowers regular, about ⅛ in. long, tubular with 4 spreading lobes, bearded within, several densely clustered in rounded heads at upper stem nodes. Smooth buttonweed can be distinguished from similar-looking but non-related water horehounds (*Lycopus*) by its possession of bristle-like stipules. IL, IN, MO, OH. **Locally common but scattered, primarily in Midwest's southern third.** *MAH*

Comandra umbellata
SANTALACEAE

bastard toadflax

Dry upland woodlands, barrens, prairies, dunes. Spring, summer.

An erect perennial, simple or branched above, glabrous, 3–18 in. tall, rhizomatous and colonial. Leaves alternate, sessile, elliptic-lanceolate, 1–2 in. long × ¼–½ in. wide. Flowers regular, ⅛–¼ in. wide, bell-shaped, petals none, with 5 lanceolate petal-like sepals, in flat-topped clusters atop stems. Bastard toadflax is hemiparasitic on roots of a variety of plants. It is an alternate host for a rust fungus that infects Jack pine (*Pinus banksiana*) and certain other pine species. The rust is generally not considered a serious problem in our region. **IA, IL, IN, MI, MN, MO, OH, WI. Widespread and locally common.** *SN*

Saururus cernuus
SAURURACEAE

lizard's tail

Floodplain forests, swamps, pond margins, vernal pools, shallow streams. Summer.

An erect perennial, branched, hairy to glabrous, 2–3 ft. tall, usually colonial. Leaves alternate, petioles sheathing at base, cordate-ovate, entire, 3–6 in. long × 2–4 in. wide. Flowers ⅛–¼ in. long, lacking petals and sepals, stamens and pistils showy, these densely arranged on 3- to 8-in.-long terminal and axillary spikes, which droop or nod at the tips. Fragrant. Both its common and scientific names allude to the shape of the inflorescence. **IL, IN, MI, MO, OH. Common south and east of a diagonal line from eastern MI to west-central MO.** *CB*

Micranthes pensylvanica
SAXIFRAGACEAE

swamp saxifrage

Swamps, marshes, sedge meadows, seeps. Spring, summer.

A densely hairy perennial, to 40 in. tall. Leaves in a basal rosette, entire or with scalloped margins, 9 in. long × 3 in. wide, short-petiolate (petiole often reddish). Flowers regular, ¼ in. wide, with 5 spreading, narrow petals (see inset), numerous in terminal panicle to 1½ ft. long on naked, solitary stem. Forbes' saxifrage (var. *forbesii*; syn. *M. forbesii*) occurs in southern part of range on moist sandstone bluffs and cliffs. IA, IL, IN, MI, MN, MO, OH, WI. Widespread through much of the Northern Lakes and adjacent Tallgrass Prairie, scattered elsewhere. *SN*

Micranthes virginiensis
SAXIFRAGACEAE

early saxifrage

Cliffs, rocky slopes, upland woods. Spring.

A glabrous to glandular-hairy perennial, 4–12 in. tall. Leaves in a basal rosette, oblong-ovate to elliptic, somewhat fleshy, with shallowly toothed margins, blades 1–2½ in. long × ½–1½ in. wide. Flowers regular, about ¼–⅓ in. wide, with 5 oblong to elliptic petals, in a compact terminal inflorescence that becomes open and lax at maturity. Early saxifrage was formerly placed in *Saxifraga* ("stone-breaker"), a genus named for its purported medicinal ability to break up kidney stones. IL, IN, MI, MN, MO, OH. Occurs almost exclusively in eastern OH, the Ozark Highlands, and counties bordering the Ohio River and Lake Superior. *BS*

Mitella diphylla
SAXIFRAGACEAE

bishop's cap, miterwort

Rich mesic forests, rock outcrops, swamps.
Spring.

An erect, unbranched perennial, hairy to
glandular-hairy, 6–12 in. tall. Leaves basal,
long-petiolate, ovate to rounded with 3–5
pointed lobes, 1–3 in. long and nearly as
wide, and a single pair of smaller, sessile,
3-lobed opposite leaves at about mid-stem.
Flowers regular, ⅛–¼ in. wide, with 5 delicately fringed petals resembling a snowflake
(see inset), several occurring on a slender
spike-like raceme. When mature the open
bowl-like capsule (the "bishop's cap," or mitre)
exposes its shiny black seeds. IA, IL, IN, MI,
MN, MO, OH, WI. Widespread and relatively
common but rare to absent in our southwestern counties. *SN/CR*

Sullivantia sullivantii
SAXIFRAGACEAE

sullivantia

Moist, shaded, mostly calcareous cliffs.
Spring, summer.

An erect to ascending perennial, stems mostly
glabrous or with scattered glandular hairs,
6–12 in. tall. Leaves predominantly in a basal
rosette, petiolate, round to kidney-shaped,
glossy, the margins shallowly lobed and
toothed, 1–4 in. long and about as wide.
Flowers regular, about ¼ in. wide, bell-shaped
with 5 small, flaring to slightly reflexed,
ovate-oblanceolate lobes, occurring on delicate stems in spreading panicles. Sullivantia
leaves look somewhat like those of alumroot
(*Heuchera*) but are quite glossy by comparison.
IA, IL, IN, MN, MO, OH, WI. Locally common in widely disjunct areas of occurrence.
It is a Midwest near-endemic. *KC*

Tiarella cordifolia
SAXIFRAGACEAE

heartleaf foamflower, false miterwort

Moist forests, swamps. Summer.

A hairy, colony-forming perennial to 15 in. tall. Leaves basal, roundish with 3–7 shallow, palmate lobes, cordate at base, broadly toothed, to 4 in. long and wide, dark green with deeply impressed veins, long-petiolate. Flowers regular, to ⅓ in. wide, with 5 spreading sepals alternating with 5 spreading petals surrounding 10 ascending stamens and single pistil, numerous, on stalks to ⅖ in. long in terminal raceme on naked, solitary stem. Fruit a pair of unequal, pointed, ovate capsules; the genus name *Tiarella* ("little tiara") references the shape of the fruit. MI, MN, OH, WI. Scattered to widespread in eastern portion of both the Eastern Forests and the Northern Lakes. *MAH*

Verbascum blattaria
SCROPHULARIACEAE

moth mullein

Fields, gravel bars, roadsides, railroads, waste areas. Spring, summer.

A biennial to 5 ft. tall. Basal leaves in rosette, pinnately lobed, somewhat wavy-margined, to 6¾ in. long × 2 in. wide. Stem leaves alternate, unlobed, coarsely toothed, sessile to clasping, to 5 in. long × 2 in. wide, smaller above, deeply veined and appearing wrinkled. Flowers slightly irregular, white or yellow with 5 spreading petals fused at purple base, to 1 in. wide; numerous in a glandular terminal raceme. Fruit a round capsule. IA, IL, IN, MI, MN, MO, OH, WI. Introduced from Eurasia. Widespread in the Ozark Highlands, Eastern Forests, eastern portion of the Tallgrass Prairie, and southern portion of the Northern Lakes, scattered to rare elsewhere. *MAH*

Datura stramonium
SOLANACEAE

jimsonweed

Barnyards, pastures, waste areas. Summer.

An erect annual, branched, glabrous, rank-smelling when bruised, 3–5 ft. tall. Leaves alternate, petiolate, ovate, irregularly lobed and toothed, 3–8 in. long × 1½–5 in. wide. Flowers regular, usually single, 2–5 in. long, funnel-shaped with deep purple throat and 5 sharply pointed lobes angled somewhat like a pinwheel, often in forking branch axils, usually opening in late afternoon and fading the next day. Fruit a prickly capsule (see inset). This species is often found where livestock are kept, avoided by them due to its highly poisonous qualities. IA, IL, IN, MI, MN, MO, OH, WI. Presumably introduced from the New World tropics. Occurs mostly in the lower Midwest. *MAH/SN*

Leucophysalis grandiflora
SOLANACEAE

large false ground-cherry, white-flowered ground-cherry

Floodplains, gravel pits, dumps, construction sites, roadsides; in sand or gravel. Summer.

A glandular-hairy, bushy annual or short-lived perennial to 3 ft. tall. Leaves simple, alternate, wavy-margined, to 5 in. long × 2½ in. wide, petiolate, ovate with pointed tip. Flowers regular, white with yellow center, pentagonal, very shallowly lobed with 5 fused corolla lobes, to 1½ in. wide, facing outward or nodding, on short stalks, in clusters of 2–4 in upper leaf axils, sometimes as many as 300 per plant. Fruit a roundish berry to ½ in., enclosed in papery sac with open tip formed by mature calyx. MI, MN, WI. Uncommon in northern portion of the Northern Lakes. *OG*

Solanum carolinense
SOLANACEAE

horse nettle

Prairies, pastures, fields, cultivated areas.
Spring, summer.

An erect to sprawling perennial, weakly
branched, hairy, spiny, rhizomatous, 1–2 ft.
tall. Leaves alternate, petiolate, lance-ovate
with a few large, angled teeth or lobes along
the margins, usually with some spines along
the main veins beneath, 3–6 in. long × ½–3 in.
wide. Flowers regular, corolla ¾–1 in. wide,
with 5 white to pale purple lobes united at
base into a star-shaped disk, several in a cluster. Fruit a berry, resembling a small yellow
(but toxic) tomato. This species is considered
native, at least in our far southern counties.
IA, IL, IN, MI, MN, MO, OH, WI. Very common in the Midwest's southern half. *MAH*

Solanum ptychanthum
SOLANACEAE

black nightshade

Disturbed ground, forest openings,
gardens, waste areas. Spring, summer.

An erect to sprawling annual, glabrous to
sparsely hairy, 1–2 ft. tall. Leaves alternate,
petiolate, blades ovate to triangular, irregularly toothed or lobed, wavy to entire, 1–4 in.
long × ½–2 in. wide. Flowers regular, corolla
¼–⅓ in. wide with 5 white to light purple and
often reflexed lobes, 2–7 in stalked umbel-like
clusters. Fruit a berry with a shiny surface.
When mature it resembles a tiny (but toxic)
black tomato. IA, IL, IN, MI, MN, MO,
OH, WI. **Common and possibly in every
county.** *MAH*

Trillium flexipes
TRILLIACEAE

drooping trillium, declined trillium

Forests, swamps, floodplains. Spring.

A glabrous perennial, unbranched, to 20 in.
tall. Leaf-like bracts 3 atop stalk, whorled,
sessile, entire, to 6½ in. long and wide. Flower
regular, solitary, white to maroon, to 2 in.
wide, stalked, nodding and often held below
bracts, with 3 recurved petals, stamens with
very short filaments. Nodding trillium (*T.
cernuum*) has filaments as long as anthers.
Stinking Benjamin (*T. erectum*) has a maroon
ovary, longer petals, and flowers held above
bracts. IA, IL, IN, MI, MN, MO, OH, WI.
Widespread but very local. *MAH*

Trillium grandiflorum
TRILLIACEAE

large-flowered trillium

Forests, swamps. Spring.

A glabrous perennial, unbranched, to 1½ ft.
tall. Leaf-like bracts 3 atop stalk, whorled,
sessile or nearly so, entire, to 6 in. long × 5 in.
wide. Flower regular, solitary, white (aging
pink) or rarely pale pink, to 4 in. wide, facing
up or to side, on ascending stalk to 3 in., with 3
recurved petals. Fruit a capsule. Seeds spread
by ants. IA, IL, IN, MI, MN, OH, WI. **Widespread in the Northern Lakes and eastern
portion of the Eastern Forests, becoming
scattered to rare westward into the Tallgrass
Prairie.** *SN*

Trillium nivale
TRILLIACEAE

snow trillium

Steep gravelly or rocky slopes, bluffs, high riverbanks, usually in calcareous soils. Spring.

A glabrous perennial, unbranched, 2–5 in. tall. Leaf-like bracts 3 atop stalk, whorled, petiolate, elliptic to lance-ovate, entire, bluish-green, 1–2½ in. long × ⅓–1¼ in. wide. Flowers regular, 1–2 in. wide, with 3 elliptic to oblong petals and 3 lanceolate sepals, solitary, mostly erect and stalked above the leaf-like bracts, reflexed in fruit. Snow trillium is the first of our region's trilliums to bloom, sometimes as early as late February. Its aboveground portion dies back just weeks after blooming. IA, IL, IN, MI, MN, MO, OH, WI. **Widespread but very local.** *MAH*

Valeriana officinalis
VALERIANACEAE

garden valerian, garden heliotrope

Fields, shores, ditches, thickets, roadsides. Summer.

A colony-forming perennial to 6 ft. tall. Leaves basal and opposite in lower half of green to red stem, deeply pinnately lobed with up to 19 entire to sparsely toothed lobes, to 8 in. long × 6 in. wide, long-petiolate below, becoming sessile above. Flowers regular, to ¼ in. long, white to tinged pink, tubular with 5 spreading lobes, stalked in branched, flat to dome-shaped clusters to 5 in. wide at top of plant. IA, IL, IN, MI, MN, OH, WI. **Introduced from Eurasia. Scattered in the Northern Lakes, becoming uncommon to rare in the Tallgrass Prairie and Eastern Forests.** *PD*

Valeriana uliginosa
VALERIANACEAE

marsh valerian, swamp valerian

Wet prairies, conifer swamps, fens. Summer.

A perennial to 3 ft. tall. Leaves basal and opposite in lower half of stem, basal leaves unlobed or with a pair of small basal lobes, to 5½ in. long × 3 in. wide, stem leaves few, deeply pinnately lobed with up to 5 entire to sparsely toothed lobes, to 6¼ in. long × 3½ in. wide, petiolate. Flowers regular, to ¼ in. long, tubular with 5 spreading lobes, stalked in branched flat cluster to 3 in. wide at top of plant. IL, IN, MI, OH, WI. Scattered to rare in eastern portion of the Northern Lakes. *SN*

Valerianella radiata
VALERIANACEAE

beaked cornsalad

Prairies, fields, glades, savannas, fens, forest openings, roadsides. Spring.

A yellow-green annual to 16 in. tall. Leaves basal and opposite, simple, to 3 in. long × 1 in. wide, with distinct midvein; basal leaves entire, short-petiolate; stem leaves entire or coarsely toothed at clasping base. Flowers regular, to ⅛ in. wide, tubular with 5 spreading lobes, in tight, flat-topped clusters to 1 in. wide, clusters surrounded by large bracts, numerous clusters in broad inflorescence at top of plant. Lewiston cornsalad (*V. locusta*), introduced, has pale blue flowers. IL, IN, MO, OH. Widespread in the Ozark Highlands, southern portion of the Tallgrass Prairie, and western portion of the Eastern Forests; rare elsewhere. *MAH*

Valerianella umbilicata
VALERIANACEAE

northern cornsalad

Fields, swamps, floodplains, seeps, roadsides. Spring, summer.

A yellow-green annual to 2 ft. tall. Leaves basal and opposite, simple, to 3 in. long × 1 in. wide, with distinct midvein; basal leaves entire, short-petiolate; stem leaves entire or coarsely toothed at sessile to clasping base. Stems with hairs on angles. Flowers regular, to ⅙ in. wide, tubular with 5 spreading lobes, in tight, flat-topped clusters to 1½ in. wide, several widely spaced clusters in broad inflorescence at top of plant. Goosefoot cornsalad (*V. chenopodiifolia*) has glabrous stems or hairs only at nodes. IL, IN, MI, OH. **Widespread in eastern portion of the Eastern Forests, becoming scattered westward; rare in the Northern Lakes and Tallgrass Prairie.** *BS*

Phyla lanceolata
VERBENACEAE

fog fruit, frog fruit

Mudflats, swamps, marshes, shorelines. Spring, summer.

A trailing perennial with ascending to erect flowering branches, rooting at nodes, hairy, 6–18 in. tall. Leaves opposite, sessile or short-petiolate, lance-elliptic with sharply toothed margins, 1–3 in. long × ½–1 in. wide. Flowers irregular, about ⅛ in. wide, tubular and 2-lipped, the lower one larger with 3 lobes, the central "eye" commonly yellow or dark pink, clustered in cone-shaped heads emerging from leaf axils. Common names probably reference the damp environment in which the plant grows. IA, IL, IN, MI, MN, MO, OH, WI. **Common and widespread except for much of the Northern Lakes.** *MAH*

Verbena urticifolia
VERBENACEAE

white vervain

Mesic lowland and upland forests, flatwoods, clearings. Summer.

An erect perennial, branched above, hairy, 2–6 ft. tall. Leaves opposite, petiolate, broadly lanceolate to ovate-lanceolate, coarsely toothed, 4–6 in. long × 1–2½ in. wide. Flowers nearly regular, about ⅛ in. wide, tubular, imperceptibly 2-lipped, with 5 tiny lobes rounded at tips. Inflorescence of long, widely spreading terminal and axillary spikes. This is the only vervain in the Midwest with consistently white flowers, and the one most likely to be encountered in forest habitats. IA, IL, IN, MI, MN, MO, OH, WI. Common, mostly throughout. *MJH*

Viola arvensis
VIOLACEAE

field pansy, wild pansy

Sandy fields, lawns, homesites, roadsides. Spring.

An annual to 14 in. tall. Leaves basal and alternate; basal leaves ovate, long-petiolate, to 1 in. long and wide, bluntly toothed; stem leaves similar but narrower, with pair of large, leaf-like, pinnately lobed stipules at bases of petioles. Flowers irregular, to ⅗ in. wide, with 5 spreading petals, white to creamy yellow with few dark purple veins and large bright yellow spot on petal with short spur on back, subtended by pointed sepals slightly longer than petals; solitary on long, erect stalks from upper leaf axils. Crushed roots emit a wintergreen odor. IA, IL, IN, MI, MN, MO, OH, WI. **Introduced from Europe. Rare to scattered throughout, most frequent to the east.** *KB*

Viola blanda
VIOLACEAE

sweet white violet

Moist forests, swamps, floodplain forests, thickets. Spring.

A perennial to 6 in. tall. Leaves basal, cordate, to 3 in. long × 2½ in. wide, long-petiolate, bluntly toothed. Flowers irregular, to ½ in. wide, white with purple veins in throat, with 5 spreading to recurved petals, 1 short-spurred; solitary on naked stalks from base, about as tall as leaves. Smooth white violet (*V. macloskeyi*) has smaller flowers that overtop the flatter-toothed leaves. Kidney-leaved violet (*V. renifolia*) has largest leaves kidney-shaped and is restricted to our northern region. IA, IL, IN, MI, MN, OH, WI. Widespread in the Northern Lakes and eastern portion of the Eastern Forests, scattered westward and south into northern portion of the Tallgrass Prairie. *SN*

Viola canadensis
VIOLACEAE

Canada violet, Canadian white violet

Rich forests. Spring, summer, fall.

An upright perennial to 14 in. tall. Leaves basal and alternate, cordate, tapering to long pointed tip, to 4 in. long × 3 in. wide (basal shorter), petiolate (short-petiolate above), bluntly toothed, with pair of narrow, entire stipules to ⅓ in. where petioles attach to stem. Flowers irregular, to ⅘ in. wide, bright white with purple veins into bright yellow throat, with 5 spreading to recurved petals purplish on back, 1 with short spur, lateral petals bearded at base, solitary on stalks from leaf axils. IA, IL, IN, MI, MN, OH, WI. Widespread to scattered in the Northern Lakes and eastern portion of the Eastern Forests, becoming rare into the Tallgrass Prairie. *SN*

Viola lanceolata
VIOLACEAE

lance-leaved violet

Wet sand prairies and fields, marshes, swamps, bogs, shores. Spring, summer.

A perennial to 6 in. tall. Leaves basal, lance-olate, ascending, to 6 in. long × ¾ in. wide, tapering to short petioles, bluntly toothed. Flowers irregular, to ⅔ in. wide, with 5 spreading petals, upper 2 petals white with yellow bases, lower 3 petals white with purple veins in throat, 1 short-spurred; solitary on long, naked stalks from base. Primrose-leaved violet (*V. primulifolia*) has more ovate leaves that are truncate at base; it is often considered a hybrid between lance-leaved violet and smooth white violet (*V. macloskeyi*). IA, IL, IN, MI, MN, MO, OH, WI. Scattered throughout, most frequent in the Northern Lakes. SN

Viola striata
VIOLACEAE

cream violet, striped white violet, pale violet

Moist to wet forests, floodplains, streambanks, thickets, disturbed sites. Spring, summer.

A mostly glabrous, sprawling, colony-forming perennial to 1 ft. tall. Leaves basal and alternate, cordate, to 2½ in. long × 2 in. wide, petiolate, bluntly toothed, with pair of ragged-toothed stipules to 1 in. where petioles attach to stem. Flowers irregular, to ¾ in. wide, solitary on stalks from leaf axils, creamy white, with 5 spreading to recurved petals, 1 with short spur and purple veins, lateral petals bearded at base. IA, IL, IN, MI, MO, OH, WI. Widespread in the Ozark Highlands, the Eastern Forests, southern portion of the Tallgrass Prairie, and southeastern portion of the Northern Lakes. MAH

PINK
TO RED
FLOWERS

Allium canadense
ALLIACEAE

wild garlic

Moist upland forests, well-drained floodplains, seeps, fields. Spring.

A glabrous perennial herb 10–20 in. tall. Ascending grass-like leaves, linear and flat, mostly basal, 5–18 in. long × ¼ in. wide. Flowers regular, ⅜–½ in. wide, with 6 tepals. Inflorescence a terminal umbel. Quite commonly only a few or no flowers are present in a cluster, often replaced by stalkless bulblets. All parts of the plant possess an onion-like odor. Compare with false garlic (*Nothoscordum bivalve*). **IA, IL, IN, MI, MN, MO, OH, WI. Widely scattered and common, sparse to absent in far northern counties.** *SN*

Allium cernuum
ALLIACEAE

nodding wild onion

Meadows, prairies, rocky woodlands. Summer.

A glabrous perennial herb with a single angled stem, 10–20 in. tall. Leaves grass-like, basal, linear, flat, 8–12 in. long × ¼–⅓ in. wide. All parts of plant possess an onion-like odor. Flowers regular, ¼ in. long, somewhat bell-shaped with 6 tepals on 1-in. stalks. Inflorescence a terminal rounded umbel 2–3 in. across that nods atop a glabrous stem. The nodding helps distinguish it from the similar-looking autumn onion (*A. stellatum*), the stem of which is erect at flowering; autumn onion occurs mostly in MN and MO. **IA, IL, IN, MI, MN, MO, OH, WI. Locally common.** *SN*

Allium vineale
ALLIACEAE

wild onion

Agricultural fields, gardens, lawns, disturbed sites. Spring, summer.

A glabrous, unbranched perennial, 1–2 ft. tall. Leaves linear, 8–12 in. long, round in cross section and hollow, positioned on lower half of flowering stalk, smelling of onion when bruised. Flowers regular, about ¼ in. long, with 6 purple to white tepals. Inflorescence a terminal ball-like umbel atop flowering stalk. Flowers commonly absent, replaced by stalkless bulblets intermixed with thread-like green leaves. This is a common weedy onion. It might be confused with wild garlic (*A. canadense*), but the latter has leaves flat in cross section. IA, IL, IN, MI, MO, OH. Introduced from Europe. Found mostly in our region's southern half. *SN*

Amaranthus tuberculatus
AMARANTHACEAE

tall waterhemp

Shores, agricultural fields, mudflats, moist to wet open areas, disturbed sites. Summer, fall.

A glabrous, dioecious annual (male plant pictured), from a few inches to 6+ ft. tall when blooming. Often with reddish stems. Leaves alternate, entire, ovate-lanceolate, shiny, to 6 in. long × 1½ in. wide. Flowers regular, lacking petals, to ⅛ in. long and wide, with bracts that give densely flowered terminal and axillary inflorescences green or reddish-pink hue. There are similar species, some of which are monoecious, have hairy leaves, and/or grow in drier areas. IA, IL, IN, MI, MN, MO, OH, WI. Widespread throughout the Midwest, but less frequent in northern portions of the Northern Lakes. *SN*

Apocynum androsaemifolium
APOCYNACEAE

spreading dogbane

Dry open woods, woodland edges, savannas, prairies, dunes, roadsides. Spring, summer.

A perennial to 5 ft. tall. Leaves opposite, entire, to 4 in. long × 2 in. wide, short-petiolate, spreading or drooping, containing milky sap. Flowers regular, bell-shaped with 5 spreading to recurved lobes, to ⅓ in. long and wide, whitish to light pink with deep pink vertical stripes inside, nodding, stalked in clusters of up to 10 at ends of branches or from axils of upper leaves. Fruit a long, skinny, dangling bean-like follicle that dries and splits to release seeds, each with long white silky hairs attached to one end. IA, IL, IN, MI, MN, MO, OH, WI. **Widespread but generally with a more northern distribution.** *SN*

Asclepias amplexicaulis
APOCYNACEAE

sand milkweed, clasping milkweed

Barrens, dunes, glades, dry savannas, gravel and sand prairies. Spring, summer.

An erect to ascending perennial, unbranched, glabrous, glaucous, 1–3 ft. tall, with milky sap. Leaves opposite, sessile and cordate-clasping, oblong and entire with wavy margins, 3–5 in. long × 1–3 in. wide. Flowers regular, about ½ in. wide, with 5 mostly green reflexed corolla lobes beneath 5 pink hoods and horns, up to 40 borne in a large umbel. Fruit a smooth follicle, with flattened seeds each bearing a tuft of long hairs. IA, IL, IN, MI, MN, MO, OH, WI. **Most occurrences are in eastern IA, northern IL, and southern WI.** *SN*

Asclepias incarnata
APOCYNACEAE

swamp milkweed

Floodplains, open swamps, wet prairies, marshes, fens, pond margins. Summer.

A perennial to 5 ft. tall. Leaves opposite, entire, glabrous, to 6 in. long × 1½ in. wide, short-petiolate, containing milky sap. Flowers regular, pink to magenta (rarely white), to ¼ in. long, with 5 reflexed-spreading corolla lobes and 5 erect corona hoods, each with protruding horn, on stalks to 2 in. long in numerous dense umbels in upper part of plant. Fruit a smooth, lanceolate follicle to 4 in. long. Fresh flowers emit bubblegum odor. Aquatic milkweed (*A. perennis*; see inset) has white flowers, roots at lower nodes, and has leaves tapering at both ends. IA, IL, IN, MI, MN, MO, OH, WI. Widespread. *SN*

Asclepias purpurascens
APOCYNACEAE

purple milkweed

Prairies, barrens, savannas, open woodlands and flatwoods. Spring, summer.

An erect, unbranched perennial, mostly glabrous to minutely hairy, 1½–3 ft. tall, with milky sap. Leaves opposite, petiolate, elliptic to oblong and entire, somewhat narrowed toward tip, mostly glabrous but undersurface hairy, 2–8 in. long × 1–4 in. wide. Flowers regular, about ⅓ in. wide, with 5 glabrous, reflexed corolla lobes beneath 5 purple hoods and shorter incurved horns, arranged in 1 to few terminal, rounded umbels of up to 50 flowers. Fruit a smooth follicle with flattened seeds each bearing a tuft of long hairs. IA, IL, IN, MI, MN, MO, OH, WI. **Most populations occur south of the Great Lakes.** *PR*

Asclepias sullivantii
APOCYNACEAE

Sullivant's milkweed, prairie milkweed

Prairies. Summer.

A glabrous, glaucous perennial to 3½ ft. tall. Leaves opposite, entire, ascending, wavy-margined, 6 in. long × 3 in. wide, with pink or cream-colored midvein, cordate at base, sessile, containing milky sap. Flowers regular, pink to rose, to ¾ in. long, with 5 reflexed-spreading corolla lobes and 5 erect corona hoods, each with incurved horn, stalked in few convex umbels. Fruit a lanceolate, smooth to warty follicle to 4 in. long. Vegetatively resembles dogbane (*Apocynum cannabinum*), which has shorter, thinner-textured leaves with less-pronounced lateral veins. IA, IL, IN, MI, MN, MO, OH, WI. Primarily in the Tallgrass Prairie, also in southern portions of the Northern Lakes and in adjacent Eastern Forests. *SN*

Asclepias syriaca
APOCYNACEAE

common milkweed

Prairies, fields, roadsides. Spring, summer.

An erect, unbranched perennial, short-hairy, 2–6 ft. tall, with milky sap. Leaves opposite, short-petiolate, entire, oblong to elliptic, densely hairy underneath, 3–12 in. long × 1–4 in. wide. Flowers regular, about ¼ in. wide, with 5 reflexed, pale pink corolla lobes, hairy beneath, and 5 whitish-pink hoods longer than the incurved horns, up to 100 in terminal and axillary umbels, fragrant. Fruit a warty follicle with flat seeds each bearing a tuft of long hairs. Common milkweed is a major food source for larvae of eastern North American monarchs. IA, IL, IN, MI, MN, MO, OH, WI. Throughout, probably in every Midwestern county. *MAH*

Carduus nutans
ASTERACEAE

musk thistle, nodding thistle

Prairies, fields, roadsides, railroads, waste areas. Summer, fall.

A biennial to 7 ft. with spiny-winged stems, white-woolly and less spiny-winged above. Leaves lobed, margins silvery-white, spiny, wavy; basal leaves in rosettes, stem leaves sessile, smaller, alternate. Flowers in composite heads; numerous disk flowers in solitary, spherical, nodding terminal heads, 3 in. wide, subtended by large, triangular, spine-tipped purplish phyllaries. Fruit with silky pappus. Plumeless thistle (*C. acanthoides*) has smaller flower heads and narrower phyllaries, spiny-winged stems throughout, and more branched arrays; its leaves lack silvery-white edges. IA, IL, IN, MI, MN, MO, OH, WI. Introduced from Eurasia. Throughout, less frequent in northern MI and northern WI. *KB*

Centaurea stoebe
ASTERACEAE

spotted knapweed

Prairies, fields, savannas, vacant lots, railroads, roadsides. Summer, fall.

A gray-green biennial or short-lived perennial to 4 ft. tall. Basal leaves to 8 in. long × 2 in. wide, stem leaves smaller, occasionally simple (at base and near inflorescence) but usually deeply lobed with linear segments. Flowers in composite heads; ray flowers absent, disk to 1 in. wide with spreading 5-lobed flowers around perimeter of smaller erect 5-lobed flowers. Syns. *C. biebersteinii*, *C. maculosa*. Several *Centaurea* species look similar, differing only in leaf shape and phyllary characteristics. IA, IL, IN, MI, MN, MO, OH, WI. Introduced from Eurasia. Throughout, less frequent in the Eastern Forests and western portion of the Tallgrass Prairie. *MJH*

Cirsium altissimum
ASTERACEAE

tall thistle

Dry-mesic forests, well-drained stream terraces, savannas, thickets. Summer, fall.

A biennial or monocarpic perennial, stem erect, branched, somewhat glabrous or with spreading hairs, to 6–9 ft. tall. Leaves alternate, entire to variously lobed, lanceolate, ovate, or elliptic, with a few spiny teeth on margins, densely white-hairy on undersurface, up to 18 in. long × 8 in. wide. Flowers in composite heads about 2 in. wide, disk flowers 100+, ray flowers absent, involucres prickly, heads occurring at tips of branches. Tall thistle is one of our least spiny thistles and is more at home in woodland environments than most thistles. IA, IL, IN, MI, MN, MO, OH, WI. Throughout the Midwest, less common northward. SN

Cirsium arvense
ASTERACEAE

Canada thistle

Pastures, fallow fields, roadsides, disturbed sites. Summer, fall.

An erect, dioecious perennial, rhizomatous and colonial, upwardly branched, glabrous to somewhat hairy stems, 1–5 ft. tall. Leaves alternate, lanceolate to oval, pinnately lobed and spiny, glabrous to densely hairy especially on the undersurface, 1–8 in. long × ½–2½ in. wide, basal leaves absent at flowering. Flowers in composite heads ½–1 in. wide, disk flowers 50–100, ray flowers absent, involucres prickly or only slightly so, numerous on branch tips. This highly invasive weed is especially notable along roadsides. IA, IL, IN, MI, MN, MO, OH, WI. Introduced from Eurasia. Mostly throughout and common. MJH

Cirsium discolor

ASTERACEAE

field thistle

Prairies, old fields, pastures. Summer, fall.

An erect biennial, branched, stems thinly hairy or nearly glabrous, to 6–8 ft. tall. Leaves alternate, variously lobed, often narrow with sinuses almost to midrib, lanceolate-elliptic, margins spiny, undersurface densely white-hairy, to 18 in. long × 8 in. wide. Flowers in 2-in.-wide composite heads, disk flowers 100+, ray flowers lacking, involucres prickly, heads on short peduncles at tips of leafy branches. Similar species include Flodman's thistle (*C. flodmanii*) in our northwest and Carolina thistle (*C. carolinianum*) in the south. The former has white hairy stems; the latter has smaller flowering heads on elongated, seemingly leafless peduncles. IA, IL, IN, MI, MN, MO, OH, WI. Widespread throughout. *MJH*

Cirsium hillii

ASTERACEAE

Hill's thistle

Prairies, savannas. Summer.

A short-lived perennial to 2 ft. tall. Leaves alternate, lobed, densely hairy, with spiny, wavy margins, to 8 in. long. Flowers in composite head; numerous disk flowers in solitary terminal head to 3 in. long; head subtended by long, narrow phyllaries with spreading to appressed spine tips. Fruiting head with silky pappus. Shorter than bull thistle (*C. vulgare*) and lacking spiny wings on stem. Sometimes considered a variety or subspecies of pasture thistle (*C. pumilum*), which is branched with longer spines and known from eastern OH. IA, IL, IN, MI, MN, MO, WI. A near-endemic of the Midwest. Rare in the Northern Lakes and northern portion of the Tallgrass Prairie. *SN*

Cirsium muticum
ASTERACEAE

swamp thistle, fen thistle

Wet prairies, sedge meadows, marshes, swamps, fens, bogs. Summer, fall.

A hairy biennial to 8 ft. tall. Stems spineless. Leaves alternate, lobed, with short spines on margins, to 10 in. long × 4 in. wide, smaller above. Flowers in composite heads; numerous disk flowers in 2-in.-long terminal heads in upper part of plant, subtended by long, narrow, sticky phyllaries with tightly appressed spines at tips (spines sometimes absent). Fruiting head with silky pappus. Marsh thistle (*C. palustre*), invasive, has more flower heads, spiny-spreading phyllaries, and conspicuously spiny-winged stems (see inset). IA, IL, IN, MI, MN, MO, OH, WI. Widespread in the Northern Lakes, scattered elsewhere. *SN*

Cirsium vulgare
ASTERACEAE

bull thistle

Pastures, fallow fields, prairies, meadows, roadsides, disturbed sites. Summer, fall.

An erect biennial, branched, stems hairy, to 7 ft. tall. Leaves alternate, deeply segmented in a pinnate fashion, prickly on upper and lower surfaces, undersurface not whitened with woolly hairs, leaf bases extending down stem producing thin ribs of spiny tissue, larger leaves 4–15 in. long × 1–6 in. wide. Flowers in composite heads 1–2 in. wide, disk flowers 100+, ray flowers absent, heads arranged on tips of branches. Involucres prickly. The somewhat similar-looking plumeless thistle (*Carduus acanthoides*) has flower heads about half the size. IA, IL, IN, MI, MN, MO, OH, WI. Introduced from Eurasia. Common. *MAH*

Echinacea pallida
ASTERACEAE

pale purple coneflower

Prairies, glades, savannas, roadsides, railroads. Spring, summer.

A hairy perennial to 3 ft. tall. Leaves basal and alternate on lower stem, to 9 in. long × 2 in. wide, entire, long-petiolate. Flowers in composite head; a deep purplish-red disk is surrounded by up to 20 reflexed lavender to pale pink ray flowers, in a single head to 3 in. wide. Narrow-leaved purple coneflower (*E. angustifolia*) has shorter ray flowers (to 1½ in. long). Glade purple coneflower (*E. simulata*) has yellow pollen (usually white in pale purple coneflower). IA, IL, IN, MI, MN, MO, WI. Widespread in southern portion of the Tallgrass Prairie and the Ozark Highlands, extending into adjacent Northern Lakes; few occurrences in the Eastern Forests. *EH*

Echinacea purpurea
ASTERACEAE

purple coneflower

Open woodlands, barrens, savannas, prairies, thickets. Summer, fall.

An erect perennial, mostly unbranched, hairy, 2–4 ft. tall. Leaves alternate, ovate to lanceolate, entire or with a few teeth, petiolate (especially lower ones), rough to the touch, 2–12 in. long × 1–5 in. wide. Flowers in composite heads 3–5 in. wide, ray flowers 10–20, disk flowers 200+, greenish-purple. This is one of our most popular cultivated native plants. The prickly pointed bracts give rise to the genus name *Echinacea*, from the Greek *echinos*, a root word shared by the class of animals (Echinoidea) that includes spiny sea urchins. IA, IL, IN, MI, MO, OH, WI. Occurs naturally primarily in the lower Midwest. *MAH*

Eutrochium maculatum
ASTERACEAE

spotted Joe-Pye-weed

Marshes, wet prairies, sedge meadows, bogs, fens, cedar swamps, ditches. Summer.

An unbranched (except inflorescence) colony-forming perennial to 8 ft. tall. Stems purple or purple-spotted. Leaves in whorls of 3 to 6, simple, short-petiolate, sharply toothed, to 9 in. long × 2½ in. wide, with impressed reticulate venation. Flowers in composite heads; ray flowers absent, 8–20 disk flowers in heads to ¼ in. wide, heads numerous, forming flattish-topped arrays to 6 in. across at top of plant. Syns. *Eupatoriadelphus maculatus*, *Eupatorium maculatum*. IA, IL, IN, MI, MN, MO, OH, WI. Widespread in the Northern Lakes and adjacent Tallgrass Prairie, occasional in the Eastern Forests. *PS*

Eutrochium purpureum
ASTERACEAE

purple-node Joe-Pye-weed

Moist forests, wooded ravines, savannas, stream corridors. Summer, fall.

An erect, unbranched perennial, mostly glabrous, stems green but commonly purple at nodes and mostly with solid pith, up to 6 ft. tall. Leaves whorled, usually 4, lanceolate or narrowly ovate, sharply toothed, 4–12 in. long × 1–3 in. wide. Flowers in composite heads ⅛–⅙ in. wide, ray flowers absent, disk flowers 4–8, in a dome-shaped array. In forest environments it is the most likely Joe-Pye-weed to be encountered. Hollow-stemmed Joe-Pye-weed (*E. fistulosum*) prefers open sites, its stem mostly purplish throughout (or streaked) and hollow, at least below. IA, IL, IN, MI, MN, MO, OH, WI. Common except for far northern counties. *SN*

Liatris aspera
ASTERACEAE

rough blazingstar, rough gayfeather

Prairies, glades, savannas, open woods, clearings, roadsides. Summer, fall.

An unbranched perennial to 5 ft. tall. Leaves alternate, simple, entire, to 12 in. long × 1 in. wide at base of plant, progressively smaller above, with pronounced midvein. Flowers in composite heads; ray flowers absent, numerous small 5-lobed disk flowers in heads to 1 in. wide, in sparse to dense terminal spike-like array to 1½ ft. long; heads sessile or nearly so, subtended by many inflated, jagged-tipped phyllaries in several series; tips of phyllaries whitish to colorless. IA, IL, IN, MI, MN, MO, OH, WI. Widespread in the Ozark Highlands, the Tallgrass Prairie, and the southern portion of the Northern Lakes, scattered elsewhere. *SN*

Liatris cylindracea
ASTERACEAE

cylindrical blazing star

Barrens, hill prairies, sand prairies, glades, blufftops, dunes. Summer, fall.

An erect perennial, unbranched, mostly glabrous, 1–2 ft. tall. Leaves alternate, entire, linear, sessile to short-petiolate, 4–10 in. long × ⅛–½ in. wide. Flowers in composite heads, about 1 in. long and wide, ray flowers lacking, disk flowers lobed and hairy within, 10–30, the cylindrical heads with tightly overlapping and appressed green to purple scale-like phyllaries. Inflorescence a spike-like raceme. It is one of our smaller blazing stars. IA, IL, IN, MI, MN, MO, OH, WI. Widespread but local in much of IN and OH as well as our far northern and western counties. *KB*

Liatris pycnostachya
ASTERACEAE

prairie blazingstar, prairie gayfeather

Prairies, glades, fields. Summer, fall.

An unbranched perennial to 5 ft. tall. Leaves numerous, alternate, entire, to 12 in. long × ½ in. wide at base of plant, progressively smaller above, with pronounced midvein. Flowers in composite heads; ray flowers absent, several small 5-lobed disk flowers in heads to ½ in. wide, in dense terminal spike-like array to 1½ ft. long; heads sessile, subtended by narrow, recurved phyllaries in several series. Marsh blazingstar (*L. spicata*) has appressed phyllaries. Dotted blazingstar (*L. punctata*) is shorter with fewer large, ciliate, appressed phyllaries. IA, IL, IN, MI, MN, MO, OH, WI. Throughout the Ozark Highlands and the Tallgrass Prairie, scattered elsewhere; introduced in MI and eastward. *SN*

Liatris scariosa
ASTERACEAE

savanna blazingstar

Prairies, glades, savannas, barrens, clearings. Summer, fall.

An unbranched perennial to 5 ft. tall. Leaves alternate, simple, entire, to 12 in. long × 1½ in. wide at base of plant, progressively smaller above, with pronounced midvein. Flowers in composite heads; ray flowers absent, disk flowers abundant, small, 5-lobed, spreading in heads to 2 in. wide, in sparse to dense terminal raceme to 2 ft. long; heads clearly stalked, subtended by many flat, entire to scalloped phyllaries in several series. Meadow blazingstar (*L. ligulistylis*) has ragged-tipped phyllaries and is more frequent northwestward. Appalachian blazingstar (*L. squarrulosa*) has heads sessile or nearly so with at least lower phyllaries spreading. IL, IN, MI, MO, OH, WI. Spotty. *SN*

Liatris squarrosa
ASTERACEAE

scaly blazing star

Barrens, glades, rocky slopes, dry
woodlands, upland prairies. Summer, fall.

An erect perennial, unbranched, glabrous
to mostly hairy, 1–2 ft. tall. Leaves alternate,
entire, linear, sessile to short-petiolate, 4–10
in. long × ⅛–½ in. wide. Flowers in composite
heads about 1 in. wide, ray flowers lacking,
disk flowers lobed and hairy within, 10–40,
grouped in a cylindrical to cup-shaped head,
outer phyllaries spreading, long-pointed
with recurved tips. Inflorescence a spike-like
raceme. Three varieties are known, separated
by differences in phyllaries and hairiness. IA,
IL, IN, MI, MO, OH. Occurs mostly in the
lower Midwest. *ST*

Palafoxia callosa
ASTERACEAE

Spanish needles, small palafox

Glades, bluffs, streambanks. Summer, fall.

An annual to 16 in. tall, hairy below, with
black, stalked glands above and in inflorescence. Leaves alternate, simple, entire, with
pointed tips, to 2 in. long × ⅛ in. wide, sessile
or short-petiolate. Flowers in composite
heads; ray flowers absent, disk flowers pink,
rose, lavender, or white (and fading to white
with age), tubular with 5 spreading lobes, to ⅖
in. long, heads to ¾ in. wide, at ends of leafless
branches in broad, open array in upper part of
plant. Leaves deciduous by flowering when in
dry conditions. MO. **Restricted to the Ozark
Highlands.** *ST*

Pluchea camphorata
ASTERACEAE

camphorweed, stinkweed, camphor pluchea, marsh-fleabane

Swamps, floodplain forests, ponds, ditches. Summer, fall.

An erect annual or short-lived perennial, unbranched, mostly hairy, 3–5 ft. tall. Leaves alternate, lanceolate to ovate-elliptic, petiolate and toothed, 2–6 in. long × 1–3 in. wide, surfaces with rounded, stalkless glands. Flowers in composite heads about ⅓–¼ in. wide, ray flowers lacking, disk flowers 30+, arranged in flat-topped inflorescences from upper leaf axils and atop stem. The leaves and stems emit a somewhat skunk-like smell upon bruising. **IL, IN, MO, OH. Found in the far southern Midwest. PR**

Cynoglossum officinale
BORAGINACEAE

common hound's tongue, gypsy flower

Fields, pastures, disturbed woods, roadsides. Spring, summer.

A densely soft-hairy biennial to 4 ft. tall. Leaves basal and alternate, oblong, with winged petioles in basal rosettes, sessile with round to clasping bases on stems, to 12 in. long × 5 in. wide, smaller along stem. Flowers regular, with 5 spreading lobes, fused at base, to ⅓ in. wide, stalked and drooping in numerous racemes at top of plant. Fruit 4-parted, ½ in. wide, with rounded nutlets covered in hooked hairs. The "sticktight" fruit attach to passing mammals or human clothing. Nutlets much larger than those of more common Virginia sticktight (*Hackelia virginiana*). **IA, IL, IN, MI, MN, MO, OH, WI. Introduced from Eurasia. Scattered throughout the Midwest. MJH**

Brasenia schreberi
CABOMBACEAE

watershield

Lakes, ponds, swamps, slow-moving streams. Spring, summer, fall.

An aquatic perennial with long stems bearing a thick coating of clear jelly-like mucilage, growing up to several feet in length, colonial. Leaves floating, alternate, broadly elliptic, petiolate and peltate, with a thick gelatinous coating underneath, glabrous and water-repellent above, 1½–5 in. long × ¾–3¼ in. wide. Flowers regular, ¾ in. wide, held just above water surface, bearing 3 linear-oblong petals and 3 similar sepals, all recurved. Individual flowers bloom for only two days, the first day as functionally female, the second as functionally male. IA, IL, IN, MI, MN, MO, OH, WI. **Most common in the Northern Lakes.** *SN*

Lobelia cardinalis
CAMPANULACEAE

cardinal flower

Floodplain forests, swamps, marshes, streambanks, ditches. Summer, fall.

An erect perennial, unbranched, glabrous or sparsely hairy, 1–4 ft. tall, with milky sap. Leaves alternate, short-petiolate, lanceolate with shallow teeth, 1–6 in. long × ¼–1½ in. wide. Flowers irregular, about 1 in. wide, tubular, 5-lobed, 3 of which are positioned lowermost and 2 spreading laterally, numerous in a tall spike-like raceme atop stem. Ruby-throated hummingbirds frequent cardinal flower in search of nectar, as do a variety of insects. IA, IL, IN, MI, MN, MO, OH, WI. **Relatively common but mostly absent in the northwest.** *MAH*

Lonicera dioica
CAPRIFOLIACEAE

limber honeysuckle, red honeysuckle, glaucous honeysuckle

Savannas, forests, thickets, swamps, sedge meadows, rock outcrops. Spring, summer.

A slender woody vine to 10 ft. long. Leaves gray-green, opposite, entire, blunt-tipped, to 3½ in. long × 2½ in. wide, short-petiolate to sessile, the upper 1–3 pairs clasping and fused around stem. Flowers irregular, to 1 in. long, red (sometimes maroon or orange-yellow), tubular, terminating in 2 lips, upper lip broad with 4 lobes, lower lip narrow, tube with protrusion at base; in tight cluster of up to 10 at ends of branches. Fruit an oblong red berry. IA, IL, IN, MI, MN, MO, OH, WI. Widespread in the Northern Lakes and adjacent portions of the Tallgrass Prairie, scattered elsewhere. *SN*

Lonicera sempervirens
CAPRIFOLIACEAE

trumpet honeysuckle

Fencerows, forest edges, flatwoods, old homesites. Spring, summer.

A perennial or woody vine, twining, to 15+ ft. long. Leaves opposite, oblong to elliptic, glabrous to short-hairy beneath and glaucous, entire, the terminal ones rounded and perfoliate, 1–3 in. long × ½–2 in. wide. Flowers nearly regular, about 2 in. long, tubular with 5 short lobes of almost equal length, in whorls at tips of new growth. This southern U.S. species is widely cultivated and has in places escaped and naturalized but rarely becomes a nuisance. May be native in our far southeastern counties. IA, IL, IN, MI, MO, OH. Widespread in lower Midwest, occasionally encountered. *SN*

Symphoricarpos albus
CAPRIFOLIACEAE

common snowberry

Barrens, dry rock outcrops, savannas, bluffs, dunes. Summer.

A small shrub, branched, glabrous to hairy, stems hollow between leaf nodes, commonly 1–2+ ft. tall. Leaves opposite, short-petiolate, elliptic to oval, 1–1½ in. long × ½–1 in. wide. Flowers regular, about ¼ in. wide, pale pink to white, bell-shaped with 5 short and pointed lobes, stamens and style not exserted, in clusters of 1–3 dangling mostly in leaf axils. Fruit white and berry-like. The flowers of western snowberry (*S. occidentalis*) have stamens and style exserted beyond the floral lobes. IA, IL, MI, MN, OH, WI. Occurs mostly in the Northern Lakes. *PR*

Triosteum perfoliatum
CAPRIFOLIACEAE

late horse gentian

Moist and well-drained to dry upland forests, open woodlands. Spring, summer.

An erect perennial, unbranched, glabrous or short-hairy, some hairs tipped with glands, to 3 ft. tall. Leaves opposite, united at their bases on lower and mid-stem, obovate to elliptic, 4–8 in. long × 1–4 in. wide. Flowers irregular, about ½ in. long, tubular, with 5 unequal lobes, reddish to yellow, hairy, 2 or 3 per leaf axil. Fruit round, reddish, topped with persistent sepals. Bases of mid-stem leaves of early horse gentian (*T. aurantiacum*) not united as is true for the southern-ranging yellow horse gentian (*T. angustifolium*). The latter differs from both by having bristly stems. IA, IL, IN, MI, MN, MO, OH, WI. Widespread but local. *EH*

Dianthus armeria
CARYOPHYLLACEAE

Deptford pink

Fields, pastures, open woodlands, roadsides, railroads, waste areas. Spring, summer, fall.

An annual to 2 ft. tall. Leaves basal and opposite, hairy (along stem) or glabrous and glossy (in basal rosette), to 3 in. long × ¼ in. wide. Flowers regular, to ½ in. wide with 5 spreading, toothed petals, sessile in few-flowered clusters at top of plant, base of each flower subtended by 3 conspicuous narrowly lanceolate green bracts. The white-spotted hot-pink petals coupled with purple anthers make this a rather attractive and conspicuous weed. IA, IL, IN, MI, MN, MO, OH, WI. Introduced from Europe. Common throughout the Midwest, with fewer occurrences in southwestern MN and northwestern IA. *SN*

Silene regia
CARYOPHYLLACEAE

royal catchfly

Prairies, glades, savannas. Spring, summer.

An erect perennial, unbranched, short-hairy and glandular, 1½–4 ft. tall. Leaves opposite, sessile, entire, lanceolate to ovate, hairy to glabrous, 2–5 in. long × 1–2 in. wide. Flowers regular, about 1 in. wide, petals 5, unnotched at tip or with a few small teeth, calyx tubular and glandular, in open panicles. In nature, this spectacular species is mostly restricted to the Midwest. Habitat destruction threatens its wild populations. Royal catchfly is often used in landscaping and prairie plantings. The brilliantly colored flowers are pollinated by hummingbirds. IL, IN, MO, OH. Widely scattered and rare, with most occurrences in the Ozark Highlands. *MAH*

Silene virginica
CARYOPHYLLACEAE

fire pink

Open woods (often rocky). Spring, summer.

A short-hairy, somewhat sticky perennial to 2 ft. tall. Leaves basal and opposite, to 6 in. long × 1 in. wide, entire, short-petiolate, becoming smaller and sessile above and glabrous with age. Flowers regular, to 1 in. long × 1½ in. wide, tubular with 5 spreading, deeply notched, narrow, laterally toothed corolla lobes, with 10 stamens, 3-parted style, and appendages at base of corolla lobes short-exserted from tube, subtended by tubular calyx nearly as long as corolla tube with 5 short triangular lobes at top; short-stalked at ends of branches at top of plant. IL, IN, MI, MO, OH, WI. Widespread in the Eastern Forests and the Ozark Highlands, scattered elsewhere. *MAH*

Streptopus lanceolatus
CONVALLARIACEAE

rose twisted-stalk

Forests, conifer swamps. Spring, summer.

An arching, colony-forming perennial to 16 in. tall with zigzag stem. Leaves sessile to clasping, simple, alternate, to 4 in. long × 1¾ in. wide, with ciliate margins and numerous parallel veins. Flowers regular, pale pink to maroon, bell-shaped with 6 tepals recurved at tips, to ⅓ in. long, 1 or 2 dangling and hidden on stalks from leaf axils. Fruit a deep red ⅓-in. round berry. Clasping-leaved twisted-stalk (*S. amplexifolius*) has whitish flowers and lacks ciliate leaf margins. Yellow fairybells (*Prosartes lanuginosa*), in eastern OH, has spreading yellow tepals and hairy leaf undersides. IA, MI, MN, OH, WI. Widespread northward. *SN*

Sedum pulchellum
CRASSULACEAE

glade stonecrop, widow's cross

Glades, cliff ledges, rocky woodlands.
Spring, summer.

An erect annual or biennial, branched above,
glabrous, succulent, 1–6 in. tall. Leaves alternate, sessile, linear, rounded in cross section,
⅛–1 in. long × 1/16–⅛ in. wide. Flowers regular,
⅓–½ in. wide with 4 linear-lanceolate petals,
arranged along the upper side of branches that
spread laterally from atop the plant (forming
something like a horizontal cross). Widow's
cross usually completes its life cycle before its
environment gets too hot and dry. Most completely expire by the middle of summer. IL,
IN, MO, OH. **In the Midwest occurs naturally
only in southern IL and southern MO.** *MAH*

Chimaphila umbellata
ERICACEAE

pipsissewa

Forests. Summer.

An evergreen perennial to 10 in. tall. Leaves
opposite to whorled, restricted to lower stem,
sharply toothed, to 2 in. long × ¾ in. wide,
glabrous, leathery, shiny, short-petiolate.
Flowers regular, to ¾ in. wide with 5 spreading
to reflexed white to pink petals and 10 pinkish
anthers around conspicuous green pistil, nodding, on stalks at top of plant. Fruit a round
capsule on erect stalk. Spotted wintergreen
(*C. maculata*) has white petals (see inset) and
dark green leaves with white middle portion
extending into some lateral veins. IA, IL,
IN, MI, MN, OH, WI. **Widespread in the
Northern Lakes and adjacent portions of the
Tallgrass Prairie, uncommon in the Eastern
Forests.** *SN*

Kalmia polifolia
ERICACEAE

bog-laurel, pale-laurel, swamp-laurel

Bogs, fens, muskegs, swales. Spring, summer.

An evergreen shrub to 2½ ft. tall. Leaves simple, opposite, glossy green above, strongly white-hairy beneath, with margins rolled under, to 1¾ in. long × ½ in. wide, sessile or nearly so. Flowers regular, pentagonal, very shallowly lobed with 5 fused corolla lobes, to ½ in. wide, stalked in terminal dome-shaped clusters of up to 13. Sheep-laurel (*K. angustifolia*) has leaves mostly in whorls of 3 and flower clusters in leaf axils. Mountain-laurel (*K. latifolia*), a larger shrub mostly in southeastern OH, has bigger, alternate leaves and larger, usually white flowers. MI, MN, WI. Widespread in the Northern Lakes. *SN*

Pyrola asarifolia
ERICACEAE

pink pyrola, pink wintergreen

Moist forests, cedar swamps, fens. Summer.

A glabrous evergreen perennial to 1 ft. tall. Leaves basal, kidney-shaped (somewhat similar in shape to wild ginger, *Asarum*), entire, to 1½ in. long and usually wider, cordate at base, shiny, dark green, thick-textured, on petioles often longer than leaves. Flowers regular, pink or white with pink margins, downward-facing, to ¾ in. wide, with 5 spreading petals surrounding exserted stamens and curved style, subtended by scale-like bract, on spreading stalks, 7–15 in raceme at top of solitary naked stem. IA, IN, MI, MN, WI. Scattered to widespread in the Northern Lakes and adjacent portions of the Tallgrass Prairie. *MJH*

Apios americana
FABACEAE

American groundnut, Indian potato, American potato bean, hopniss, cinnamon vine

Streambanks, wet meadows, marshes, seeps, thickets, moist woodlands. Summer, fall.

A perennial herbaceous vine to 20 ft. Leaves alternate, pinnately compound, with usually 5 or 7 leaflets (sometimes 3 on younger stems); leaflets stalked (terminal stalk longest), to 4 in. long × 2½ in. wide, margins entire. Flowers irregular with typical pea family shape, to ¾ in. wide, in densely flowered racemes to 6 in. long on stalks from leaf axils. Fruit a legume to 4 in. long. IA, IL, IN, MI, MN, MO, OH, WI. Widespread in most of the Midwest, less common in western MN, western IA, northern WI, and northern MI (including Upper Peninsula). *SN*

Lathyrus latifolius
FABACEAE

everlasting pea, perennial pea

Fields, woodland margins, vacant lots, homesites, roadsides, railroads, waste areas. Spring, summer, fall.

A gray-green perennial herbaceous vine to 6 ft. long, sprawling or climbing by tendrils. Stems broadly winged. Leaves alternate, on winged petioles with 2 leafy stipules at base, compound with 2 entire leaflets to 3 in. long × 1 in. wide and terminating in branched tendril. Flowers irregular with typical pea family shape, showy, to 1 in. wide, in clusters of up to 10 from leaf axils. Fruit a flattened legume to 2 in. long. *Lathyrus sylvestris* (same common names) has longer, narrower leaflets. IA, IL, IN, MI, MN, MO, OH, WI. Introduced from Europe. Scattered, less frequent in northwestern portion of the Midwest. *DT*

Mimosa nuttallii
FABACEAE

Nuttall's sensitive-brier, cat-claw

Prairies, glades, open woods, thickets, roadsides. Spring, summer, fall.

A sprawling, prickly perennial to 4 ft. long, reaching 2 ft. tall. Leaves twice pinnately compound, alternate, to 4¾ in. long with numerous leaflets to ¼ in. long, sensitive to touch (folding in half lengthwise). Flowers regular, tubular, with 5 inconspicuous lobes surrounding 5–10 long-exserted stamens with yellow anther and pink filament, very numerous in dense spherical clusters to ¾ in. wide, reminiscent of a Koosh ball, on stalks to 2¾ in. from leaf axils. Syns. *M. quadrivalvis, Schrankia nuttallii*. IA, IL, MI, MO, WI. Widespread in MO, scarce elsewhere; introduced in MI and WI. *MJH*

Phaseolus polystachios
FABACEAE

wild bean

Thickets, open woodlands, forest edges, rocky slopes. Summer.

A perennial twining and climbing vine, branched, glabrous to hairy, 6–15 ft. long. Leaves alternate, petiolate, compound with 3 pointed leaflets, broadly ovate with the lateral ones somewhat asymmetric, entire, 1½–5 in. long × 1–3 in. wide. Flowers irregular with typical pea family shape, ⅓–½ in. long, the keel strongly coiled. Inflorescence of branching axillary racemes. Wild bean is in the same genus as cultivated beans. It is being researched for possible use in crossbreeding with cultivated species for disease resistance. IL, IN, MI, MO, OH. Occurs mostly in the southern third of the Midwest. *CB*

Securigera varia
FABACEAE

crownvetch

Roadsides, fields, ditches, prairies, woodlands, waste areas. Spring, summer.

An ascending and sprawling perennial, somewhat branched to almost bushy, mostly glabrous, 1–3 ft. tall. Leaves alternate, sessile, pinnately compound, 2–6 in. long with 11–25 oblong leaflets ½–¾ in. long × ¼ in. wide. Flowers irregular with typical pea family shape, about ½ in. long, the banner petal usually darker pink than others. Inflorescence a long-stalked umbel from a leaf axil. Fruit a linear 4-angled legume. Crownvetch is invasive and difficult to control. Syn. *Coronilla varia*. IA, IL, IN, MI, MN, MO, OH, WI. **Introduced from Eurasia. Widespread and common throughout.** *MAH*

Strophostyles helvola
FABACEAE

annual wild bean

Mesic and dry-mesic woodlands, savannas, thickets. Summer.

An annual vine, twining and climbing, branched, glabrous to sparsely hairy, 3–6 ft. long. Leaves alternate, petiolate, compound with 3 ovate and often bluntly lobed leaflets, 1–2½ in. long × ½–1 in. wide. Flowers irregular with typical pea family shape, ⅓–½ in. wide, the keel strongly curled upward into a beak, calyx mostly glabrous, occurring in few-flowered clusters terminating long axillary stalks. Fruit a narrow, finely hairy legume with seed surfaces dull. Perennial wild bean (*S. umbellata*) is more southern with slightly smaller and typically unlobed leaflets. IA, IL, IN, MI, MN, MO, OH, WI. **Common except in Northern Lakes.** *MAH*

Strophostyles leiosperma
FABACEAE

small-flowered wild bean

Sand prairies and barrens, acidic glades, sandy fields, eroded slopes. Summer.

An annual trailing and twining vine, branched, hairy, 1–4 ft. long. Leaves alternate, petiolate, compound with 3 linear to narrowly oblong-elliptic and unlobed leaflets 1–2 in. long × ¼–¾ in. wide. Flowers irregular with typical pea family shape, about ¼ in. wide, the keel strongly curled upward into a beak, the calyx mostly hairy. Inflorescence few-flowered clusters terminating long axillary stalks. Fruit a narrow, often densely hairy legume with seeds normally shiny. IA, IL, IN, MN, MO, OH, WI. Occurs primarily in western half of the Midwest. *PD*

Tephrosia virginiana
FABACEAE

goat's rue

Dry woodlands, sand savannas, acidic glades, dunes. Spring, summer.

An erect perennial, normally unbranched, hairy, 1–2 ft. tall. Leaves alternate, petiolate, stipulate, pinnately compound, up to 5 in. long, with 9–27 narrowly elliptic to oblong leaflets ½–1¼ in. long × ¼ in. wide. Flowers irregular with typical pea family shape, about ¾ in. wide, bicolored with a cream-colored banner petal and pink wings and keel, occurring in terminal racemes. Fruit a hairy legume. IA, IL, IN, MI, MN, MO, OH, WI. Locally common and widespread but mostly absent from the far north and northwest as well as IN and OH till plains. *PR*

Trifolium arvense
FABACEAE

rabbitfoot clover, hare's-foot clover

Sandy fields, roadsides, railroads, waste areas. Spring, summer, fall.

A hairy, branching, gray-green annual to 16 in. tall. Leaves alternate, short-petiolate, compound with 3 entire to sparsely toothed, sessile (or nearly so) leaflets to 1 in. long × ⅓ in. wide; a pair of pointed leaf-like stipules at base of each leaf. Flowers irregular with typical pea family shape, individually inconspicuous, white to pale pink, to ¼ in. long, subtended by longer, greenish-red, feathery calyx, numerous in tight, fuzzy, pink-tinged, cylindrical clusters to 1½ in. long × ½ in. wide at ends of stems and branches. IA, IL, IN, MI, MN, MO, OH, WI. **Introduced from Eurasia. Scattered throughout, most frequent in the Northern Lakes and Ozark Highlands.** *SN*

Trifolium hybridum
FABACEAE

alsike clover

Fields, roadsides, pastures, meadows, disturbed sites. Spring, summer.

An ascending perennial, branched, glabrous, 6–18 in. tall. Leaves alternate, petiolate with lanceolate stipules at base, compound with 3 rounded to oblanceolate to elliptic leaflets, these finely toothed ½–1½ in. long × ¼–1 in. wide. Flowers irregular with typical pea family shape, tubular, calyx lobes slightly longer than tube, pink and white flowers mixed in heads about ¾ in. wide on stalks from upper leaf axils. Fruit a legume, hidden by withered flower. Ignore its specific epithet: it is not a hybrid. IA, IL, IN, MI, MN, MO, OH, WI. **Introduced from Eurasia. Common, most likely in every county.** *MAH*

Trifolium pratense
FABACEAE

red clover

Pastures, old fields, meadows, roadsides, disturbed sites. Spring, summer.

An erect or ascending perennial, branched, hairy, 6–24 in. tall. Leaves alternate, petiolate with oblong, bristle-tipped stipules at base, compound with 3 broadly elliptic to ovate leaflets, these ½–2½ in. long × ⅓–1 in. wide, entire to finely toothed and often with a light green chevron patch. Flowers irregular with typical pea flower shape, tubular. Inflorescence of numerous flowers in rounded, mostly sessile heads 1–1½ in. wide positioned above a pair of opposite leaves. Fruit a legume, hidden by withered flower. IA, IL, IN, MI, MN, MO, OH, WI. Introduced from Europe. Common, probably in every county. *MAH*

Centaurium pulchellum
GENTIANACEAE

branching centaury, showy centaury

Fields, lawns, roadsides, railroads. Summer, fall.

A glabrous annual to 8 in. tall. Stems 4-angled. Leaves opposite, entire, somewhat clasping, to 1 in. long × ¾ in. wide. Flowers regular, pink with yellow throat, ¼ in. wide, tubular with 5 spreading lobes, tube mostly concealed by sepals, stalked in branched clusters. Forking centaury (*C. erythraea*), introduced, is less frequent, known from near Lake Michigan and Lake Erie; it has a basal rosette and is slightly larger with sessile flowers with calyces shorter than corolla tubes. IA, IL, IN, MI, MN, OH, WI. Introduced from Eurasia and northern Africa. Mostly in the Northern Lakes, primarily distributed near the Great Lakes. *SN*

Sabatia angularis

GENTIANACEAE

rosepink, rose gentian

Dry to moist prairies, glades, savannas,
old fields, swales, roadsides. Summer.

An erect annual or biennial, branched above,
glabrous, stems 4-angled, 1–2 ft. tall. Leaves
opposite, sessile and clasping, ovate, entire,
½–1½ in. long × 1 in. wide. Flowers regular,
1–1½ in. wide, short-tubular with 5 elliptic to
oblanceolate lobes, these pink to occasionally white, the bases of which are commonly
yellow-green bordered by red, the inflorescence of terminal, oppositely branching
clusters. Western rose gentian (*S. campestris*)
is mostly in open, dry sites in southern IL and
MO; its inflorescence has alternate branching.
**IL, IN, MI, MO, OH. Common, mostly in the
lower Midwest.** *SN*

Hypericum fraseri

HYPERICACEAE

Fraser's marsh St. John's-wort

Bogs, fens, marshes, sedge meadows,
swamps. Summer.

A glabrous gray-green perennial, to 2 ft. tall.
Leaves opposite, entire, clasping, to 2½ in.
long × 1¼ in. wide. Flowers (see inset) regular,
usually closed, to ¾ in. wide, with 5 petals,
3 groups of 3 stamens, and 3 styles to ¹⁄₁₆ in.
in clusters in upper part of plant. Fruit a
3-sectioned, ½-in.-long red capsule. Virginia
marsh St. John's-wort (*H. virginicum*) has
longer styles. Greater marsh St. John's-wort
(*H. walteri*) has short-petiolate leaves. These
all have translucent glands on leaves; glands
absent on lesser marsh St. John's-wort (*H. tub-
ulosum*). Pink-flowered St. John's-worts sometimes placed in *Triadenum*. **IA, IL, IN, MI,
MN, OH, WI. Widespread northward.** *PD*

Monarda bradburiana
LAMIACEAE

eastern beebalm

Dry upland woods, rocky slopes, usually acidic soils. Spring, summer.

An erect perennial, stems 4-angled, glabrous to hairy, 1–1½ ft. tall. Leaves opposite, sessile or nearly so, ovate-lanceolate with somewhat cordate base, toothed, 2–4 in. long × 1–2 in. wide. Flowers irregular, about 1 in. long, purple to pink, tubular and 2-lipped, the narrow upper lip upright to slightly arching forward over the wider lower lip, lower lip 3-lobed and spotted. Inflorescence a dome-shaped head. Eastern beebalm is distinct in possessing only 1 flower cluster per stem tip in combination with sessile or nearly sessile leaves. IL, IN, MO. Relatively common in the Ozark Highlands and southern IL. *MAH*

Monarda fistulosa
LAMIACEAE

wild bergamot, beebalm

Wet to dry grasslands, open woodlands, roadsides. Summer.

An erect perennial, branched, hairy, stems 4-angled, 2–4 ft. tall. Leaves petiolate, opposite, lanceolate with toothed margins, 2–5 in. long and up to 2 in. wide. Flowers 1–1½ in. long, irregular, tubular and 2-lipped, the upper one narrow and unlobed, the lower one 3-lobed, lavender to pink, subtended by commonly pink-tinged leafy bracts, in dome-like heads terminating branch tips. Oswego tea (*M. didyma*) and basil beebalm (*M. clinopodia*) have scarlet and white flowers, respectively; the latter inhabits IL, IN, MI, and OH, and native populations of Oswego tea occur mostly in eastern OH. IA, IL, IN, MI, MN, MO, OH, WI. Common throughout. *MAH*

Physostegia virginiana
LAMIACEAE

obedient plant, false dragonhead

Prairies, wet fields, glades, swamps, floodplains, wet thickets, shores. Spring, summer, fall.

A square-stemmed perennial to 4 ft. tall. Leaves opposite, simple, sharply toothed, shiny, sessile, to 5 in. long × 1½ in. wide. Flowers irregular, white to purplish-pink, dotted or striped with darker pink-purple internally, to 1 in. long, tubular, upper lip broad and hood-like, lower lip 3-lobed with middle lobe largest, in vertical rows within dense spikes to 10 in. long at top of plant, remaining in place when manually rotated horizontally. Variable, with several named varieties sometimes treated as species. Narrowleaf obedient plant (*P. angustifolia*) has leaves to ½ in. wide. IA, IL, IN, MI, MN, MO, OH, WI. Scattered to widespread. *MAH*

Stachys hyssopifolia
LAMIACEAE

hyssop hedgenettle

Dry sand prairies, fields, sandy or mucky shores and flats. Summer.

A slender, square-stemmed perennial to 20 in. tall. Leaves opposite, entire to sparsely toothed, to 2½ in. long × ½ in. wide, sessile or very short-petiolate, becoming smaller bracts in inflorescence. Flowers irregular, pale pink with darker lavender-pink spots, to ⅓ in. long, tubular with hood-like upper lip and lower lip with 2 lateral lobes and shallowly cleft basal lobe, with 5-lobed calyx covering tube, opposite or in whorls in several separated clusters at top of stem. Rough hedgenettle (*S. aspera*; syn. *S. hyssopifolia* var. *ambigua*) has deeply impressed reticulate leaf venation and sharply toothed leaf margins. IN, MI. In zone around Lake Michigan. *SN*

Stachys pilosa
LAMIACEAE

marsh hedgenettle, hairy hedgenettle, woundwort

Wet prairies, marshes, shores, floodplains, ditches. Summer.

A square-stemmed perennial to 3 ft. tall with stem hairy throughout. Leaves opposite, sharply toothed, with deep reticulate venation, to 4 in. long × 1¾ in. wide, sessile or very short-petiolate, becoming smaller bracts in inflorescence. Flowers irregular, pale pink with darker lavender-pink spots, to ½ in. long, tubular with hood-like upper lip and lower lip with 2 lateral lobes and unlobed to shallowly cleft basal lobe, with 5-lobed calyx covering tube, in whorls in several separated clusters at top of stem. IA, IL, IN, MI, MN, MO, OH, WI. Mostly northwestern, primarily in the Northern Lakes and the Tallgrass Prairie, scattered in the Ozark Highlands and the Eastern Forests. *PD*

Stachys tenuifolia
LAMIACEAE

smooth hedge nettle

Wet woods, swamps, riverbanks, marshes. Summer.

An erect perennial, stems square, mostly glabrous or with few downward hairs on angles, 1–3 ft. tall. Leaves opposite, petiolate, blades lance-ovate to elliptic, shallowly toothed, mostly glabrous, 2–6 in. long × ½–1½ in. wide. Flowers irregular, ⅓–½ in. long, 2-lipped, the upper forming hood, lower 3-lobed with middle lobe longest, calyx 5-lobed with lobes as long as tube. Rough hedge nettle (*S. aspera*) has narrower, mostly sessile leaves, and heart-leaved hedge nettle (*S. cordata*) grows in more upland habitats with typically ovate-cordate leaves and calyx lobes half as long as tube. IA, IL, IN, MI, MN, MO, OH, WI. Mostly widespread and common. *MAH*

Linnaea borealis
LINNAEACEAE

twinflower

Moist forests and swamps, openings and edges, often with conifers. Summer.

An herb-like, colony-forming dwarf shrub to 6 in. tall, rooting at nodes. Leaves evergreen, opposite, confined to lower part of plant, short-petiolate, toothed in upper half, to ⅔ in. long × ½ in. wide. Flowers 2, regular, funnel-shaped, delicately nodding atop hairy, forked stem, to ½ in. long, with 5 ascending lobes. The genus name honors Carl Linnaeus, the father of modern taxonomy; it was said to be his favorite plant. IA, IL, IN, MI, MN, OH, WI. Mostly in the Northern Lakes, less frequent in adjacent portions of the Tallgrass Prairie. Presumed extirpated from southern-most areas, more common northward. *PS*

Spigelia marilandica
LOGANIACEAE

Indian pink, pinkroot

Moist upland forests, bottomlands, stream terraces, flatwoods. Spring, summer.

An erect perennial, unbranched, mostly glabrous, 1–2 ft. tall. Leaves opposite, sessile, entire, ovate-lanceolate, up to 4 in. long × 1–2 in. wide. Flowers regular, about 2 in. long, tubular, scarlet-red externally and yellow within, bearing 5 flaring, pointed lobes and a long-exserted style. Inflorescence a 1-sided arcing spike atop the main stem, with lowest flowers blooming first. Fruit a capsule. This attractive species is slowly becoming a popular garden ornamental. IL, IN, MO. Confined mostly to the southeastern Ozark Highlands and southwestern Eastern Forests. *MAH*

Ammannia coccinea

LYTHRACEAE

valley redstem, scarlet toothcup

Pond and lake borders, shallow mudflats, open flood-prone areas. Summer, fall.

An annual, single-stemmed or branched, glabrous, 6–18 in. tall. Leaves opposite, linear, entire, sessile with slight flaring at base where clasping the stem, up to 3 in. long × ¼–½ in. wide. Flowers regular, about ¼ in. wide, with 4 rounded petals, usually 2–5 on short stalks ¹⁄₁₆–⅛ in. long in leaf axils. Grand redstem (*A. robusta*) is similar, but its flowers and fruits are sessile; its flower color may also be less intense. IA, IL, IN, MN, MO, OH. Mostly in middle and southwestern Midwest. *MAH*

Callirhoë triangulata

MALVACEAE

clustered poppy-mallow

Sand prairies, barrens. Summer.

An ascending to weakly erect perennial, branched, with numerous stellate hairs, 1–2 ft. tall. Leaves alternate, petiolate, basal and lower leaves triangular with rounded teeth or a few lobes and somewhat cordate base, 2–6 in. long × 1–4 in. wide. Flowers regular, 1–2 in. wide, somewhat cup-shaped, the 5 broad petals red with basal white spot, in clusters at stem tips or stalked in leaf axils. Four other *Callirhoë* species occur in our region, but this is the only one with triangular lower leaves. IA, IL, IN, MN, MO, WI. Occurs mostly in northern IL and southwestern WI. *BS*

Hibiscus laevis
MALVACEAE

halberd-leaf rosemallow

Swamps, marshes, floodplains, sloughs, ponds, ditches. Summer, fall.

A glabrous perennial to 6 ft. tall. Leaves simple, finely toothed, alternate, to 6 in. long × 4 in. wide, with conspicuous lobe on each side at base, long-petiolate. Flowers regular, to 4 in. wide, with 5 rounded red-based white to pink petals, with stigma-tipped column covered with stamens slightly exserted, numerous on stalks from upper leaf axils; each flower typically opens for only one day. Fruit a dry, oblong capsule opening from tip into 5 segments. Seeds hairy. Commonly included in restoration seed mixes. Syn. *H. militaris*. IA, IL, IN, MI, MN, MO, OH, WI. Throughout, but rare in MI, MN, and WI. *PR*

Hibiscus moscheutos
MALVACEAE

swamp rosemallow

Marshes, floodplains, open swamps, ponds. Summer, fall.

A perennial to 7 ft. tall. Leaves simple, toothed, alternate, to 8 in. long × 4 in. wide, ovate with pointed tip, unlobed or with pair of lobes near middle, densely short-hairy beneath, petiolate. Flowers regular, to 6 in. wide, with 5 rounded white (with red base) or pink petals, with stigma-tipped column covered with stamens slightly exserted, several stalked from upper leaf axils. Fruit a dry, oblong capsule opening from tip into 5 segments. Seeds glabrous. Sometimes split into subsp. *moscheutos* and subsp. *lasiocarpos*. Syn. *H. palustris*. IL, IN, MI, MO, OH, WI. Widespread in the southern and eastern portions of Midwest and around the Great Lakes. *SN*

Rhexia virginica
MELASTOMATACEAE

Virginia meadow beauty

Acid wet prairies, meadows, marshes, sand/
muck flats, sandstone glades. Summer.

An erect perennial, branched above, stem
4-angled, conspicuously winged, hairy, 1–2 ft.
tall. Leaves opposite, sessile, ovate to elliptic,
toothed, covered with perpendicular and
moderately glandular hairs, 1–3 in. long × ½–1
in. wide. Flowers regular, 1–1½ in. wide, with 4
rounded to obovate petals and 8 bright yellow,
curving stamens, together arising from summit
of an urn- or vase-shaped floral tube. Flowers
clustered on terminal and axillary branches.
The less-common Maryland meadow beauty
(R. mariana) has stems unwinged or only
slightly so, and petals pale pink or white. IA,
IL, IN, MI, MO, OH, WI. Locally common but
distribution is patchy. SN

Mirabilis nyctaginea
NYCTAGINACEAE

wild four-o'clock

Prairies, fallow fields, roadsides, railroads,
waste areas. Summer.

An erect perennial, branched, glabrous, 2–3
ft. tall. Leaves opposite, short-petiolate,
triangular-ovate with cordate base, entire, 1–4
in. long × ¾–3 in. wide. Flowers regular, about
½ in. wide, petals absent, sepals 5, notched at
tip and petal-like, extending from a greenish
cup (involucre) in clusters at stem tips. The
common name references late afternoon, the
time when its flowers begin to open. Related
is the garden four o'clock (M. jalapa), native to
South America. Mirabilis means "wonderful."
IA, IL, IN, MI, MN, MO, OH, WI. Common
throughout, perhaps native only in our western counties. NP

Gaura biennis
ONAGRACEAE

biennial gaura

Prairies, glades, fields, thickets, riverbanks, roadsides, waste areas. Summer, fall.

A hairy biennial to 7 ft. tall. Leaves in basal rosette first year, then alternate, 7 in. long × 1 in. wide, smaller above, sessile to short-petiolate, often infused with red. Flowers irregular, ½ in. wide, with 4 pale pink to white upwardly spreading petals, 8 long anthers and solitary style spreading downward, numerous, sessile in upper part of plant. Syn. *Oenothera gaura*. Small-flowered gaura (*G. parviflora*) has smaller flowers. Large-flowered gaura (*G. longiflora*) has appressed or incurved (versus spreading) stem hairs. IA, IL, IN, MI, MN, OH, WI. Widespread in eastern portion of the Eastern Forests and the Tallgrass Prairie and the adjacent Northern Lakes, scattered elsewhere. *MAH*

Oenothera speciosa
ONAGRACEAE

showy evening primrose

Prairies, glades, roadsides, railroads, waste areas. Spring, summer.

An erect to ascending or reclining perennial, branched, hairy, 4–24 in. long. Leaves alternate, sessile to short-petiolate, linear to oblanceolate, entire to irregularly toothed, 1–3 in. long × ½–1 in. wide. Flowers regular, 2–3 in. wide, with 4 rounded petals shallowly notched at tips, white to pink, sepals reflexed, occurring in upper leaf axils. Fruit an elliptically shaped capsule. Nuttall's evening primrose (*O. nuttallii*), found mostly in MN and WI, has mostly linear leaves and whitish, glabrous stems. IA, IL, IN, MO, OH. Perhaps native only in western parts of IA and MO. *DT*

Arethusa bulbosa
ORCHIDACEAE

dragon's mouth

Sphagnum bogs, conifer swamps, fens.
Spring, summer.

An erect, unbranched perennial, glabrous,
5–8 in. tall. Leaf solitary, entire, lanceolate,
emerging after flowering, 1½–3 in. long × ¼–½
in. wide. Flowers irregular, about 1 in. long,
single or rarely 2, with 3 petals, the lip obovate
and reflexed with yellow bristles, larger than
other 2 petals, which are narrowly oblong and
arching forward, sepals 3, petal-like, narrowly
oblong, upright. Due to habitat degradation
it is becoming less common, especially in the
southern part of its range. **IL, IN, MI, MN,
OH, WI. Occurs mostly within the Northern
Lakes with a few outliers elsewhere.** *AG*

Calopogon tuberosus
ORCHIDACEAE

tuberous grass pink

Bogs, fens, wet meadows, wet sand prairies.
Summer.

An erect perennial, unbranched, glabrous,
1–2 ft. tall. Leaves typically 1, basal, grass-like,
linear-lanceolate, entire, 6–9 in. long × ⅓–1
in. wide. Flowers irregular, 1–2 in. wide, lip
uppermost, with an expanded triangular
tip bearing yellow club-shaped bristles,
petals ovate-lanceolate to oblong, sepals 3
and petal-like, ovate-lanceolate, in a loosely
flowered raceme positioned above the leaf.
Calopogon is quite unusual as its lip petal is
uppermost; in almost all other orchids it is lowermost. Oklahoma grass pink (*C. oklahomensis*)
is similar and extremely rare with us, its flowering stem usually not taller than the leaf. **IA, IL,
IN, MI, MN, MO, OH, WI. Located predominantly in the Northern Lakes.** *PG*

Calypso bulbosa
ORCHIDACEAE

fairy slipper, calypso, Venus' slipper

Moist coniferous and mixed forests.
Spring, summer.

A perennial to 6 in. tall. Single glabrous,
entire, ovate basal leaf to 2¼ in. long × 1¼ in.
wide forms in fall, lying nearly flat on ground,
with wrinkled surface and parallel veins,
present through winter, withering soon after
flower develops. Flower irregular, solitary
(rarely 2), 1¼ in. long × 1 in. wide, with 3 pink
sepals and 2 pink petals erect to spreading,
and downward-pointing pink to white lip with
prominent red-spotted yellow beard in center
forming "slipper" with pink to red vertical
striping or spotting. MI, MN, WI. Rare and
restricted to northern portion of the Northern Lakes. *BS*

Corallorhiza maculata
ORCHIDACEAE

spotted coral-root

Mesic upland forests, swamps,
dune forests. Summer.

An erect, perennial mycoheterotroph,
unbranched, glabrous, fleshy, 6–18 in. tall.
Leaves reduced to sheaths, floral bracts
minute. Flowers irregular, about ½ in. wide,
the lip ovate, purple-spotted, lobed at base,
surrounded by 2 petals and 3 purplish and
spreading petal-like sepals. Inflorescence with
10–15 flowers in a spike-like raceme. Rhizomes
resemble branching coral, hence the genus
and common names. In var. *occidentalis* the
lip is widest toward tip; in var. *maculata* the
lip width is uniform. Variety *occidentalis* does
not occur as far south in the Midwest as var.
maculata. IA, IL, IN, MI, MN, OH, WI. Occurs
principally in the Northern Lakes. *PR*

Corallorhiza odontorhiza
ORCHIDACEAE

autumn coral-root

Mesic to dry upland forests. Summer, fall.

An erect, perennial mycoheterotroph, unbranched, purplish and fleshy, glabrous, swollen at base with branching rhizomes, 4–8 in. tall. Leaves reduced to sheaths, floral bracts minute. Flowers irregular, ⅛–¼ in. wide, the lip ovate-obovate, purple-spotted, unlobed at base, and surrounded by 3 purplish petal-like sepals that along with 2 petals form a hood. Inflorescence a spike-like raceme. Capsules are often greenish, indicating presence of some chlorophyll. Most plants have cleistogamous (closed) flowers (var. *odontorhiza*); those with chasmogamous (open) flowers are less common (var. *pringlei*). IA, IL, IN, MI, MN, MO, OH, WI. Widespread but mostly absent in far northern counties. *CB*

Corallorhiza striata
ORCHIDACEAE

striped coralroot, hooded coralroot

Moist forests, swamps. Spring.

A perennial with pink to red-purple stem to 1½ ft. tall. Leaves inconspicuous and scale-like, reduced to yellowish, alternate scales at base of plant. Flowers irregular, to ½ in. wide, with 3 red-purple and white striped sepals, 2 similar-looking lateral petals, and a solid to striped red-purple lip petal, the sepals and 2 lateral petals forming an arching hood over the lip; flowers to 25 in spike-like inflorescence. Fruit a capsule. Obtains nutrients from subterranean fungi rather than through photosynthesis (mycoheterotrophic). MI, MN WI. Restricted to northern portion of the Northern Lakes. *MJH*

Corallorhiza wisteriana
ORCHIDACEAE

spring coral-root, Wister's coral-root

Mesic to dry upland forests. Spring.

An erect, perennial mycoheterotroph, unbranched, yellowish to purplish and fleshy, glabrous, 6–12 in. tall with branching rhizomes. Leaves reduced to sheaths, floral bracts minute. Flowers irregular, about ¼ in. wide, the lip petal ovate, entire, purple-spotted, and surrounded by 3 purplish petal-like sepals and 2 petals that together form a hood. Inflorescence a spike-like raceme. This and showy orchis (*Galearis spectabilis*) are often the earliest native orchids to bloom in our region, sometimes as early as late March–early April in the south. IL, IN, MO, OH. Mostly in the Midwest's southern third. *CB*

Cypripedium acaule
ORCHIDACEAE

pink lady's slipper, moccasin flower

Acid bogs, swamps, well-drained upland forests. Spring, summer.

An erect perennial, unbranched, hairy, 8–18 in. tall. Leaves normally 2, basal, elliptic, entire, 4–8 in. long × 3–5 in. wide. Flowers irregular, bearing 3 petals, the lowermost a pink pouch-shaped lip with a center cleft, 1½–3 in. long, the other petals purplish, hairy, ovate-lanceolate and slightly twisted, sepals 3 (effectively 2; lateral ones fused), hairy. This orchid often grows in saturated boggy soils but also in well-drained upland soils, especially in the far northern counties and those of the unglaciated southeast. IL, IN, MI, MN, OH, WI. Mostly in the Northern Lakes and unglaciated OH. *AB*

Cypripedium reginae
ORCHIDACEAE

showy lady's-slipper

Fens, seep springs, thickets, wet limestone cliffs. Spring, summer.

An erect perennial, unbranched, hairy, 1½–3 ft. tall. Leaves alternate, sheathing, elliptic to ovate-lanceolate, entire, 5–8 in. long × 3–5 in. wide. Flowers irregular, bearing 3 petals, the lowermost a pink and pouch-shaped lip, about 2½ in. long, the other 2 white, ovate-lanceolate, not twisted, sepals 3 (effectively 2, the lateral sepals united) and white. This is the largest of all native Midwestern orchids. Its preferred habitat is where the ground is deep muck and mosquitoes rule. IA, IL, IN, MI, MN, MO, OH, WI. Mostly in the Northern Lakes, scattered elsewhere, including, amazingly, the Ozark Highlands. *AG*

Pogonia ophioglossoides
ORCHIDACEAE

rose pogonia, snakemouth orchid

Bogs, fens, calcareous interdunal wetlands, conifer swamps. Summer.

A glabrous perennial to 14 in. tall. Leaf solitary halfway up stem (rarely second leaf from base of plant), to 4 in. long × 1 in. wide, entire, sheathing. Flowers irregular, subtended by a bract, pink (rarely white), solitary (rarely 2) at top of stem, to 1½ in. wide, with upper sepal erect to ascending, lateral petals forming arching hood, lateral sepals spreading, and fringed and bearded lip petal spreading to horizontal with yellow spot toward base. IA, IL, IN, MI, MN, MO, OH, WI. Primarily in the Northern Lakes, where widespread. *BS*

Capnoides sempervirens
PAPAVERACEAE

pink corydalis, pale corydalis, rock harlequin

Rocky ledges, gravel shores, savannas, recently burned areas. Spring, summer, fall.

A glaucous biennial to 2½ ft. tall. Leaves to 1¼ in. long, twice compound into deeply lobed round-tipped segments. Flowers irregular, tubular, pink with yellow tips and rounded pink spur, to ¾ in. long, stalked and dangling in clusters at top of plant. Thrives with disturbance (such as fire) and persists in areas with little competition (such as cracks in boulders). Syn. *Corydalis sempervirens*. IA, IL, IN, MI, MN, OH, WI. **Widespread throughout much of the Northern Lakes (especially in MN and WI), rare in adjacent Tallgrass Prairie; uncommon in the Eastern Forests.** *SN*

Chelone obliqua
PLANTAGINACEAE

pink turtlehead, rose turtlehead, red turtlehead

Swamps, floodplain forests, seeps, fens. Summer, fall.

A glabrous perennial to 3 ft. tall. Leaves opposite, to 6 in. long × 2¼ in. wide, shiny, dark green, deeply veined, abundantly sharply toothed, on petioles to ½ in. Flowers irregular, pink to rose, to 1½ in. long, tubular with broad, hood-like upper lip and slightly longer, shallowly 3-lobed lower lip, with middle lobe somewhat elevated, spreading in densely flowered terminal spikes to 3½ in. long. Individual flowers look somewhat like a turtle's head. IA, IL, IN, MI, MN, MO. **Scattered to rare, mostly in the southeastern portion of the Tallgrass Prairie and adjacent Eastern Forests.** *BS*

Phlox glaberrima
POLEMONIACEAE

smooth phlox

Moist to wet prairies, fens, openings in wet woods. Spring, summer.

A glabrous perennial to 2½ ft. tall. Leaves opposite, entire, lanceolate, shiny, sessile, with strongly impressed midvein, to 4 in. long × ⅖ in. wide. Flowers regular, to ¾ in. wide, tubular with 5 flaring lobes at apex of ⅗-in.-long tube, with stamens and pistil included in floral tube, in showy, slightly dome-shaped cluster at top of plant. Fruit a capsule. IL, IN, MO, OH, WI. **Widespread in eastern portion of the Tallgrass Prairie, barely ranging north into the Northern Lakes and southwest into the Ozark Highlands; also scattered in the Eastern Forests.** *SN*

Phlox maculata
POLEMONIACEAE

wild sweetwilliam, speckled phlox

Sedge meadows, wet prairies, flatwoods, fens. Summer.

A perennial to 3 ft. tall with purplish-red-spotted or -streaked stem. Leaves opposite, entire, ovate-lanceolate, usually shiny, sessile or nearly so, with strongly impressed midvein, to 4 in. long × ¾ in. wide. Flowers regular, purplish-pink (rarely white), to ¾ in. wide, tubular with 5 flaring lobes at apex of 1-in.-long tube, with stamens and pistil included in floral tube, short-stalked, in showy, cylindrical cluster to 12 in. long at top of plant. IA, IL, IN, MI, MN, MO, OH. **Scattered in the Eastern Forests, the Tallgrass Prairie, and the Ozark Highlands, rare in southern portion of the Northern Lakes.** *MAH*

Phlox paniculata
POLEMONIACEAE

garden phlox, summer phlox

Rich and well-drained floodplain forests, streambanks. Summer.

An erect perennial, glabrous to hairy, 2–3+ ft. tall. Leaves opposite, short-petiolate to sessile, lanceolate to narrowly elliptic, with minutely barbed margins, 2–4 in. long × 1–1½ in. wide. Flowers regular, about ¾ in. wide, tubular, pinkish to sometime white, with 5 broad obovate and sometimes shallowly notched lobes, in a compact, pyramidal to rounded cluster atop the stem. This species is quite popular in landscaping. IA, IL, IN, MI, MN, MO, OH, WI. Common and native mostly in the lower Midwest; upper Midwest occurrences are likely garden escapes. *MAH*

Phlox pilosa
POLEMONIACEAE

prairie phlox, downy phlox

Prairies, glades, savannas, fens, dry open woods. Spring, summer, fall.

A multi-stemmed perennial to 2 ft. tall, usually hairy. Leaves opposite, entire, sessile, with strongly impressed midvein, to 3 in. long × ½ in. wide. Flowers regular, purplish-pink (rarely white), to ¾ in. wide, tubular with 5 flaring lobes at apex of ⅗-in.-long tube, with stamens and pistil included in floral tube, in slightly dome-shaped terminal cluster to 3 in. wide. Several varieties differ in hair length and presence or absence of glands in inflorescence, leaf shape, and flower color. IA, IL, IN, MI, MN, MO, OH, WI. Widespread in the Ozark Highlands, the Tallgrass Prairie, and southern portion of the Northern Lakes, scattered elsewhere. *SN*

Polygala incarnata
POLYGALACEAE

pink milkwort, procession flower

Prairies, glades. Spring, summer, fall.

A glaucous annual to 2 ft. tall. Leaves few, alternate, inconspicuous and scale-like with pointed tip, often absent below and deciduous above, to ½ in. long. Flowers irregular, pinkish-purple (rarely white), ascending, to ¼ in. long × ⅛ in. wide, 3 petals and 3 sepals forming a tube and 2 petal-like sepals (wings) enlarged, lower petal with 6 lobes that are fringed and lobed, in dense terminal raceme to 1 ⅗ in. long × ⅗ in. wide. IA, IL, IN, MI, MO, OH, WI. Most frequent in the Ozark Highlands and southern portion of the Tallgrass Prairie, scattered elsewhere. *DT*

Polygala paucifolia
POLYGALACEAE

fringed polygala, gaywings

Rich woods, calcareous shores. Spring.

A colony-forming perennial to 6 in. tall, emerging from along underground stems. Leaves few, alternate, scale-like below, clustered and appearing whorled at top of stem, dark green, ovate, to 1½ in. long × ¾ in. wide, simple, entire. Flowers irregular, pinkish-purple (rarely white), to ¾ in. wide, with 3 sepals and 3 petals forming tube and 2 sepals enlarged and petal-like (wings), lower petal whitish with delicate pinkish-purple fringe at tip; 1–4 showy flowers from upper leaf axils. Also produces highly reduced, self-fertilizing subterranean flowers along rhizome. Syn. *Polygaloides paucifolia*. IL, IN, MI, MN, OH, WI. Widespread throughout the Northern Lakes. *SN*

Polygala polygama
POLYGALACEAE

racemed milkwort, purple milkwort

Sandy barrens, savannas, prairies, open woods. Summer.

A single- to several-stemmed glabrous biennial to 1 ft. tall. Leaves alternate, entire, sessile, to 1 in. long × ⅙ in. wide. Flowers irregular, pinkish-purple, to ¼ in. wide, with 3 petals and 3 sepals forming tube and 2 sepals enlarged and petal-like (wings), lower petal with fringed tip at tube opening (see inset), in loose, spike-like, 3-in.-long terminal raceme. Also produces highly reduced, self-fertilizing subterranean flowers, a likely adaptation to grazing pressures and/or fire. IA, IL, IN, MI, MN, OH, WI. Widespread throughout the Northern Lakes, scattered in eastern portion of the Tallgrass Prairie, rare in the Eastern Forests. *SN*

Polygala sanguinea
POLYGALACEAE

field milkwort, purple milkwort

Prairies, barrens, savannas, glades, fallow fields. Summer.

An erect annual, usually branched above, glabrous, 3–12 in. tall. Leaves alternate, sessile, linear, ½–1½ in. long × 1/16–⅙ in. wide. Flowers irregular, ⅛–¼ in. long with 3 small petals, the lower one keel-shaped, and 5 sepals, 2 of which are enlarged and petal-like (wings), crowded in compact, cylindrical racemes about 1 in. long × ⅓ in. wide. Cross-leaved milkwort (*P. cruciata*), found mostly in counties radiating from southern Lake Michigan, has whorled leaves. IA, IL, IN, MI, MN, MO, OH, WI. Common except in far northern and northwestern counties. *MAH*

Persicaria amphibia
POLYGONACEAE

water smartweed

Lakes and ponds, sloughs, marshes, quiet waters. Summer, fall.

An aquatic or wetland perennial, erect to sprawling, simple or branched, glabrous or hairy, rhizomatous, 1–4 ft. tall. Leaves alternate, petiolate, summit of ocreae (nodal sheaths) with a flared collar (var. *stipulacea*) or not (var. *emersa*), blades lanceolate to elliptic, 1–6 in. long × ½–3 in. wide. Leaves floating (var. *stipulacea*; pictured) or usually not (var. *emersa*). Flowers regular, ⅛–¼ in. long, tepals 5, in 1 or 2 spike-like terminal racemes, compact and egg-shaped (var. *stipulacea*) or longer and cylindrical (var. *emersa*). Intermediates exist. IA, IL, IN, MI, MN, MO, OH, WI. Scattered throughout the Midwest, more common northward. *PS*

Persicaria careyi
POLYGONACEAE

Carey's smartweed, Carey's heartsease

Wet sand prairies, swamps, bogs, streambanks, thickets. Summer, fall.

An annual to 4 ft. tall. Leaves alternate, simple, entire, to 7 in. long × 1 ⅓ in. wide, sessile or short-petiolate, with ocreae (nodal sheaths) with summits fringed. Flowers regular, pale pink to rose, to ⅛ in. long, with 5 tepals, appearing closed, numerous and tightly packed in upright to nodding spike-like racemes to 4 in. long in upper part of plant. Stem beneath spike with spreading gland-tipped hairs. IL, IN, MI, MN, OH, WI. Uncommon in scattered pockets in the Northern Lakes, adjacent parts of the Tallgrass Prairie, and the Eastern Forests. *SN*

Persicaria longiseta
POLYGONACEAE

creeping smartweed, oriental lady's-thumb

Moist disturbed ground, floodplains, trails. Spring, summer, fall.

A sprawling, colony-forming annual to 3 ft. tall. Leaves alternate, simple, entire, to 3 in. long × 1¾ in. wide, broadest at middle, sessile or nearly so, with ocreae (nodal sheaths) with summits fringed (fringes to ½ in. long). Flowers regular, whitish to deep rose (sometimes mixed), to ⅛ in. long with 5 nearly closed tepals, numerous in erect spike-like racemes to 1½ in. long in upper part of plant and in leaf axils, individual flowers subtended by sheaths topped with bristles over 1/12 in. long. IA, IL, IN, MI, MN, MO, OH, WI. Introduced from Asia. Scattered throughout; increasing. *SN*

Persicaria maculosa
POLYGONACEAE

lady's-thumb

Moist fields, marsh borders, ponds, shores, ditches, gardens, cultivated ground, roadsides. Spring, summer, fall.

A somewhat sprawling annual to 3 ft. tall, often colony-forming. Leaves alternate, simple, entire, to 7 in. long × 1 in. wide, sessile to short-petiolate, with ocreae (nodal sheaths) with summits fringed. Flowers regular, whitish to rosy pink (often mixed), nearly closed, to ⅛ in. long with 5 tepals, numerous and tightly packed in erect spike-like racemes to 2 in. long in upper part of plant and in leaf axils. Stem beneath spike glabrous. As with many other *Persicaria* species, leaves usually have dark "lady's thumb" print near middle. IA, IL, IN, MI, MN, MO, OH, WI. Introduced from Europe. Common and widespread. *SN*

Persicaria pensylvanica
POLYGONACEAE

pinkweed, Pennsylvania smartweed

Moist fields, marshes, sloughs, ponds, shores, streambanks, ditches. Spring, summer, fall.

An annual to 4 ft. tall, often colony-forming. Leaves alternate, simple, entire, to 7 in. long × 2½ in. wide, on petioles to 1 in., with ocreae (nodal sheaths) with summits entire. Flowers regular, white to rosy pink, to just over ⅛ in. long with 5 erect tepals, numerous and tightly packed in erect to slightly nodding spike-like racemes to 2½ in. long in upper part of plant. Stem immediately beneath spike with spreading gland-tipped hairs. IA, IL, IN, MI, MN, MO, OH, WI. **Common and widespread nearly throughout, with fewer occurrences north.** *SN*

Rumex acetosella
POLYGONACEAE

sheep sorrel

Sand barrens, acid dry woods, savannas, old fields, roadsides, waste areas. Spring, summer.

An erect, dioecious perennial, branched above, glabrous, stems 4-angled, 5–18 in. tall, colonial. Leaves basal and alternate on stems, petiolate, summit of ocreae (nodal sheaths) silvery-brown and torn, blades usually 3-lobed, terminal lobe longest, lanceolate-elliptic, base with 2 small triangular lobes, 1–2½ in. long × ¼–¾ in. wide. Flowers regular, about ⅛ in. wide, with 6 persistent tepals, scattered along spreading branches, color varies from green to red or purple. Sour-tasting leaves inspired an alternative common name, sour grass. IA, IL, IN, MI, MN, MO, OH, WI. **Introduced from Eurasia. Possibly in every county.** *PR*

Phemeranthus parviflorus
PORTULACACEAE

**prairie fameflower,
few-flowered flower-of-an-hour**

Acidic glades and cliff tops. Summer.

A glabrous succulent perennial, 3–6 in. tall.
Leaves alternate, sessile and crowded atop
stem, cylindrical and quill-like, 1–2 in. long
× ⅛ in. wide. Flowers regular, to about ½ in.
wide, with 5 elliptic petals and 4–8 stamens,
on a few-flowered terminal inflorescence.
Seed surfaces smooth. The similar largeflower
fameflower (*P. calycinus*) and rough-seeded
fameflower (*P. rugospermus*) differ by having
larger flowers bearing 20+ stamens, and seeds
with rough and wrinkled surfaces, respectively. IA, IL, MN, MO. Occurs mostly in the
Ozark Highlands and southwestern MN. *DT*

Aquilegia canadensis
RANUNCULACEAE

wild columbine

Rock outcrops, dunes, steep slopes,
well-drained creek banks, seeps. Spring.

A multi-branched perennial with glabrous
stems, 15–30 in. tall. Leaves basal as well as
alternate on stems, once or twice compound,
the leaflets up to 2 in. long and wide with
rounded teeth. Flowers regular, about 1½
in. long with 5 petals, yellow with red to
salmon-colored straight spurs and red sepals,
nodding from stem tips. Wild columbine is
the only native columbine in eastern North
America. The purple-flowered European columbine (*A. vulgaris*) sometimes escapes from
cultivation. IA, IL, IN, MI, MN, MO, OH, WI.
Occurs in most of the region. *MAH*

Comarum palustre
ROSACEAE

marsh cinquefoil, purple marshlocks

Marshes, swamps, bogs, swales, ditches. Summer.

A gray-green, sprawling, branched perennial to 2 ft. tall. Leaves alternate, petiolate, pinnately compound with 3–7 narrowly elliptic, coarsely toothed leaflets to 3 in. long × 1 in. wide. Flowers regular, with 5–8 small petals alternating between larger, pointed sepals, to 1 in. wide, with numerous protruding stamens and pistils; in stalked clusters of up to 3 at tips of branches at top of plant. The flowers, although a distinctive maroon, are easily overlooked. Syn. *Potentilla palustris*. IA, IL, IN, MI, MN, OH, WI. **Widespread in the Northern Lakes with fewer occurrences in adjacent portions of the Tallgrass Prairie and Eastern Forests.** *SN*

Filipendula rubra
ROSACEAE

queen-of-the-prairie

Wet prairies, fens, sedge meadows. Summer.

An unbranched (except inflorescence) perennial to 6 ft. tall. Leaves alternate, petiolate, to 2 ft. long, shorter up stem, yellow-green, pinnately compound with palmately lobed, jagged-margined leaflets. Flowers regular, pink to rose, to ⅓ in. wide, with 5 small petals, numerous in a showy, oddly shaped, plume-like, densely flowered panicle to 8 in. across at top of plant. An indicator of a natural community but often planted in native landscaping and restorations. Queen-of-the-meadow (*F. ulmaria*), introduced and uncommon, has white flowers and is primarily in the Northern Lakes. IA, IL, IN, MI, MN, MO, OH, WI. Scattered. *SN*

Geum triflorum
ROSACEAE

prairie smoke

Prairies, savannas, limestone outcrops.
Spring, summer.

A hairy perennial to 1 ft. tall. Leaves in basal
rosette, petiolate, to 5 in. long × 2 in. wide,
variable, pinnately compound with irregularly
lobed and coarsely toothed leaflets of variable
size. Flowers regular, with 5 white to pale pink
petals shorter than 5 red sepals, to 1 in. wide,
nodding, subtended by 5 red, narrowly linear
pointed bracts, few on stalks at top of plant.
Fruiting heads to 2 in. wide, upright, roundish, feathery-hairy, giving the appearance of
smoke when seen en masse from a distance.
**IA, IL, MI, MN, MO, WI. Throughout the
southern portion of the western Northern
Lakes and the northern portion of the Tallgrass Prairie.** *CR*

Rosa carolina
ROSACEAE

Carolina rose, pasture rose

Upland woodlands, barrens, glades,
savannas. Summer.

A low, sparingly branched shrub, prickly,
rhizomatous, 1–2½ ft. tall. Leaves alternate,
petiolate with widened stipules, compound,
leaflets 5–7 (usually 5), these oblong or
elliptic-ovate, serrated, glabrous to sparsely
hairy, 1–2 in. long × ½–1¼ in. wide. Flowers
regular, 2–3 in. wide, with 5 petals and sepals
atop the hypanthium (floral cup), the latter
along with lower surface of sepals normally
covered with stalked glands. The hypanthium
matures into a pulpy hip (fruit). Prairie rose
(*R. arkansana*) has 9–11 leaflets and a glabrous
hypanthium. **IA, IL, IN, MI, MO, OH, WI.
Common except for the far northern counties.** *MAH*

Spiraea tomentosa
ROSACEAE

steeplebush, hardhack

Swamps, bogs, sand prairies, marshes, flatwoods, thickets. Summer.

An erect but mostly low-growing shrub, sparsely branched, younger twigs finely hairy becoming glabrous with age, 1–4 ft. tall. Leaves alternate, short-petiolate, ovate to oblong, toothed, thick and heavily veined, undersurface densely white-hairy, 1–3 in. long × ½–1 in. wide. Flowers regular, ⅛–¼ in. wide, 5-petaled, numerous and showy, arranged in terminal and densely flowered pyramidal inflorescence. The cultivated Japanese spiraea (*S. japonica*) normally has similar-colored flowers and occasionally escapes from cultivation; its inflorescence is rounded, as broad or broader than long. IL, IN, MI, MN, MO, OH, WI. Occurs mostly in the Northern Lakes. *MAH*

Diodia teres
RUBIACEAE

rough buttonweed, poorjoe

Prairies, fields, glades, gravel bars, railroads, disturbed sites, often in sand or gravel. Summer, fall.

A hairy annual to 1 ft. tall. Leaves opposite, entire, to 1½ in. long × ¼ in. wide, with long-fringed stipules at base. Flowers regular, to ¼ in. wide, with 4 spreading whitish, pink, or bluish lobes, sessile, 1–3 in leaf axils. Virginia buttonweed (*D. virginiana*) has pure white flowers and broader leaves. Sometimes placed in *Diodella*. IA, IL, IN, MI, MO, OH, WI. Widespread in the Ozark Highlands and southern portion of the Tallgrass Prairie, less frequent in the Eastern Forests, rare in the Northern Lakes. *MAH*

Valeriana pauciflora
VALERIANACEAE

large-flowered valerian, pink valerian

Rich forested stream terraces, ravine bottoms. Spring.

An erect perennial with basal and stem leaves, plus runners, mostly unbranched, glabrous, colonial, 1–2½ ft. tall. Leaves opposite, basal ones petiolate and cordate with wavy margins, about 2–3 in. long and wide, stem leaves pinnately compound with 3–7 leaflets, sessile, the larger terminal ones about 2 in. long and wide. Flowers nearly regular, ½–¾ in. long, tubular, pinkish to white, like miniature herald trumpets and tipped with 5 small flaring lobes in a rounded cluster atop stem. Fruit an achene bearing feathery bristles. IL, IN, OH. Confined mostly to the southern counties. Relatively common. *SN*

ORANGE FLOWERS

Asclepias tuberosa
APOCYNACEAE

butterfly milkweed, butterfly weed, orange milkweed, pleurisy root

Prairies, glades, savannas, open woodlands, fields, roadsides. Spring, summer, fall.

A hairy perennial to 3 ft. tall. Leaves mostly alternate, entire, shiny, deeply veined, to 4 in. long × 1 in. wide, sessile or short-petiolate, containing clear sap. Flowers regular, orange (ranging from yellowish to reddish), to ½ in. long, with 5 reflexed-spreading corolla lobes and 5 erect corona hoods, each with incurved horn, on stalks to 1½ in. in numerous small umbels. Fruit a narrowly lanceolate, hairy follicle to 6 in. long with seeds bearing tufts of hairs. IA, IL, IN, MI, MN, MO, OH, WI. Widespread, with fewer occurrences in northern portion of the Northern Lakes. *SN*

Hieracium aurantiacum
ASTERACEAE

orange hawkweed, devil's paintbrush

Fields, pastures, woods, swamps, roadsides. Summer, fall.

A long-hairy (often with some black hairs) perennial to 2 ft. tall. Leaves entire, simple, basal or with few reduced alternate leaves on lower stem, to 5 in. long × 1 in. wide, sessile, containing milky sap. Flowers in composite heads; numerous orange ray flowers in 1-in.-wide heads, disk flowers lacking; few to numerous stalked heads at top of plant. Fruit an achene with silky pappus. Syn. *Pilosella aurantiaca*. IA, IL, IN, MI, MN, OH, WI. **Introduced from Europe. Abundant and widespread in the Northern Lakes, adjacent portions of the Tallgrass Prairie, and eastern portion of the Eastern Forests, scarce elsewhere.** *SN*

Impatiens capensis
BALSAMINACEAE

jewelweed, spotted touch-me-not

Seeps, creek bottoms, lake borders, ditches, other wetland types. Summer, fall.

An erect annual, branched, hollow-stemmed, glabrous, somewhat glaucous, juicy, 2–4 ft. tall. Leaves alternate, petiolate, ovate to elliptic, wavy or bluntly toothed, 1–5 in. long and up to 3 in. wide. Flowers irregular, about 1 in. long, spotted, corolla 5-lobed, the uppermost one broadest, sepals 3, the lower cone-shaped, open in front then narrowing to a small spur. Flowers numerous, dangling from upper leaf axils. Fruit a capsule that explodes when disturbed. The leaves have a silvery appearance when submersed in water. IA, IL, IN, MI, MN, MO, OH, WI. Widespread, in almost every county. *MAH*

Campis radicans
BIGNONIACEAE

trumpet creeper

Bottomland forests, shallow swamps, forest edges, fencerows. Summer.

A woody vine with aerial roots, climbing into trees and up poles, mostly glabrous, to 50+ ft. Leaves opposite, petiolate, pinnately compound, with 7–11 ovate, toothed leaflets, leaves 1–1½ ft. long × 5–8 in. wide. Flowers nearly regular, about 3 in. long, tubular with 5 flaring lobes, somewhat waxy, occurring in clusters at branch tips. The flowers are attractive to ruby-throated hummingbirds, their main pollinator. Leaves of the related crossvine (*Bignonia capreolata*), found in our far southern counties, bear only 2 leaflets. IA, IL, IN, MI, MO, OH, WI. Occurs mostly in the southern Midwest, introduced northward. *ST*

Hemerocallis fulva
HEMEROCALLIDACEAE
orange daylily
Old homesites, cemeteries, thickets, roadsides. Summer.

A glabrous, waxy, colony-forming perennial with stems to 5 ft. tall. Leaves basal, simple, entire, to 3 ft. long × 1¼ in. wide, linear, tapering to point, flaccid; stem leaves few, reduced to scales on flowering branches. Flowers regular, to 3½ in. wide, upright, with 3 spreading to recurved orange petals with yellow-orange bases and 3 similar sepals, 6 stamens and single pistil exserted, in branching clusters of up to 9 at top of solitary to few stems. IA, IL, IN, MI, MN, MO, OH, WI. Introduced from Asia. Scattered throughout, with fewer occurrences in MN and northern IA. *SN*

Belamcanda chinensis
IRIDACEAE
blackberry lily
Old homesites, roadsides, disturbed sites. Summer.

An erect perennial, branched above, glabrous, 1–3 ft. tall. Leaves alternate, mostly basal, sword-shaped, arranged fan-like, glaucous, 1–2 ft. long and up to 1 in. wide. Flowers regular, about 2 in. wide with 6 spreading elliptic tepals sitting atop a green ovary, the tepals splotched with red or purple. Fruit a capsule that when ripe opens to reveal a conglomeration of fleshy black seeds that resembles a blackberry. Blackberry lily is neither a lily nor a blackberry. Syn. *Iris domestica.* IA, IL, IN, MI, MN, MO, OH, WI. Introduced from Asia. Occurs mostly in our southern counties. *CB*

Lilium michiganense
LILIACEAE

Michigan lily

Wet prairies, fens, thickets, savannas, wet woods, swamps. Summer.

An unbranched perennial to 6 ft. tall with glaucous stem. Leaves sessile or nearly so in whorls of 3–7 (occasionally opposite or alternate above), simple, entire, to 5 in. long × ¾ in. wide. Flowers regular, to 3 in. long and wide, nodding, with 6 recurved tepals, each orange with reddish-orange spots, with 6 stamens and 1 pistil long-exserted, numbering 1–8 on stalks at top of plant. Fruit an erect, 6-parted capsule. Canada lily (*L. canadense*), found only in the extreme eastern portion of Midwest, has spreading tepals and barely exserted stamens and pistil. IA, IL, IN, MI, MN, MO, OH, WI. Throughout. *SN*

Lilium philadelphicum
LILIACEAE

wood lily, prairie lily

Prairies, fens, thin woods, dunes, rock outcrops. Summer.

An unbranched perennial to 3 ft. tall with glaucous stem. Leaves sessile, simple, entire, alternate below (in var. *andinum*), in ascending whorls of 3–11 above, to 4 in. long × 1 in. wide. Flowers regular, to 3 in. long and wide, erect, with 6 ascending tepals, each orange-red narrowed to yellowish-orange (or dark) stalk-like base spotted with purplish-brown, with stamens and pistil erect and only slightly exserted, numbering 1–5 on stalks at top of plant. Fruit an erect 6-parted capsule. IA, IL, IN, MI, MN, MO, OH, WI. Scattered in the Northern Lakes and northern portion of the Tallgrass Prairie, scarce in the Eastern Forests. *SN*

Anagallis arvensis
MYRSINACEAE

scarlet pimpernel, common pimpernel, poor man's barometer

Lawns, gardens, pastures, trails, roadsides. Spring, summer, fall.

A glabrous, square-stemmed, branching annual to 12 in. tall (usually shorter). Leaves simple, opposite, entire, to ⅗ in. long × ⅖ in. wide, sessile, often somewhat clasping. Flowers regular, to ¼ in. wide, orange (blue in forma *caerulea* or sometimes white) with 5 spreading petals with minutely toothed or glandular margins, with single pistil and 5 purple filaments topped by yellow anthers, solitary on stalks to ⅘ in. from leaf axils. Fruit a round capsule. Flowers close when cloudy. Syn. *Lysimachia arvensis*. IA, IL, IN, MI, MN, MO, OH, WI. Introduced from Eurasia. Scattered throughout, more common south. *MAH*

Platanthera ciliaris
ORCHIDACEAE

orange fringed orchid

Sphagnum bogs, acid sandy meadows, savannas. Summer.

An erect perennial, unbranched, glabrous, 1–2½ ft. tall. Leaves alternate, lanceolate, sheathing, 3–6 in. long × 1–2 in. wide. Flowers irregular, about 1 in. long, with 3 petals and 3 ovate sepals, the lip petal lowermost, oblong-elliptic, deeply and finely fringed, with a basal spur about 1 in. long, other petals linear with fringed tips. Inflorescence of 20–30 flowers in a loose raceme. The similar white fringed orchid (*P. blephariglottis*) has white flowers and resides with us in lower MI almost exclusively. IL, IN, MI, MO, OH. Mostly in southern MI and northern IN. *PG*

Castilleja coccinea
OROBANCHACEAE

Indian paintbrush

Prairies, glades, savannas, cedar swamps, fens, interdunal flats. Spring, summer.

A biennial to 2 ft. tall. Basal rosette often absent when flowering; stem leaves 3-lobed, sessile, to 3 in. long. Flowers in terminal cluster, irregular, tubular, green, ¾ in. long, subtended by bright red, orange, or yellow calyx and bracts. Calyx tubular, shorter than corolla, bracts leafy, several-lobed. Hemi-parasitic. Downy painted-cup (*C. sessiliflora*) has pale flowers longer than green bracts (see inset). IA, IL, IN, MI, MN, MO, OH, WI. Frequent in the Northern Lakes, Ozark Highlands, and eastern portion of the Tallgrass Prairie; scarce in western portion of the Tallgrass Prairie and the Eastern Forests. *SN*

YELLOW FLOWERS

Acorus americanus
ACORACEAE

American sweetflag

Marshes, fens, seep springs, wet prairies.
Spring, summer.

An erect, unbranched perennial, glabrous, colonial, 2–5 ft. tall. Leaves 2–4 ft. long × ⅜–¾ in. wide, sword-shaped, aromatic, with 2–6 prominent, equally raised parallel veins. Flowers minute, tightly clustered in a club-like spike, the spike 2–3 in. long and about ¾ in. wide and projecting outwardly from the base of leaf-like spathe. Plants fertile and producing fruits. The introduced European sweetflag (*A. calamus*) has 1 vein prominently raised above the others, it offset from middle of blade. The plant is sterile, reproducing by vegetative means only. **IA, IL, IN, MI, MN, OH, WI. Mostly in the Northern Lakes with a few populations farther south.** *SN*

Narcissus pseudonarcissus
AMARYLLIDACEAE

daffodil

Homesites, woods, roadsides.
Winter, spring.

A glabrous perennial to 18 in. tall, from a bulb. Leaves basal, linear, entire, to 18 in. long × ⅗ in. wide, flat in cross section, blue-green. Flower regular, to 2 in. long, with tubular, wavy-topped corona surrounded by and about as long as 6 spreading tepals, subtended by papery bract, solitary and horizontal or nodding at top of single stem to 18 in. tall. Poet's narcissus (*N. poeticus*) has white tepals and a red-tipped yellow corona much shorter than the tepals. A commonly used spring-blooming ornamental; numerous cultivars available in the nursery trade. **IL, IN, MI, MO, OH. Introduced from Europe. Scattered throughout.** *EH*

Pastinaca sativa
APIACEAE

wild parsnip

Roadsides, fields, fencerows, pastures, ditch banks. Spring, summer.

An erect biennial, glabrous to sparsely hairy, stems strongly ridged, 3–5 ft. tall. Leaves alternate, petiolate, oblong and pinnately compound, coarsely toothed or lobed, up to 1½ ft. long × 3–7 in. wide. Flowers regular, up to ⅛ in. wide, with 5 curled to spreading petals, sepals absent, in flat, compound umbels up to 8 in. across terminating branch tips, 25+ per umbellet. Wild parsnip is the ancestor of garden parsnip. Its phototoxic sap can cause serious blistering of the skin. IA, IL, IN, MI, MN, MO, OH, WI. Introduced from Europe. Generally common throughout our range. *MAH*

Polytaenia nuttallii
APIACEAE

prairie parsley

Dry prairies, barrens, open woodlands. Spring, summer.

An erect perennial, glabrous to minutely hairy, 1½–3 ft. tall. Leaves alternate, petiolate, ovate-triangular, 2 times pinnately compound with a few teeth on leaflet margins, relatively thick and shiny, up to 7 in. long × 5 in. wide. Flowers regular, about ⅛ in. wide, with 5 oblong and shallowly notched petals, in somewhat rounded compound umbels of 25+ flowers per umbellet. Prairie parsley is so named because its foliage looks somewhat like flat-leaved parsley (*Petroselinum crispum* var. *neapolitanum*). IA, IL, IN, MI, MN, MO, WI. Mostly in the Tallgrass Prairie, especially in IA, IL, and MO. *SN*

Sanicula odorata
APIACEAE

clustered black snakeroot

Moist and floodplain forests, seeps, thickets. Spring, summer.

A glabrous perennial to 3 ft. tall. Leaves mostly basal or alternate, to 5 in. long and wide, on petioles to 6 in. long basally, becoming sessile above, palmately compound with 3–5 shallowly lobed and sharply toothed deep-veined leaflets. Flowers regular, tiny, with 5 petals, in tight globose ½-in.-wide clusters, with 1–3 sessile (or nearly so) perfect flowers and numerous stalked male flowers per cluster, perfect flowers with straight to arching styles longer than ovary bristles, clusters several in umbels at ends of branches. Fruit spherical, to ⅓ in. wide, covered in hooked bristles. IA, IL, IN, MI, MN, MO, OH, WI. **Common and widespread.** *MAH*

Taenidia integerrima
APIACEAE

yellow pimpernel

Rocky wooded hillsides, limestone cliffs, glades, savannas, dunes. Spring, summer.

An erect perennial, glabrous and glaucous, 1–2½ ft. tall. Leaves alternate, petiolate, ovate, 1–10 in. long and wide (or wider), commonly 3 times compound, leaflets ovate to elliptic, entire. Flowers regular, with 5 ovate petals, ⅛ in. wide, in large rounded compound umbels of 10–15 flowers per umbellet. This is our only yellow-flowered carrot family member whose leaflets have entire margins. The spreading inflorescence provides an appearance reminiscent of an aerial fireworks display. IA, IL, IN, MI, MN, MO, OH, WI. **Widespread but rare to absent in most of MN and upper WI.** *JM*

Thaspium barbinode
APIACEAE

bearded meadow parsnip

Well-drained stream terraces, moist cliffs, mesic forests, thickets. Spring, summer.

An erect perennial, branched above, stems glabrous to short-hairy (at least upper nodes hairy), 1–3 ft. tall. Leaves alternate, petiolate, ovate, 2 or 3 times compound, leaflets mostly glabrous on undersurface, sharply toothed, leaves 3–8 in. long and wide (or wider). Flowers regular, ⅛ in. wide, 5-petaled, golden yellow, in somewhat rounded umbels with 20+ flowers per umbellet, all flowers/fruits on short stalks. Chapman's meadow parsnip (*T. chapmanii*) is similar, but its leaflets are hairy on the undersurface and petals are cream to pale yellow. IA, IL, IN, MN, MO, OH, WI. Occurs primarily in the lower Midwest. *MAH*

Thaspium trifoliatum
APIACEAE

meadow parsnip

Dry-mesic forests, upland wooded slopes, savannas, prairies. Spring, summer.

An erect perennial, mostly unbranched, glabrous, 1–3 ft. tall. Leaves on stem alternate, petiolate, with 3 ovate-lanceolate and toothed leaflets with thickened whitish margins, the lower 2 leaflets sometimes 2- or 3-lobed, basal leaves simple, ovate or cordate, or like the stem leaves, 1–3 in. long and about as wide. Flowers regular, about ⅛ in. wide, 5-petaled, yellow (var. *flavum*) or purple (var. *trifoliatum*), in spreading, somewhat rounded umbels with 6–15 flowers per umbellet. All flowers/fruits on short stalks. Variety *flavum* is more common in our region. IA, IL, IN, MI, MO, OH, WI. Occurs primarily in the lower Midwest. *EH*

Zizia aptera
APIACEAE

heart-leaved golden alexanders

Dry rocky woodland slopes, glades, prairies.
Spring, summer.

An erect perennial, mostly glabrous, somewhat branched, 1–2 ft. tall. Stem leaves alternate, short- or long-petiolate, with 3 ovate,
somewhat blunt-tipped, toothed and glossy
leaflets having thickened whitish margins,
basal leaves similar but usually all simple,
cordate, and long-petiolate, 1–6 in. long and
up to 3 in. wide. Flowers regular, about ⅛ in.
wide, 5-petaled, in umbels of 10–25 stalked
flowers per umbellet with the central flower/
fruit sessile. Compare with meadow parsnip
(*Thaspium trifoliatum*). IA, IL, IN, MI, MN,
MO, OH, WI. Most populations occur in
the Midwest's northwestern quarter and
in MO. *SN*

Zizia aurea
APIACEAE

golden alexanders

Streambanks, glades, savannas, prairies,
fens. Spring, summer.

An erect perennial, somewhat branched, glabrous, 1–3 ft. tall, often colonial. Stem leaves
alternate, short-petiolate, ovate to lanceolate,
sharply toothed, 2 times ternately compound
becoming just lobed on the upper stem. Basal
leaves similar but regularly 2 times ternately
compound, 2–6 in. long and about as wide.
Flowers about ⅛ in. wide, petals 5, in umbels
with 10–25 flowers per umbellet, the central
flower sessile. The common name appears
to reference its presumed similarity to alex-
anders (*Smyrnium olusatrum*), a European
relative. IA, IL, IN, MI, MN, MO, OH, WI.
Widespread and common but somewhat less
so in the Eastern Forests. *SN*

Bidens aristosa
ASTERACEAE

tickseed sunflower

Marshes, meadows, prairies, abandoned fields. Summer, fall.

An erect annual, glabrous to sparsely hairy, 1–4 ft. tall. Leaves opposite, petiolate, pinnately compound into 3–7 lanceolate to linear segments with serrated margins, 2–8 in. long × 2–4 in. wide. Flowers in composite heads 2–3 in. wide with 7–10 ray flowers and 20+ disk flowers, outer phyllaries 8–10, heads in an open, branched inflorescence. Fruit a flat, convex-sided seed-like achene with 2 barbed awns. Ozark tickseed sunflower (*B. polylepis*) has 11+ outer phyllaries. Northern tickseed sunflower (*B. trichosperma*) differs from both in having straight-sided fruits. IA, IL, IN, MI, MO, OH, WI. Widespread and common except for in the Northern Lakes. *MAH*

Bidens beckii
ASTERACEAE

water marigold

Lakes, ponds, rivers, streams. Summer, fall.

An aquatic perennial in water to 13 ft. deep with emerged shoots to 6 in. above water. Leaves opposite. Emergent leaves simple, sessile, entire to toothed, to 1½ in. long. Submerged leaves deeply dissected into thread-like segments, appearing whorled, round to fan-shaped in outline, to 2 in. long. Flowers in composite head; approximately 8 ray flowers surround disk in solitary emergent head to 1½ in. wide at top of plant. Syn. *Megalodonta beckii*. Considered an indicator of good water quality. IA, IL, IN, MI, MN, MO, OH, WI. Almost entirely restricted to the Northern Lakes, rare elsewhere. *AG*

Bidens bipinnata
ASTERACEAE

Spanish needles

Clearings in upland forests, old fields, riverbanks, roadsides, waste areas. Summer, fall.

An erect annual, glabrous or sparsely hairy, 1–3 ft. tall. Leaves opposite, 2 or 3 times pinnately divided with ovate-lanceolate leaflets triangular in outline, petiolate, about 7 in. long × 4 in. wide. Flowers in composite heads about ¼ in. wide with 0–5 ray flowers and 10–25 disk flowers, arranged in a widely spreading inflorescence. Fruit a narrow needle-like achene. Awns of the fruits allow them to attach to clothing and fur, not to mention being a pain when jabbing the skin! IA, IL, IN, MI, MO, OH. **Widespread and common in lower Midwest.** *MAH*

Bidens cernua
ASTERACEAE

nodding bur-marigold, nodding beggarticks

Mudflats, fens, swamps, bogs, seeps, marshes, wet meadows, floodplains, pond, lake, and river margins. Summer, fall.

An annual to 3 ft. tall with glabrous to rough-hairy stems. Leaves opposite, simple, clasping, coarsely toothed, glabrous, to 5 in. long × 1 in. wide, becoming purplish. Flowers in composite heads; disk surrounded by approximately 8 ray flowers, or ray flowers absent; heads to 2 in. wide, several in upper part of plant, facing outward and nodding with age. Fruit an achene with 4 barbed awns that stick in fur of passing animals and human clothing. Often forms large colonies in areas that were inundated early in year. IA, IL, IN, MI, MN, MO, OH, WI. **Widespread.** *PS*

Bidens frondosa
ASTERACEAE

devil's beggarticks

Lake and pond borders, streambanks, ditches, fallow agricultural fields. Summer, fall.

An erect annual, mostly glabrous, branching, 3+ ft. tall. Leaves opposite, pinnately compound with 3–5 serrated, lanceolate to narrowly ovate leaflets 4 in. long × 1 in. wide. Flowers in composite heads ¼–⅓ in. long, ray flowers few if present, disk flowers 15–50, with 6–8 outer phyllaries much longer than the heads. Fruit a 2-awned achene. Swamp beggarticks (*B. discoidea*) and tall beggarticks (*B. vulgata*; see inset) are quite similar but have up to 4 and 10+ outer phyllaries per head, respectively. IA, IL, IN, MI, MN, MO, OH, WI. **Widespread and common throughout.** *PR/SN*

Bidens tripartita
ASTERACEAE

three-lobed beggarticks, swamp beggarticks

Wet open to shaded ground. Summer, fall.

An annual to 4½ ft. tall. Leaves opposite, simple, sometimes 2- to 3-lobed, sessile to petiolate, toothed, sometimes becoming purplish. Flowers in composite heads; disk surrounded by few ray flowers, or ray flowers absent, in several heads to 1¼ in. wide. Fruit an achene with 2 or 3 barbed awns that stick in animal fur and human clothing. Some apply the name *B. tripartita* strictly to the European species with dense white hairs on outer (green) phyllary margins. We consider *B. tripartita* a native complex that includes *B. comosa* (pale yellow 4-lobed disk flowers) and *B. connata* (orange-yellow 5-lobed disk flowers). IA, IL, IN, MI, MN, MO, OH, WI. **Widespread.** *KC*

Coreopsis lanceolata
ASTERACEAE

sand coreopsis, lanceleaf coreopsis

Prairies, savannas, glades, open woods, dunes, roadsides. Spring, summer.

A perennial to 2½ ft. tall. Basal leaves in rosette, simple, strap-like, sometimes with 1 or 2 deep lobes, to 6 in. long × 1 in. wide, entire; stem leaves similar, opposite, up to 3 pairs along lower half of stem, sometimes lobed. Flowers in composite head; a ½-in. disk is surrounded by coarsely and irregularly toothed ray flowers in terminal solitary head to 3 in. wide. Bigflower coreopsis (*C. grandiflora*) and star tickseed (*C. pubescens*) have at least 4 pairs of stem leaves, some above middle of stem; the leaves of latter usually unlobed and more broadly elliptic. IA, IL, IN, MI, MN, MO, OH, WI. Scattered. *SN*

Coreopsis palmata
ASTERACEAE

prairie coreopsis, finger coreopsis, stiff tickseed

Prairies, savannas, glades, open woods. Spring, summer.

A colony-forming perennial to 2½ ft. tall. Leaves opposite, hairy where connected to stem, deeply 3- to 5-lobed, dark green with prominent light green vein running to tip of each lobe, to 3 in. long × 1½ in. wide, entire. Flowers in composite heads; a ⅓-in. disk is surrounded by 8–12 ray flowers in 1 to few 2-in.-wide heads at top of plant. Leaves usually have shape of a bird's foot but occasionally, especially lower on stem, are unlobed. IA, IL, IN, MI, MN, MO, WI. Widespread in the Ozark Highlands, Tallgrass Prairie, and adjacent portions of the Northern Lakes. *SN*

Coreopsis tinctoria
ASTERACEAE

plains coreopsis, golden coreopsis, calliopsis

Prairies, glades, roadsides, railroads, waste areas. Summer, fall.

A glabrous annual to 3 ft. tall. Leaves opposite, sessile or short-petiolate, once or twice pinnately compound with linear-elliptic to thread-like leaflets, to 5 in. long × 4 in. wide. Flowers in composite heads; a red-purple ⅓-in. disk is surrounded by 6–12 ray flowers that are deep reddish-purple at base and yellow in outer half (sometimes entirely red-purple) with 3 or 4 coarse, irregular lobes at tip; several 2-in.-wide heads at ends of upper branches. IA, IL, IN, MI, MN, MO, OH, WI. Scattered but native only in the Ozark Highlands and parts of the Tallgrass Prairie; most frequent in southern MO. *ST*

Coreopsis tripteris
ASTERACEAE

tall tickseed, tall coreopsis

Prairies, glades, barrens, savannas, dry upland forests, thickets. Summer.

An erect perennial, branched above, on glabrous stems, 2–6 ft. tall. Leaves opposite, petiolate, glabrous to hairy, usually divided into 3 narrowly elliptic leaflets 1–5 in. long × ¼–1½ in. wide Flowers in composite heads 1–2 in. wide, ray flowers 5–8, the purplish or dark brown fertile disk flowers number 40–75, on narrow terminal branches forming a somewhat flat-topped inflorescence. The name *Coreopsis* translates as "bug appearance," alluding to its small, dark achenes. IA, IL, IN, MI, MO, OH. Widespread in southern two-thirds of Midwest. *MAH*

Dyssodia papposa
ASTERACEAE

fetid marigold

Thin or compacted soil in glades, trails, pastures, roadsides. Spring, summer, fall.

A multi-branched annual, sparsely hairy to glabrous, malodorous, 4–16 in. tall. Leaves opposite, deeply pinnately divided with linear segments, dotted with yellowish oil glands, ½–1½ in. long and up to ½ in. wide. Flowers in composite heads about ¼ in. wide bearing 4–8 ray flowers and 10–20 disk flowers. Fetid marigold is native in our western counties but has spread eastward, especially along the shoulders of interstate highways. In the Great Plains it is a common plant of many prairie dog towns. IA, IL, IN, MI, MN, MO, OH, WI. **Occurs mostly in IA, IL, and MO.** *SN*

Echinacea paradoxa
ASTERACEAE

yellow coneflower, Ozark coneflower

Prairies, glades. Spring, summer.

A perennial to 3 ft. tall. Leaves basal and alternate on lower stem, to 12 in. long × ¾ in. wide, entire, petiolate below, becoming smaller and sessile up stem. Flowers in composite head; a deep reddish-brown disk is surrounded by up to 30 reflexed yellow ray flowers (in var. *paradoxa*), in a single head to 2½ in. wide. Ray flowers pinkish to white in var. *neglecta*, found just outside the Midwest. Flower heads of yellow coneflower look similar to those of *Rudbeckia* species but have stiff bracts within the disks, which are large, as in other *Echinacea* species. **MO. Restricted to the Ozark Highlands.** *MJH*

Euthamia graminifolia
ASTERACEAE

flat-top goldentop, grass-leaved goldenrod

Moist to wet prairies, fields, fens, conifer swamps, ditches. Summer, fall.

A branched perennial to 4 ft. tall. Leaves alternate, linear, entire, to 4 in. long × ½ in. wide, with at least 5 parallel veins. Flowers in composite heads; disks surrounded by short, spreading ray flowers in heads to ⅛ in. wide, heads numerous, forming flat-topped terminal arrays. Plants with hairy stems and underside veins are var. *nuttallii*. Slender goldentop (*E. caroliniana*) and Great Plains goldentop (*E. gymnospermoides*) have narrower leaves with 1–3 veins; the former has glistening upper leaf surfaces, the latter has fewer flowers per head. IA, IL, IN, MI, MN, MO, OH, WI. **Widespread but absent in the Ozark Highlands and uncommon in MO.** *SN*

Grindelia squarrosa
ASTERACEAE

curlytop gumweed

Prairies, old fields, roadsides, railroads, waste areas. Summer, fall.

A sticky biennial to 3 ft. tall. Leaves in basal rosette first year, alternate second year, clasping, to 3 in. long × ¾ in. wide, with numerous short, sharp teeth on margins (except at top of plant). Flowers in composite heads; a ½-in.-wide disk is surrounded by numerous ½-in.-long ray flowers in 3–20 heads in upper part of plant. Heads subtended by abundant narrow, conspicuously curved, sticky phyllaries. Narrowleaf gumweed (*G. lanceolata*), found mostly in prairies and glades in MO, has ascending phyllaries and spine-tipped teeth on leaf margins. IA, IL, IN, MI, MN, MO, OH, WI. **Scattered, with most occurrences in MN and WI.** *PR*

Helenium amarum

ASTERACEAE

bitterweed, yellowdicks

Prairies, glades, sandy clearings, roadsides, railroads. Summer, fall.

An annual to 3 ft. tall (often much shorter). Leaves alternate, numerous, simple, sessile, entire, narrowly linear, to 3 in. long × ⅛ in. wide, often appearing whorled or compound due to presence of axillary clusters. Flowers in composite heads; 5–10 ray flowers, 3- or 4-lobed, surround domed disk in several stalked heads to ¾ in. wide in upper part of plant. IA, IL, IN, MI, MO, OH, WI. **Widespread in southernmost portion of Midwest (mostly in MO), spotty elsewhere.** *ST*

Helenium autumnale

ASTERACEAE

common sneezeweed

Wet prairies, moist old-fields, marshes, swamps, fens, floodplains, thickets. Summer, fall.

A wing-stemmed perennial to 5 ft. tall. Leaves alternate, sessile, usually sparsely shallowly toothed, to 5 in. long × 1½ in. wide. Flowers in composite heads; up to 15 spreading to reflexed 3- or 4-lobed ray flowers surround globose yellow disk; heads in upper part of plant, numerous, stalked, to 1 in. wide. Common name comes from historical use of dried leaves in snuff to induce sneezing and rid one of evil spirits. Wingstem (*Verbesina alternifolia*) also has conspicuously winged stems but has fewer, narrower ray flowers, more sparsely flowered disks, and wider leaves. IA, IL, IN, MI, MN, MO, OH, WI. **Widespread, with fewer occurrences in northern MI.** *SN*

Helenium flexuosum

ASTERACEAE

purple-headed sneezeweed

Wet fields and prairies, depressions in glades, swamps, ditches, roadsides. Summer, fall.

A wing-stemmed perennial to 3 ft. tall, often covered in short hairs. Leaves alternate, simple, sessile, entire, to 3 in. long × 1 in. wide. Flowers in composite heads; up to 14 spreading to reflexed 3- or 4-lobed yellow ray flowers surround globose purple-brown disk, stalked heads to 1 in. wide and numerous in upper part of plant. Heads somewhat resemble black-eyed Susan (*Rudbeckia hirta*) but are much smaller. Syn. *H. nudiflorum*. IL, IN, MI, MN, MO, OH, WI. Widespread in southern portion of the Midwest, widely scattered north. *EH*

Helianthus annuus

ASTERACEAE

common sunflower, annual sunflower

Prairies, fields, roadsides, railroads, waste areas. Summer, fall.

An upright, hairy annual to 10 ft. tall. Leaves opposite below, alternate above, simple, triangular, rough-hairy, toothed or wavy on margins, to 8 in. long × 6 in. wide, on petioles to 8 in. long. Flowers in composite heads; up to 40 spreading yellow, conspicuously veined ray flowers surround yellow to brown disk in stalked heads to 6 in. wide at ends of branches. The seeds, removed from achenes, make a delicious field snack! Prairie sunflower (*H. petiolaris*) is shorter with smaller leaves and heads with proportionally smaller disks. IA, IL, IN, MI, MN, MO, OH, WI. Introduced from western North America. Scattered throughout, more frequent in western portion of Midwest. *KY*

Helianthus grosseserratus
ASTERACEAE

sawtooth sunflower

Prairies, marshes, fens, roadsides, ditches. Summer, fall.

An erect colonial perennial, stems glabrous below, purplish and glaucous, 3–10 ft. tall. Leaves alternate above, petiolate, lanceolate to narrowly ovate, margins usually sharply toothed, blades mostly flat, the upper surface rough, undersurface soft hairy. Flowers in composite heads to 4 in. wide, ray flowers 10–20, disk flowers 100+, arrayed in upper branches. Phyllaries lanceolate, pointed, spreading to recurved. Giant sunflower (*H. giganteus*) and Maximilian's sunflower (*H. maximiliani*) have hairy or roughened stems. The former has flat leaves; those of the latter are arcing and folded longitudinally. IA, IL, IN, MI, MN, MO, OH, WI. **Widespread, most prevalent in the Tallgrass Prairie.** *PR*

Helianthus hirsutus
ASTERACEAE

hairy sunflower

Prairies, glades, woods, savannas, thickets, roadsides, railroads. Summer, fall.

A rough-hairy perennial to 5 ft. tall. Leaves opposite (sometimes alternate in inflorescence), simple, shallowly toothed, to 6 in. long × 3 in. wide, on ½-in. petioles. Flowers in composite 3-in.-wide heads; up to 15 spreading, conspicuously veined ray flowers surround disk in heads at ends of upper branches. Phyllaries lanceolate, sharp-pointed. Rough sunflower (*H. divaricatus*), woodland sunflower (*H. strumosus*), thin-leaved sunflower (*H. decapetalus*), and small woodland sunflower (*H. microcephalus*) are very similar, differing in flower head size, leaf margins, and other characters. IA, IL, IN, MI, MN, MO, WI. **Patchy distribution throughout.** *MAH*

Helianthus mollis
ASTERACEAE

ashy sunflower

Prairies, glades, barrens, thickets.
Summer, fall.

An erect perennial, unbranched with a densely
hairy stem, 2–5 ft. tall, colonial. Leaves
opposite, sessile, ovate with cordate bases
clasping, entire or serrated, with dense grayish
hairs, 2–5 in. long × 1–3 in. wide. Flowers in
composite heads 2–4 in. across, ray flowers
15–25, disk flowers 75+. Phyllaries lanceolate,
sharp-pointed and appressed. The grayish
appearance of stem and foliage confirms the
appropriateness of the common name. This
species is generally a good indicator of remnant natural grasslands but is also regularly
included in prairie plantings. IA, IL, IN, MI,
MO, OH, WI. Mostly in the Tallgrass Prairie
and Ozark Highlands. *EH*

Helianthus occidentalis
ASTERACEAE

western sunflower

Barrens, dunes, glades, savannas,
prairies. Summer, fall.

An erect perennial, unbranched, glabrous
to hairy, 2–4 ft. tall, colonial. Leaves mostly
basal, surfaces hairy and rough, petiolate, with
a few smaller pairs of mostly sessile, opposite
leaves on flowering stem, ovate to elliptical,
entire or finely toothed, 2–6 in. long × 1–3
in. wide. Flowers in composite heads about 2
in. wide, ray flowers 10–20, disk flowers 50+.
Phyllaries lanceolate, pointed and appressed.
Where it and ashy sunflower (*H. mollis*) occur
together, look for *H. ×cinereus*, their hybrid
progeny. IA, IL, IN, MI, MN, MO, OH, WI.
**Occurs mostly in Ozark Highlands and the
western Great Lakes area.** *SN*

Helianthus tuberosus
ASTERACEAE

Jerusalem artichoke

Prairies, fields, floodplains, sloughs, thickets, roadsides, railroads, waste areas. Summer, fall.

A rough-hairy, colony-forming perennial to 10 ft. tall. Leaves opposite below, alternate above, simple, toothed or entire, to 9 in. long × 4 in. wide, on 2½-in. petioles. Flowers in composite heads; up to 20 spreading, conspicuously veined ray flowers surround disk in heads to 3½ in. wide at ends of upper branches. Phyllaries lanceolate, sharp-pointed. Often produces edible tubers. Stiff sunflower (*H. pauciflorus*) has reddish-brown disk flowers and blunt phyllaries. Cheerful sunflower (*H. ×laetiflorus*) is their intermediate hybrid. IA, IL, IN, MI, MN, MO, OH, WI. Widespread throughout; fewer occurrences in northern MI and northern WI. *MAH*

Heliopsis helianthoides
ASTERACEAE

false sunflower, smooth oxeye

Prairies, glades, wet meadows, open woods, thickets, roadsides, railroads. Spring, summer, fall.

A rough-hairy perennial to 5 ft. tall. Leaves opposite, simple, sharply toothed, to 5 in. long × 3½ in. wide, on petioles to 1¼ in. Flowers in composite heads; up to 20 spreading, conspicuously veined ray flowers surround disk in heads to 3 in. wide at ends of upper branches. Phyllaries blunt, spreading to somewhat recurved. Differs from rosinweeds (*Silphium*) in having upper leaves petiolate and toothed. Differs from sunflowers (*Helianthus*) in phyllary shape and by having fertile ray and disk flowers (ray flowers sterile in sunflowers). IA, IL, IN, MI, MN, MO, OH, WI. Widespread throughout; less common in MI. *MAH*

Heterotheca camporum
ASTERACEAE

prairie goldenaster, lemon-yellow false goldenaster

Prairies, glades, sandy fields, forest openings. Summer, fall.

A hairy, colony-forming perennial to 4 ft. tall. Leaves alternate, simple, sparsely toothed, to 2¾ in. long × ¾ in. wide, smaller up stem, sessile. Flowers in composite heads; up to 35 spreading ray flowers surround disk in heads to 1 in. wide at ends of upper branches, forming flat-topped to dome-shaped array. Phyllaries appressed in several series. Hairy goldenaster (*H. villosa*) is more branched, to 20 in. tall, with entire leaves. Camphorweed (*H. subaxillaris*) is short-gray-hairy and has clasping, somewhat sticky-hairy leaves. Syn. *Chrysopsis camporum*. IA, IL, IN, MI, MO, OH. Scattered in the Ozark Highlands and eastern portion of the Tallgrass Prairie, rare elsewhere. *ST*

Hieracium caespitosum
ASTERACEAE

king devil, yellow hawkweed, field hawkweed

Prairies, fields, pastures, open woods, roadsides. Spring, summer.

A long-hairy (often with some black hairs) creeping perennial to 2½ ft. tall. Leaves entire, simple, basal or with few reduced alternate leaves on lower stem, to 6 in. long × 1 in. wide, sessile, containing milky sap. Flowers in composite heads; numerous yellow ray flowers in ¾-in.-wide heads, disk flowers lacking; heads few to numerous in dense cluster at top of plant. Fruit with silky pappus. Syns. *H. pratense, Pilosella caespitosa. Hieracium piloselloides* (aka king devil) has glaucous leaves. IL, IN, MI, MN, MO, OH, WI. Introduced from Europe. Mostly in the Northern Lakes and Eastern Forests (more common east). *SN*

Hieracium paniculatum
ASTERACEAE

panicled hawkweed, Allegheny hawkweed

Forests, especially in sand. Summer.

A perennial to 3 ft. tall. Leaves simple, alternate, sparsely toothed, thin-textured, to 6 in. long × 1½ in. wide, short-petiolate below, sessile above, containing milky sap. Flowers in composite heads; numerous yellow ray flowers in ½-in.-wide heads, disk flowers lacking; heads numerous (up to 50) on slender stalks in large, open paniculate array at top of plant. Fruit with golden brown, silky pappus. **IN, MI, OH. In the eastern portion of the Midwest, particularly in the Eastern Forests and southern portion of the Northern Lakes. *THK***

Hieracium scabrum
ASTERACEAE

rough hawkweed

Dry upland forests, sand savannas, rock outcrops. Summer.

An unbranched perennial 1–3 ft. tall with ¼-in.-long spreading hairs on lower stem, some glandular upward. Leaves alternate, entire, oblanceolate to elliptic, sessile, hairy, basal leaves usually withered by flowering, 2–8 in. long × ½–2 in. wide. Flowers in composite heads ½–1 in. wide, ray flowers 40–100, disk flowers lacking, involucres glandular-hairy, in a spreading, flat-topped array. Basal leaves of beaked hawkweed (*H. gronovii*) usually persist. Lower leaves of veiny hawkweed (*H. venosum*) bear conspicuous purple veins. Long-haired hawkweed (*H. longipilum*) has lower stem hairs up to ¾ in. long. **IA, IL, IN, MI, MN, MO, OH, WI. Scattered throughout except for western Tallgrass Prairie. *JD***

Hieracium umbellatum
ASTERACEAE

Canada hawkweed, northern hawkweed

Forests, thickets, often in sandy or rocky soils. Summer, fall.

A perennial to 4 ft. tall. Leaves alternate, sparsely toothed, to 4 in. long × 1¼ in. wide, shorter above, sessile or very short-petiolate, containing milky sap. Flowers in composite heads; numerous ray flowers in heads to 1¼ in. wide, disk flowers lacking, phyllaries reflexed; up to 30 heads in often large paniculate array at top of plant. Fruit with light brown, silky pappus. Syns. *H. canadense*, *H. kalmii*. **IA, IL, IN, MI, MN, MO, OH, WI. Widespread in the Northern Lakes and adjacent Tallgrass Prairie.** *SN*

Hypochaeris radicata
ASTERACEAE

rough cat's-ear

Fields, lawns, vacant lots, roadsides. Spring, summer, fall.

A perennial to 2 ft. tall. Leaves mostly basal, shallow-lobed, hairy, to 6 in. long × 2 in. wide, short-petiolate, with pale midvein and milky sap; stem leaves few, alternate, scale-like. Flowers in composite heads; numerous ray flowers, no disk flowers, in ¾-in.-wide stalked terminal heads (1 to few). Phyllaries purple-tipped, appressed, in several series. Fruit beaked with silky feather-like terminal pappus, forming globose heads similar to those of common dandelion (*Taraxacum officinale*). Hawksbeard (*Crepis tectorum*), annual, and fiddle-leaf hawksbeard (*C. runcinata*), perennial, have barbed (but not feather-like) pappus hairs. **IL, IN, MI, MO, OH, WI. Introduced from Eurasia. Scattered.** *RS*

Inula helenium
ASTERACEAE

elecampane

Fields, forest openings, roadsides.
Summer, fall.

A hairy, colony-forming perennial to 8 ft. tall.
Leaves basal and alternate, to 16 in. long × 8
in. wide, long-petiolate, becoming smaller,
sessile, and clasping above, simple, abundantly toothed and wavy on margins, with
strong white midvein. Flowers in composite
heads; up to 100 narrow, spreading ray flowers
surround disk (turning brown with age) in
stalked heads to 4 in. wide at top of plant.
Phyllaries large, broad, spreading, in several
series. Has been used medicinally and as
food/drink in a variety of ways. Superficially
resembles *Silphium*. IA, IL, IN, MI, MN, MO,
OH, WI. Introduced from Eurasia. Scattered,
more common eastward. *SN*

Krigia biflora
ASTERACEAE

false dandelion

Dry-mesic forests, rocky slopes, savannas,
prairies. Spring, summer.

An erect perennial, branched in inflorescence, glabrous to sparsely hairy, 6–24 in. tall.
Leaves mostly basal, petiolate, but also with
a few clasping upper-stem leaves, alternate
and entire or with wavy or toothed margins,
oblanceolate to lanceolate, with milky sap, 2–5
in. long × 1–2 in. wide. Flowers in composite
heads about 1 in. wide, ray flowers 20–50,
notched at tip, disk flowers lacking, heads 1–3
in a terminal cluster. Dwarf dandelion (*K.
dandelion*) of the lower Midwest has only basal
leaves and a single flowering head. IA, IL,
IN, MI, MN, MO, OH, WI. Well distributed
throughout except for our far western counties. *MAH*

Krigia virginica
ASTERACEAE

Virginia dwarf dandelion

Prairies, fields, glades, sandy savannas
and barrens. Spring, summer.

An annual with 1 to few stems to 1½ ft. tall.
Leaves gray-green, basal, shallowly lobed, to 3
in. long × ¾ in. wide, sessile or nearly so, containing milky sap. Flowers in composite heads;
numerous ray flowers in ½-in.-wide heads,
disk flowers absent; heads solitary, terminal,
with appressed green phyllaries with purplish
tips in a single series. Fruit with hairs at tip,
forming globose heads. Opposite-leaved dwarf
dandelion (*K. cespitosa*) has a pair of opposite
to subopposite stem leaves beneath 2 or more
flower heads and fruit lacks pappus. IA, IL,
IN, MI, MO, OH, WI. **Spotty throughout, less
frequent in the Eastern Forests.** *SN*

Lactuca canadensis
ASTERACEAE

Canada wild lettuce, tall lettuce

Prairies, glades, fields, savannas, forest
openings and borders, swamps, roadsides.
Summer, fall.

An unbranched biennial with glaucous stem
to 10 ft. tall. Leaves alternate, unlobed to
toothed and deeply lobed, clasping stem, to 10
in. long × 3 in. wide, smaller above, containing
(sometimes faint) salmon-colored sap. Flowers in composite heads; numerous yellow to
reddish-orange ray flowers in heads to ⅓ in.
wide, disk flowers absent, phyllaries appressed
in several series within bulbous-based involucre, heads numerous in tight to open
panicle-like terminal array to 2 ft. long. Fruit
with white pappus on beaked achenes, in globose heads. Variable in many respects. IA, IL,
IN, MI, MN, MO, OH, WI. **Widespread and
common.** *CR*

Lactuca serriola
ASTERACEAE

prickly lettuce

Fields, gardens, roadsides, waste areas.
Summer, fall.

A gray-green biennial to 7 ft. tall, prickly at
base. Leaves alternate, prickly-toothed and
pinnately lobed, clasping stem with pointed
basal lobes, with prickles along midrib beneath,
to 12 in. long × 4 in. wide, smaller above, with
milky sap. Flowers in composite heads; numerous ray flowers in ½-in. heads, no disk flowers,
phyllaries appressed in several series within
bulbous-based involucre, heads numerous in
leafy open panicle-like array. Fruit with white
pappus on beaked achenes, in globose heads.
Willow-leaved lettuce (*L. saligna*) has long,
narrow, unlobed, entire leaves mostly lacking
prickles. **IA, IL, IN, MI, MN, MO, OH, WI.
Introduced from Europe. Widespread, less
frequent northward. *DT***

Lapsana communis
ASTERACEAE

common nipplewort

Woodland trails, gravel bars, gardens,
roadsides, disturbed sites. Summer, fall.

A colony-forming annual or biennial to 4 ft.
tall. Leaves toothed, compound or deeply
3-lobed, with a large, triangular terminal lobe
and 2 small lateral lobes, alternate, to 6 in.
long × 2½ in. wide, on long, winged petioles,
containing milky sap; smaller, narrower, and
becoming unlobed above. Flowers in composite heads; numerous ray flowers in ½-in.-wide
heads, disk flowers absent; heads numerous
in large open panicle-like array in upper
part of plant. Fruit an achene lacking fluffy
pappus. Flower head in bud said to resemble
a nipple. **IA, IL, IN, MI, MN, MO, OH, WI.
Introduced from Eurasia. Widely scattered,
increasing. *SN***

Matricaria discoidea
ASTERACEAE

pineapple-weed

Gardens, lawns, fields, gravel lots, pavement cracks, gravel pits, roadsides, railroads. Spring, summer, fall.

A branching annual to 1 ft. tall (usually much shorter). Leaves to 2 in. long × ¾ in. wide, alternate, pinnately dissected into numerous narrow, linear segments. Flowers in composite heads; few to numerous ⅓-in.-wide conical, button-like flower heads are made up of abundant tiny disk flowers, ray flowers absent, on ascending stalks to 1½ in. from upper leaf axils. Fruit lacks pappus. Crushed leaves and especially flower heads smell like pineapple, and flower heads somewhat resemble miniature pineapples. Syn. M. *matricarioides*. IA, IL, IN, MI, MN, MO, OH, WI. Introduced from western North America. Scattered to widespread. *SN*

Nothocalais cuspidata
ASTERACEAE

prairie false dandelion

Prairies, glades. Spring, summer.

A hairy perennial to 1 ft. tall. Leaves in basal rosette, simple, unlobed, entire, wavy-margined and fringed with dense white hairs, to 12 in. long × ¾ in. wide, sessile, with broad, pale midvein, folded or rolled lengthwise, containing milky sap. Flowers in composite head; numerous ray flowers in terminal, solitary head to 2 in. wide, disk flowers absent, with appressed purplish phyllaries in single series to 1 in. long subtending flower head. Fruit an achene with silky white hairs at top, forming globose head similar to those of common dandelion (*Taraxacum officinale*). IA, IL, MN, MO, WI. Primarily scattered in the Tallgrass Prairie. *SN*

Packera aurea
ASTERACEAE

golden ragwort, heart-leaved groundsel

Stream corridors, floodplain forests, thickets, fens, swamps. Spring.

An erect perennial, glabrous (though hairy when young), 8–24 in. tall, colonial. Leaves basal, cordate at base with long petioles and rounded teeth, up to 5 in. long × 4 in. wide, stem leaves alternate smaller, narrower, sessile and deeply pinnately lobed. Flowers in composite heads about ¾ in. wide, ray flowers 8–15, disk flowers 50–75, arranged in a somewhat flat-topped inflorescence atop stem. Western golden ragwort (*P. pseudaurea*) is similar, with smaller, less cordate basal leaves; it occurs mostly in our western counties. IA, IL, IN, MI, MN, MO, OH, WI. Common. *SN*

Packera glabella
ASTERACEAE

butterweed

Floodplain forests, swamps, agricultural fields, ditches. Spring.

An erect annual, unbranched, glabrous, stem hollow and rather weak, from a few inches to 4 ft. tall. Leaves alternate, elliptic-obovate, margins toothed and pinnately divided, terminal leaflet the largest, 2–6 in. long × 1–2 in. wide. Flowers in composite heads about ¾ in. wide, ray flowers 8–15, disk flowers 50–75, arranged in a somewhat flat-topped inflorescence atop stem. No-till agriculture has contributed to the marked increase of this species. Butterweed does not appear to create a serious impact on crop farming. IA, IL, IN, MI, MO, OH. Common in much of the southern and central Midwest. *MAH*

Packera obovata

ASTERACEAE

roundleaf ragwort

Moist to dry-mesic upland forests, cliff
ledges, ravine slopes, stream terraces.
Spring.

An erect perennial, mostly glabrous, 8–24
in. tall, colonial. Stem leaves sessile, pinnately lobed and alternate, basal leaves
ovate-obovate, toothed, usually tapering to
a long, winged petiole, 2–5 in. long (includes
petiole) × ½–2 in. wide. Flowers in composite heads about ¾ in. wide, ray flowers 7–15,
disk flowers 50–75, arranged in a somewhat
flat-topped inflorescence atop stem. This
species is primarily woodland-dwelling and
seems partial to alkaline bedrock, gravel, and
associated soil. IL, IN, MI, MO, OH. Most
occurrences in the Eastern Forests and Ozark
Highlands. *MAH*

Packera paupercula

ASTERACEAE

balsam ragwort, balsam groundsel

Moist prairies, meadows, fens, calcareous
shores, cedar swamps, alvars. Spring,
summer.

A perennial to 1½ ft. tall. Leaves in basal rosette,
to 3 in. long × ¾ in. wide, with toothed margins,
simple or rarely with few lobes, petiolate. Stem
leaves few, alternate, smaller above, sessile,
pinnately lobed, coarsely toothed. Flowers in
composite heads about ¾ in. wide; up to 13
ray flowers surround disk in few to numerous
stalked heads, forming nearly flat-topped array.
Fruit with white pappus. Several varieties exist.
Prairie ragwort (*P. plattensis*), more western, is
hairier with lobed basal leaves. IA, IL, IN, MI,
MN, MO, OH, WI. Primarily in the Northern
Lakes and the Tallgrass Prairie, scattered to
rare elsewhere. *SN*

Pyrrhopappus carolinianus
ASTERACEAE

false dandelion

Barrens, prairies, old fields, pastures.
Spring, summer, fall.

An erect annual, branched, mostly glabrous,
1–2 ft. tall. Leaves basal and alternate on stem,
sessile or short-petiolate, entire to pinnately
lobed, linear to lanceolate, 2–8 in. long and
up to 2 in. wide. Flowers in composite heads
up to 2 in. wide, ray flowers 50–100+, disk
flowers lacking, in a few solitary heads atop
stem branches. The flowering head looks like
common dandelion (*Taraxacum officinale*)
and even has the rounded "puffball" fruiting
head. Leaves, some of which are along stems,
are usually narrower in *Pyrrhopappus* than in
Taraxacum, which has all leaves basal. IA, IL,
IN, MO. **Mainly throughout MO, southern
IL, and southwestern IN.** *ST*

Ratibida columnifera
ASTERACEAE

Mexican hat, long-headed coneflower

Prairies, roadsides, railroads, disturbed
sites. Summer.

A perennial to 3 ft. tall. Leaves to 6 in. long ×
2 in. wide, alternate, sessile or short-petiolate,
very deeply pinnately lobed with up to 15
narrow, linear segments, each segment
sometimes deeply lobed, entire. Flowers in
composite heads; a brown, cylindrical, 1¾-in.-
long columnar disk is surrounded by 4–12
deflexed, shallowly notched, 1¼-in.-long ray
flowers that are yellow, maroon, or yellow with
maroon spot at base; solitary at ends of naked
branches in upper part of plant. IA, IL, IN,
MI, MN, MO, OH, WI. **Scattered throughout
but native and more frequent in western portion of the Midwest.** *SN*

Ratibida pinnata
ASTERACEAE

gray-headed coneflower,
yellow coneflower

Prairies, glades, barrens, open woodlands,
savannas. Spring, summer.

An erect, branching perennial, short-hairy,
2–5 ft. tall. Leaves alternate, lower ones
long-petiolate, the petioles becoming progressively shorter up the stem, deeply pinnately
divided, commonly toothed, 2–10 in. long ×
1–6 in. wide. Flowers in composite heads held
singly atop stem branches, ray flowers 5–10,
about 2 in. long and drooping, disk flowers
100+, greenish to dark purple-brown on a conical disk about ¾ in. wide. **IA, IL, IN, MI, MN,
MO, OH, WI. Mainly in the Tallgrass Prairie
and Ozark Highlands.** *MAH*

Rudbeckia fulgida
ASTERACEAE

orange coneflower

Barrens, dry woodlands, fens, marshes.
Summer.

An erect perennial, unbranched, glabrous to
softly hairy, 1–3 ft. tall, colonial. Basal and
lower leaves lanceolate to cordate and petiolate, upper leaves narrow and sessile, entire
to lobed and toothed, 2–10 in. long × 1–4
in. wide. Flowers in composite heads 2–3 in.
wide, ray flowers 8–15, disk flowers 100+, style
branches short and blunt-tipped. At least 7
varieties have been named. Variety *sullivantii*
'Goldsturm' is very popular in the nursery
trade. **IL, IN, MI, MO, OH, WI. Occurs
mainly in the Eastern Forests and Ozark
Highlands.** *MAH*

Rudbeckia hirta
ASTERACEAE

black-eyed Susan

Prairies, fields, glades, savannas, fens, woodland openings, roadsides, railroads. Spring, summer, fall.

A densely bristly-hairy gray-green annual or perennial to 2½ ft. tall. Leaves alternate, simple, petiolate at base, sessile to clasping above, entire or with a few teeth, to 7 in. long × 2 in. wide. Flowers in composite heads; a brown, flattened-conical, ¾-in.-wide disk is surrounded by up to 20 spreading, shallowly 3-toothed, yellow ray flowers (rarely with reddish bases) to 1¾ in. long in heads to 3 in. wide, style branches somewhat sharp-tipped. Heads solitary at top of plant or few on stalks from upper leaf axils. IA, IL, IN, MI, MN, MO, OH, WI. **Common and widespread.** *SN*

Rudbeckia laciniata
ASTERACEAE

cutleaf coneflower, wild goldenglow

Floodplain forests, stream terraces, forested wetlands. Summer, fall.

An erect perennial, unbranched, glabrous, with glaucous stems, 5–8 ft. tall, colonial. Leaves alternate, petiolate, broadly ovate, with 3–5 deeply cut lobes, these toothed, 6–18 in. long and wide. Flowers in composite heads 2–3 in. wide, ray flowers 6–12, drooping, disk flowers greenish-yellow, 100+, arranged on widely spreading branches atop plant. Of the several named varieties, only one cutleaf coneflower occurs here, var. *laciniata*. Its cultivar 'Hortensia', treasured in gardens since the early 1600s, is double-flowered, with only ray flowers. IA, IL, IN, MI, MN, MO, OH, WI. **Common.** *MAH*

Rudbeckia subtomentosa
ASTERACEAE

sweet coneflower

Moist prairies, savannas, woodlands, fens, meadows, thickets. Summer, fall.

An erect perennial, branched, hairy on often red-striped upper stem and branches, 3–5 ft. tall. Leaves alternate, petiolate, lanceolate to ovate, thick and firm, simple to deeply 3- to 5-lobed, short-hairy with toothed margins, gland-dotted, 2–10 in. long × 1–5 in. wide. Flowers in composite heads about 3 in. wide, ray flowers 10–20, disk flowers 200+, in a spreading array atop floral branches. The flower heads thought to emit a mild scent of anise, especially when crushed (hence the common name). IA, IL, IN, MO, WI. Mostly in the Tallgrass Prairie and Ozark Highlands. *MAH*

Rudbeckia triloba
ASTERACEAE

brown-eyed Susan

Low wet woods, swamp and pond borders, prairies, thickets. Summer, fall.

An annual, biennial, or short-lived perennial, erect, branched, moderately to coarsely hairy, 3–5 ft. tall. Leaves alternate, petiolate or sessile, elliptic-ovate, simple to deeply 2- or 3-lobed, thin, toothed, 4–10 in. long × 2–4 in. wide. Flowers in composite heads, 1–2 in. wide, ray flowers 8–15, disk flowers 100+, in a loosely spreading array of up to 30 heads, comparatively more than our other *Rudbeckia* species. Although the specific epithet is *triloba* ("3-lobed"), many (sometimes all) of its leaves may be un-lobed. IA, IL, IN, MI, MN, MO, OH, WI. Widespread and common throughout except in the far north. *MAH*

Senecio vulgaris
ASTERACEAE

common groundsel

Disturbed ground, urban areas, gardens, nursery plots, flower beds. Spring, summer, fall.

An erect annual, glabrous but with cobwebby hairs when young, 4–12 in. tall. Leaves alternate, shallowly pinnately lobed with blunt teeth, narrowly elliptic, sessile, somewhat fleshy, 1–4 in. long × ¼–1 in. wide. Flowers in composite heads about ¼ in. wide, ray flowers absent, disk flowers barely exceeding the involucre, 50+ in a somewhat flat-topped inflorescence terminating the stem. IA, IL, IN, MI, MN, MO, OH, WI. Introduced from Europe. Found mostly in the upper Midwest. *MJH*

Silphium asteriscus
ASTERACEAE

starry rosinweed, southern rosinweed

Glades, barrens, open blufftops, open woodlands. Summer.

An erect perennial, unbranched, short-hairy and rough, 3–5 ft. tall. Leaves alternate, whorled, or less commonly, opposite, sessile to petiolate, elliptic, ovate-lanceolate, entire or coarsely toothed, 5–8 in. long × ½–2 in. wide. Flowers in composite heads 2–3 in. wide, ray flowers 8–20, disk flowers 40+, few atop stem. Phyllaries broad, spreading. At least 5 varieties, including whorled rosinweed (var. *trifoliatum*; syn. *S. trifoliatum*), have been described. IL, IN, MO, OH. Occurs in a patchy distribution mostly in the Eastern Forests and Ozark Highlands. *DT*

Silphium integrifolium
ASTERACEAE

rosinweed

Prairies, prairie fens, glades, savannas, open woods. Summer, fall.

A stout perennial to 6 ft. tall. Leaves opposite (sometimes alternate above), to 5 in. long × 2½ in. wide, sessile to somewhat clasping, simple, ascending, entire or with few minute teeth, often rough-textured on top surface. Flowers in composite heads; up to 25 spreading ray flowers surround ¾-in.-wide disk in stalked heads to 3 in. wide in upper part of plant. Phyllaries large, broad, spreading, in several series. Varieties described based on phyllary and leaf underside pubescence. IA, IL, IN, MI, MO, WI. Widespread in the Ozark Highlands and southern portion of the Tallgrass Prairie, scattered in adjacent parts of the Northern Lakes and the Eastern Forests. *MAH*

Silphium laciniatum
ASTERACEAE

compass plant

Moist and dry-mesic prairies, glades, savannas. Summer.

An erect perennial, unbranched, short-hairy and rough, 6–9 ft. tall. Leaves basal and alternate on stem, ovate, deeply pinnately lobed, up to 2 ft. long × 1 ft. wide. Flowers in composite heads about 4 in. wide, ray flowers 20–30, disk flowers 100+, aligned along the stem in short-branched panicle-like arrays. Phyllaries broad, spreading. A classic prairie species, compass plant is so named because of the tendency of its basal leaves to align in a north-south position. IA, IL, IN, MI, MN, MO, OH, WI. Primarily a plant of the Tallgrass Prairie. *SN*

Silphium perfoliatum
ASTERACEAE

cup plant

Prairies, floodplain meadows, fens, thickets, riverbanks. Summer.

A stout, square-stemmed perennial to 10 ft. tall. Leaves opposite, simple, ascending, to 10 in. long × 6 in. wide, clasping and attached across stem, forming "cup" that collects precipitation, coarsely and sharply toothed, rough-textured, with strong white midvein. Flowers in composite heads; up to 40 spreading shallowly notched ray flowers surround ¾-in.-wide disk in up to 30 stalked heads to 4 in. wide at top of plant and from upper leaf axils. Phyllaries large, broad, with spreading tips, in several series. IA, IL, IN, MI, MN, MO, OH, WI. Widespread in the Ozark Highlands, the Eastern Forests, and the Tallgrass Prairie, scattered in southern portion of the Northern Lakes. *MAH*

Silphium terebinthinaceum
ASTERACEAE

prairie dock

Prairies, savannas, glades, fens, roadsides. Summer, fall.

A stout perennial to 10 ft. tall. Leaves mostly basal (few alternate, much reduced stem leaves), simple (deeply lobed in var. *pinnatifidum*), upright, to 18 in. long × 12 in. wide, with pointed tip and shallowly cordate base, long-petiolate, coarsely toothed, with strong white midvein. Flowers in composite heads; up to 30 spreading ray flowers surround 1-in.-wide disk in numerous stalked heads to 2¾ in. wide at top of naked stem. Phyllaries large, broad, with spreading tips, in several series. IA, IL, IN, MI, MO, OH, WI. Widespread in the Ozark Highlands, eastern portion of the Tallgrass Prairie, and southern portion of the Northern Lakes; scattered in the Eastern Forests. *SN*

Smallanthus uvedalia
ASTERACEAE

yellow-flowered leafcup, bear's-foot

Rich upland forests, rocky slopes, moist thickets. Summer.

An erect perennial, branched, glabrous to moderately hairy (some hairs may be glandular), 3–5 ft. tall. Leaves opposite, petiolate, broadly ovate, somewhat palmately lobed with scattered teeth, to 1 ft. long and wide. Flowers in composite heads about 2 in. wide, ray flowers 8–12, disk flowers 50+, arranged in clusters at branch tips. The plump fruits are in various shades of purple. The large, irregularly shaped leaves gave rise to the common name bear's-foot. IL, IN, MI, MO, OH. Occurs mostly in the lower Midwest. *CR*

Solidago altissima
ASTERACEAE

tall goldenrod

Fields, prairies, disturbed sites. Summer, fall.

An erect perennial, unbranched, hairy, rhizomatous, 2–6 ft. tall. Leaves alternate, elliptic to lanceolate with scattered small teeth on lower leaves, upper ones nearly entire, 3-nerved, the upper surface commonly rough, 2–6 in. long × ¼–1 in. wide. Flowers in composite heads ¼–⅓ in. wide, with 10–15 ray flowers and 3–8 disk flowers, phyllaries appressed, involucre ⅛–⅙ in. long, in a spreading pyramidal array atop stem. Smooth goldenrod (*S. gigantea*) and common goldenrod (*S. canadensis*) differ in having stems glabrous below the inflorescence, and obviously toothed leaves (plus an involucre often less than ⅛ in. long), respectively. IA, IL, IN, MI, MN, MO, OH, WI. **Throughout, probably in every county.** *SN*

Solidago caesia
ASTERACEAE

bluestem goldenrod, wreath goldenrod

Forests, woodland edges. Summer, fall.

An arching, glabrous, glaucous-stemmed perennial to 3 ft. tall. Leaves alternate, simple, sharply toothed, to 5 in. long × 1 in. wide, smaller above, with distinct midvein, sessile. Flowers in composite heads; up to 5 ray flowers surround several disk flowers in heads to ⅙ in. wide, each head subtended by appressed phyllaries, in clusters of up to 12, short-stalked to sessile, in mid to upper leaf axils and at top of plant. IA, IL, IN, MI, MO, OH, WI. **Widespread in the Eastern Forests, southern portion of the Ozark Highlands, southeastern portion of the Northern Lakes, and eastern portion of the Tallgrass Prairie, rare elsewhere.** *SN*

Solidago flexicaulis
ASTERACEAE

zigzag goldenrod

Moist forested ravines, slopes, mesic floodplain forests. Summer, fall.

An erect, glabrous to short-hairy perennial with unbranched zigzag stems, rhizomatous, 1½–2 ft. tall. Leaves alternate, oval with coarse teeth, 4–6 in. long × 2–3 in. wide, reducing in size from lower stem to upper. Flowers in composite heads about ¼ in. wide with 4–8 disk flowers and 3–5 ray flowers, phyllaries appressed, clustered in leaf axils and atop stem. This goldenrod is an easier one to identify, with its broad, sharply toothed leaves, zigzag stem, colonial habit, and woodland habitat. IA, IL, IN, MI, MN, MO, OH, WI. **Mostly throughout the Midwest.** *KB*

Solidago gattingeri
ASTERACEAE

Gattinger's goldenrod

Rocky calcareous glades and outcrops.
Summer, fall.

An erect perennial, non-rhizomatous, 1–2½ ft.
tall, glabrous to short-hairy. Leaves alternate,
with mostly entire margins, middle leaves
linear to narrowly oblanceolate, 1–2½ in. long
× ½–1 in. wide. Flowers in composite heads
up to ¼ in. wide, with 3–8 ray flowers and 3–9
disk flowers, phyllaries appressed, arranged in
a broader-than-tall, pyramidal inflorescence
atop stem. The somewhat similar Missouri
goldenrod (*S. missouriensis*) usually has more
ray flowers per head and possesses creeping
rhizomes. **MO. Confined to the Ozark Highlands.** *DH*

Solidago hispida
ASTERACEAE

hairy goldenrod

Open dry woods, acidic rock outcrops,
dunes. Summer, fall.

An erect perennial, unbranched, 1–3 ft. tall,
hairy. Leaves basal and alternate, petiolate,
elliptic to obovate, toothed, 2–6 in. long ×
1½–2½ in. wide. Flowers in composite heads
up to ¼ in. wide, 5–14 yellow ray flowers and
6–12 disk flowers, phyllaries appressed, clustered in leaf axils on a usually unbranched
wand. Compared to hairy goldenrod, silverrod
(*S. bicolor*) has white ray flowers; slender goldenrod (*S. erecta*) is glabrous below the inflorescence; and bog goldenrod (*S. uliginosa*) grows
in wet habitats. **IA, IL, IN, MI, MN, MO, OH,
WI. Most occurrences in the Ozark Highlands and Northern Lakes.** *PD*

Solidago juncea
ASTERACEAE

early goldenrod

Prairies, fields, savannas, open forests, roadsides. Summer.

A mostly glabrous, colony-forming perennial to 4 ft. tall. Leaves in basal rosettes and alternate along stem, entire to shallowly toothed, with midvein but lacking parallel veins, to 8 in. long × 1½ in. wide, conspicuously smaller above and in axillary fascicles, sessile. Flowers in composite heads; up to 12 ray flowers surround small disk in heads to ¼ in. wide, with appressed phyllaries, stalked in arching, branching, broad-based terminal array. Missouri goldenrod (*S. missouriensis*) has narrower basal and lower leaves that have 3 parallel veins. IA, IL, IN, MI, MN, MO, OH, WI. Widespread through much of the Midwest, scattered to absent in western portion of the Tallgrass Prairie. *SN*

Solidago nemoralis
ASTERACEAE

gray goldenrod, old field goldenrod

Dry poor soil, prairies, old fields, open woodlands, roadsides. Summer, fall.

An erect or commonly arching perennial, unbranched, densely covered with minute hairs, colonial, ½–2½ ft. tall. Leaves basal and alternate, basal and lower leaves oblanceolate with few teeth, 1–4 in. long × ½–¾ in. wide. Flowers in composite heads ⅛–¼ in. wide, with 5–9 ray flowers and 3–8 disk flowers, phyllaries appressed, arranged in an arching panicle. Subspecies *decemflora* is mostly in the western Tallgrass Prairie; it has larger flower heads and narrower basal leaves than the widespread subsp. *nemoralis*. IA, IL, IN, MI, MN, MO, OH, WI. Common throughout. *EH*

Solidago ohioensis
ASTERACEAE

Ohio goldenrod

Wet prairies, sedge meadows, fens, shores, interdunal swales. Summer.

A glabrous perennial to 3½ ft. tall. Leaves basal (long-petiolate) and alternate (sessile), minutely bluntly toothed, flat, with strong midvein, blunt at tip, to 6 in. long × 1¾ in. wide, conspicuously smaller above. Flowers in composite heads; up to 8 ray flowers surround disk in heads to ⅓ in. wide, each head subtended by appressed phyllaries, stalked in branched, flattish-topped cluster at top of stem. Syn. *Oligoneuron ohioense*. IL, IN, MI, OH, WI. **Mostly in proximity to the Great Lakes in the Northern Lakes, rare in the Eastern Forests and the Tallgrass Prairie. A Midwest near-endemic.** *MJH*

Solidago patula
ASTERACEAE

swamp goldenrod

Sedge meadows, fens, seeps, interdunal wetlands, swamps. Summer, fall.

A perennial to 6 ft. tall. Leaves basal (petiolate) and alternate (becoming sessile above), shallowly toothed, with midvein but lacking parallel veins, sandpapery rough-textured, to 12 in. long × 4 in. wide, smaller above. Flowers in composite heads; up to 12 ray flowers surround small disk in heads to ⅛ in. wide, subtended by appressed phyllaries, stalked in branching, spreading, narrow, leafy arrays at top of stem. Sharp-leaved goldenrod (*S. arguta*) has leaves more sharply toothed and often smooth or nearly so. IA, IL, IN, MI, MO, OH, WI. **Scattered in eastern half of the Midwest, more frequent in southern Great Lakes.** *SN*

Solidago petiolaris
ASTERACEAE

downy goldenrod, downy ragged goldenrod

Open woodlands, glades, and rock outcrops, typically with acidic soils. Summer, fall.

An erect perennial, unbranched, short-hairy, 1½–3 ft. tall. Leaves alternate, mostly lanceolate to elliptic, thick-textured, upper surface glabrous or somewhat roughened, entire or with few scattered teeth, 1–4 in. long × ½–1½ in. wide. Flowers in composite heads approximately ¼ in. wide, 6–10 ray flowers and 10–15 disk flowers, the phyllaries glandular and commonly with somewhat reflexed tips, clustered in upper leaf axils and terminal spreading branches. A close relative with similar range is Buckley's goldenrod (*S. buckleyi*); it has broader, thinner leaves with sharply toothed margins. IL, MO. Principally in the Ozark Highlands. *ST*

Solidago radula
ASTERACEAE

western rough goldenrod

Glades, rocky calcareous bluffs and dry upland forests. Spring, summer, fall.

An erect perennial, unbranched, 1–3 ft. tall, rough with dense short hairs. Leaves alternate, numerous, thickened, stiff, petiolate (upper ones sessile), entire or with fine teeth on margins, lanceolate to elliptic, rough to the touch, commonly 3-nerved, 1–3 in. long × ⅓–1½ in. wide. Flowers in composite heads ⅛–¼ in. wide, ray and disk flowers 4–7 per head, phyllaries appressed, in a mostly dense pyramidal array. Leaves of the similar Drummond's goldenrod (*S. drummondii*) are larger, ovate, and smooth, with prominent sharp teeth; its range is similar to *S. radula*. IL, MO. Principally in the Ozark Highlands and western IL counties. *EH*

Solidago riddellii
ASTERACEAE

Riddell's goldenrod

Fens, seeps, wet meadows, moist prairies. Summer, fall.

An erect perennial, mostly glabrous, unbranched, 1½–3 ft. tall. Leaves alternate, entire, linear and somewhat folded longitudinally and curved, clasping at base, 3-nerved, 2–7 in. long × ¼–½ in. wide. Flowers in composite heads up to ¼ in. wide, 5–10 ray flowers and disk flowers, phyllaries appressed, on somewhat hairy branches in a crowded flat-topped array. Syn. *Oligoneuron riddellii*. In similar habitats, Ohio goldenrod (*S. ohioensis*) has unfolded, 1-nerved leaves and glabrous floral branches; it occurs mostly in the eastern Midwest. IA, IL, IN, MI, MN, MO, OH, WI. Principally in the Northern Lakes and Tallgrass Prairie. *SN*

Solidago rigida
ASTERACEAE

stiff goldenrod

Prairies, glades, savannas, fields, roadsides. Summer, fall.

A gray-hairy, colony-forming perennial to 5 ft. tall. Leaves basal (long-petiolate) and alternate (clasping), bluntly toothed to entire and sometimes wavy, stiff and thick-textured with strong midvein, blunt at tip, to 10 in. long × 4 in. wide, conspicuously smaller above. Flowers in composite heads; up to 13 ray flowers surround disk in heads to ½ in. wide, each head subtended by appressed phyllaries, stalked in branched, flattish-topped array to 5 in. wide at top of stem. Syn. *Oligoneuron rigidum*. IA, IL, IN, MI, MN, MO, OH, WI. Widespread in the Ozark Highlands, the Tallgrass Prairie, and adjacent portions of the Northern Lakes, scattered elsewhere. *MJH*

Solidago rigidiuscula
ASTERACEAE

stiff-leaved showy goldenrod

Prairies, open woodlands, savannas.
Summer, fall.

An erect perennial, unbranched, glabrous or
minutely hairy, 1½–3 ft. tall. Leaves alternate,
entire, middle stem leaves firm and crowded,
lanceolate to elliptic-ovate, to 1¼ in. wide,
lower leaves commonly absent during flowering. Flowers in composite heads up to ¼
in. wide, ray flowers 3–9, disk flowers 5–15,
phyllaries appressed, in pyramidal branches
bearing up to 250 crowded heads. The similar
showy goldenrod (*S. speciosa*) is a slightly taller
plant with middle stem leaves to 2 in. wide;
its lower leaves are typically present during
flowering. IA, IL, IN, MI, MN, MO, OH, WI.
Mostly throughout except for the far northern and southeastern counties. *KB*

Solidago rugosa
ASTERACEAE

wrinkleleaf goldenrod

Wet prairies, fields, dunes, thickets, forests,
swamps. Summer, fall.

A usually hairy, colony-forming perennial
to 5 ft. tall. Leaves alternate, toothed, with
midvein but lacking parallel veins, also with
deeply impressed reticulate veins giving wrinkled appearance, to 4 in. long × 1½ in. wide,
smaller above, sessile or nearly so. Flowers in
composite heads; up to 8 ray flowers surround
small disk in heads to ⅛ in. wide, subtended
by appressed phyllaries, stalked in branching,
leafy, variably shaped array (sometimes widely
spreading, sometimes more erect) at top of
stem. Compare with elm-leaved goldenrod
(*S. ulmifolia*). IL, IN, MI, MO, OH, WI. Widespread in scattered pockets throughout the
Midwest, less frequent northwestward. *MJH*

Solidago simplex
ASTERACEAE

dune goldenrod

Savannas, dunes, beaches, crevices in rocky shores. Summer, fall.

A mostly glabrous, ascending perennial to 2½ ft. tall. Leaves basal and alternate, sharply toothed, with midvein but lacking parallel veins, to 6 in. long × 1 in. wide and petiolate below, becoming smaller and sessile above. Flowers in composite heads; up to 10 ray flowers surround disk in heads to ⅙ in. wide, each head subtended by appressed, sticky phyllaries, stalked in narrow, branched, terminal, leafy array to 10 in. long × 2½ in. wide. Several varieties documented in our region, varying mostly in plant height, leaf dimensions, and involucre length. IN, MI, WI. Restricted to the Northern Lakes, particularly in proximity to the Great Lakes. *SN*

Solidago uliginosa
ASTERACEAE

bog goldenrod

Sedge meadows, marshes, pannes, bogs, fens, conifer swamps, rocky shores. Summer, fall.

A glabrous (except in inflorescence) perennial to 5 ft. tall. Leaves basal and alternate, entire or with scattered teeth, with midvein but lacking parallel veins, to 9 in. long × 1½ in. wide and petiolate below, becoming smaller and sessile above, somewhat sheathing. Flowers in composite heads; up to 8 ray flowers surround small disk in heads to ¼ in. wide, each head subtended by appressed phyllaries, stalked in narrow, upright-branched terminal array. IA, IL, IN, MI, MN, OH, WI. Widespread through much of the Northern Lakes and adjacent portions of the Tallgrass Prairie, scattered in the Eastern Forests. *SN*

Solidago ulmifolia
ASTERACEAE

elm-leaved goldenrod

Dry-mesic forests, sand savannas, bluffs. Summer, fall.

An erect to arching perennial, unbranched, glabrous to hairy, 1½–3 ft. tall. Leaves alternate, lanceolate to elliptic, toothed, 3–5 in. long × 1–2 in. wide, lower stem leaves often absent at flowering. Flowers in composite heads up to ¼ in. wide containing 3–6 ray flowers and 4–6 disk flowers, phyllaries appressed, in a widely spreading and nodding pyramidal array. Rough-leaved goldenrod (*S. rugosa*) usually grows by long, creeping rhizomes in moist soils. Autumn goldenrod (*S. sphacelata*) has wide, ovate-cordate basal and lower leaves. IA, IL, IN, MI, MN, MO, OH, WI. Common throughout except for the far northern counties. *MJH*

Sonchus arvensis
ASTERACEAE

perennial sowthistle, field sowthistle

Fields, gardens, roadsides, railroads, disturbed sites. Summer, fall.

A glabrous perennial to 4 ft. tall. Leaves alternate, mostly in lower half of plant, shallowly lobed to unlobed, prickly-toothed, clasping with ear-like bases, with white midvein, to 12 in. long × 3½ in. wide, smaller above, containing milky sap. Flowers in composite heads; numerous ray flowers, ¾-in.-long involucre with appressed phyllaries in several unequal series; heads to 1½ in. wide, solitary at ends of several long, naked stalks. Fruit with white pappus; achenes beakless. Stalked glands in inflorescence in subsp. *arvensis*, lacking in subsp. *uliginosus*. IA, IL, IN, MI, MN, MO, OH, WI. Introduced from Europe. Widespread northward, scattered to absent southward. *SN*

Sonchus asper
ASTERACEAE

spiny-leaved sow thistle

Pastures, agricultural fields, disturbed sites. Summer, fall.

An erect annual, stems hollow and rather weak, mostly glabrous, with milky sap, 1–3 ft. tall. Leaves alternate, sessile, oblong to lanceolate, shiny above, wavy-margined with prickles and rounded basal lobes (auricles) that partially clasp stem, 3–12 in. long × 1–5 in. wide. Flowers in composite heads about 1 in. wide, ray flowers 80+, disk flowers lacking, in a somewhat flat-topped inflorescence at stem tips. Its rounded, curling leaf bases resemble the shape of a nautilus, whereas common sow thistle (*S. oleraceus*) has pointed leaf bases. IA, IL, IN, MI, MN, MO, OH, WI. Introduced from Europe. Widespread and common. *MAH*

Tanacetum vulgare
ASTERACEAE

common tansy

Fields, shores, ditches, homesites, vacant lots, roadsides, waste areas. Summer, fall.

A colony-forming glabrous (or nearly so) perennial to 4 ft. tall. Leaves alternate, to 12 in. long × 4 in. wide, smaller up stem, 2 or 3 times deeply pinnately lobed and also toothed, appearing fern-like, short-petiolate. Flowers in composite heads; numerous tiny, yellow disk flowers in ⅓-in.-wide button-shaped heads, ray flowers absent; up to 200 heads in a flattish array at top of plant. Lake Huron tansy (*T. bipinnatum*), native and restricted to northern MI and WI, has hairy leaves and flower heads over ½ in. wide. IA, IL, IN, MI, MN, MO, OH, WI. Introduced from Eurasia. Scattered, most frequent northward. *SN*

Taraxacum officinale
ASTERACEAE

common dandelion

Lawns, roadsides, pastures, waste areas.
Year-round, mostly spring.

An erect perennial, unbranched, glabrous
to sparsely hairy, 2–15 in. tall. Leaves basal,
somewhat pinnately lobed, oblanceolate to
elliptic with deep triangular lobes, toothed,
4–10 in. long × 1½–3 in. wide. Flowers in composite heads 1–2 in. wide, ray flowers 50–100+,
disk flowers absent, displayed singly atop
stem. Fruit a light olive-brown to tan achene
in globose heads. Red-seeded dandelion (*T.
erythrospermum*) is quite similar but has reddish achenes. IA, IL, IN, MI, MN, MO, OH,
WI. Introduced from Europe. Widespread
and abundant. *MAH*

Tragopogon dubius
ASTERACEAE

western goat's-beard, yellow salsify

Fields, roadsides, railroads, waste areas.
Spring, summer.

A gray-green biennial to 3 ft. tall. Leaves alternate, entire, sheathing, to 12 in. long × ¾ in.
wide, smaller above, containing milky sap.
Flowers in composite heads, closing daily;
numerous spreading ray flowers in heads to 2
in. wide, disk flowers lacking, involucre conspicuously longer than ray flowers and made up
of about 13 spreading phyllaries; heads solitary
at top of stems, stem expanded and hollow just
below head. Fruit with pale brownish pappus
on long-beaked achenes, forming spherical
cluster to 3 in. wide. Salsify (*T. porrifolius*) has
lavender ray flowers. IA, IL, IN, MI, MN, MO,
OH, WI. Introduced from Eurasia. Widespread and common, especially westward. *SN*

Tragopogon pratensis
ASTERACEAE

meadow goat's-beard, meadow salsify

Prairies, fields, roadsides, railroads, waste areas. Spring, summer.

A gray-green biennial to 3 ft. tall. Leaves alternate, entire, sheathing, to 12 in. long × 1 in. wide, smaller above, containing milky sap. Flowers in composite heads, closing daily; numerous spreading ray flowers in heads to 2 in. wide, disk flowers lacking, involucre as long as ray flowers or shorter and made up of about 8 spreading phyllaries; heads solitary at top of stems. Fruit with pale brownish pappus on long-beaked achenes, forming spherical cluster to 3 in. wide. IA, IL, IN, MI, MN, MO, OH, WI. Introduced from Eurasia. Rare in the Ozark Highlands, scattered elsewhere, more common northward. *PD*

Tussilago farfara
ASTERACEAE

coltsfoot

Moist forests, seeps, rocky slopes, roadsides, streambanks, disturbed sites. Spring.

An erect perennial, unbranched, stems reddish with clasping scaly bracts, colonial, 2–12 in. tall. Leaves basal, petiolate, round or kidney-shaped and cordate, shallowly lobed with scalloped edges, mostly glabrous above, hairy below with webbing of felt-like white hairs, 2–8 in. long and wide. Flowers in composite heads about 1 in. across, ray flowers 100+, disk flowers 20–40, head solitary atop stalk. It often starts blooming before its leaves emerge. IL, IN, MI, MN, OH, WI. Introduced from Eurasia. Occurs mostly in eastern OH but spreading elsewhere. *SN*

Verbesina alternifolia
ASTERACEAE

wingstem

Floodplain forests, stream terraces, moist thickets. Summer, fall.

An erect perennial, upwardly branching, hairy, with winged ridges, 2–8+ ft. tall. Leaves alternate (a few may be opposite), with short and decurrent leaf bases, lanceolate to elliptic, toothed or entire, rough to the touch, 4–10 in. long × 1–4 in. wide. Flowers in composite heads 1–3 in. wide, ray flowers 4–8, drooping, disk flowers greenish-yellow, 40–60, heads 10–50+ arranged in panicles atop stems. The common name refers to the thin green tissue that runs from leaf petioles down the stem. IA, IL, IN, MI, MO, OH. Occurs mostly in the lower Midwest, common. *MAH*

Verbesina helianthoides
ASTERACEAE

yellow crownbeard

Dry woodlands, barrens, savannas, glades. Spring, summer, fall.

An erect perennial, stems unbranched, hairy, narrowly winged, 2–4 ft. tall, colonial by rhizomes. Leaves alternate, lanceolate to elliptic, finely toothed, 3–5 in. long × 1–2 in. wide. Flowers in composite heads about 2 in. wide, ray flowers 8–15, disk flowers 40+, 2–8 heads in a rather tight, somewhat flat-topped inflorescence atop the main stem. Compare with wingstem (*V. alternifolia*). IL, IN, MO, OH. **Confined to the lower Midwest.** *EH*

Impatiens pallida
BALSAMINACEAE

pale touch-me-not, yellow touch-me-not, pale jewelweed, yellow jewelweed

Moist to damp forests, swamps, streambanks, ditches. Spring, summer, fall.

A glabrous, glaucous-stemmed, branched annual to 6 ft. tall. Stems translucent. Leaves mostly alternate, ovate, to 5 in. long × 3 in. wide, petiolate, with numerous blunt teeth on margins (more than jewelweed, *I. capensis*). Flowers irregular, to 1¼ in. long, trumpet-shaped with 5 flaring corolla lobes (2 lateral very small), often red-spotted in throat, 1 of the sepals inflated and narrowed to hooked, nectar-bearing spur at base, numerous, stalked, dangling, in clusters of 1–3 from upper leaf axils. Fruit an inflated capsule that explodes along sutures when touched. IA, IL, IN, MI, MN, MO, OH, WI. Widespread, less common northward. *MAH*

Caulophyllum thalictroides
BERBERIDACEAE

blue cohosh

Rich mesic forests, ravines, stream terraces. Spring.

An erect perennial, glabrous, 1–2 ft. tall. Leaves 2 per plant when flowering, upper smaller than lower, the latter 3 or 4 times ternately compound, greenish-yellow, glaucous, usually fully expanded when in bloom, obovate in outline as are leaflets, the latter commonly divided with rounded teeth, 1–3 in. long × ¾–4 in. wide. Flowers regular, ½ in. wide, petal-like sepals 6, greenish-yellow or yellow and purple-tinged, petals obscure, the style less than 1/16 in. long. Inflorescence of 5–15+ flowers in open clusters. The berry-like seeds (see inset) are fleshy and poisonous. IA, IL, IN, MI, MN, MO, OH, WI. Common throughout. *SN*

Lithospermum canescens
BORAGINACEAE

hoary puccoon

Prairies, savannas, open rocky woods, glades. Spring, summer.

A densely soft-hairy perennial to 1½ ft. tall. Leaves alternate, entire, sessile, to 2 in. long × ½ in. wide, rounded at tip, with strong midvein. Flowers regular, with 5 spreading lobes, fused at base, to ½ in. wide, with stamens and pistil included in floral tube, short-stalked in flattish cluster at top of plant. Fruit a hard, white, stone-like nutlet to ⅛ in. long. IA, IL, IN, MI, MN, MO, OH, WI. Widespread in the Ozark Highlands and the Tallgrass Prairie, less frequent in the Northern Lakes and the Eastern Forests, primarily in western portion of Midwest. *MAH*

Lithospermum caroliniense
BORAGINACEAE

hairy puccoon

Prairies, savannas, dunes, open rocky woods, glades. Spring, summer.

A rough-hairy perennial to 2 ft. tall. Leaves alternate, entire, sessile, to 2 in. long × ½ in. wide, pointed (but blunt) at tip, with strong midvein. Flowers regular, with 5 spreading yellow to orange-yellow lobes, fused at base, to ¾ in. wide, with stamens and pistil included in floral tube, short-stalked in flattish cluster at top of plant. Fruit a hard, white, stone-like nutlet to ⅛ in. long. Midwestern plants are var. *croceum*. IA, IL, IN, MI, MN, MO, OH, WI. Primarily in the Northern Lakes and northern portion of the Tallgrass Prairie, also in western portion of the Eastern Forests, absent from the Ozark Highlands. *SN*

Lithospermum incisum
BORAGINACEAE

fringed puccoon

Prairies, savannas, dunes, glades.
Spring, summer.

A densely appressed-hairy perennial to 16 in.
tall. Leaves alternate, entire, sessile, to 2½ in.
long × ¼ in. wide (largest above), pointed (but
blunt) at tip, with strong midvein. Flowers regular, to ¾ in. wide, with 5 spreading to ascending pale to lemon yellow lobes fringed and wavy
along margins, fused at base, with stamens and
pistil included in floral tube, short-stalked in
flattish terminal cluster, mostly sterile; fertile
self-pollinating flowers later produced in leaf
axils. Fruit a hard, white, pitted, stone-like nutlet about ⅛ in. long. IA, IL, IN, MI, MN, MO,
WI. Primarily in the Tallgrass Prairie, also
frequent in portions of the Ozark Highlands,
scattered elsewhere. *SN*

Lithospermum latifolium
BORAGINACEAE

American gromwell

Rocky (especially alkaline) slopes,
woodlands, mesic upland forests. Spring.

An erect perennial, branched above, hairy,
1–2½ ft. tall. Leaves alternate, lanceolate
to narrowly ovate, sessile, sparsely hairy to
densely so underneath, prominently veined,
1–6 in. long × ½–2½ in. wide. Flowers regular,
about ¼ in. wide, with 5 rounded lobes and
tubular base, occurring singularly in upper
leaf axils. Fruit a shiny white oval nutlet.
Lithospermum literally means "stone seed," an
apt name for plants of this genus given the
hardness of the nutlets they produce. IA, IL,
IN, MI, MN, MO, OH, WI. Widespread, in all
states but local. *MAH*

Barbarea vulgaris
BRASSICACEAE

yellow rocket, winter cress

Roadsides, meadows, ditch banks, disturbed sites. Spring.

An erect biennial, branched above, glabrous, 1–2½ ft. tall. Leaves basal and alternate on stems, lower ones petiolate, upper ones sessile, ovate to oblanceolate, pinnately lobed and toothed with the terminal lobe largest, 1–8 in. long × ½–3 in. wide. Flowers regular, about ¼–⅓ in. wide, 4-petaled, in compact clusters terminating the branches. Fruit a silique, erect or slightly spreading. American yellow rocket (*B. orthoceras*) occurs in our far northern counties; its petals are smaller than those of yellow rocket. IA, IL, IN, MI, MN, MO, OH, WI. Introduced from Eurasia. Common and widespread. *SN*

Brassica nigra
BRASSICACEAE

black mustard

Fields, streambanks, roadsides, waste areas. Spring, summer, fall.

An annual to 8 ft. tall. Stems glaucous, glabrous above, scattered stiff hairs below. Leaves alternate, to 10 in. long × 4 in. wide, petiolate, coarsely toothed, pinnately lobed below, becoming smaller, unlobed, entire, and sessile above. Flowers regular, to ⅓ in. wide, with 4 petals (see inset), in many-flowered racemes. Fruit a knobby ⅔-in.-long silique. Brown mustard (*B. juncea*) is glabrous with fruit over 1 in. long. Turnip (*B. rapa*) has clasping leaves. Charlock (*Sinapis arvensis*) has upper leaves toothed to lobed. IA, IL, IN, MI, MN, MO, OH, WI. Introduced from Eurasia. Widespread in central portion of Midwest, scattered to rare north and south. *PR*

Camelina microcarpa
BRASSICACEAE

small-seeded false flax, littlepod false flax

Fields, pastures, roadsides, railroads, disturbed sites. Spring, summer.

An annual to 2 ft. tall. Stems rough-hairy below, glabrous in inflorescence. Leaves basal and alternate, rough-hairy, entire to minutely toothed, to 2 in. long × ¾ in. wide, smaller above, short-petiolate at base, sessile and clasping with pointed basal lobes above, often held erect. Flowers regular, to ⅙ in. wide, with 4 petals, on spreading-ascending stalks in racemes that elongate with age at top of plant and ends of branches. Fruit a roundish (often broader above) silicle to ⅙ in., tapering at base and with persistent beak. IA, IL, IN, MI, MN, MO, OH, WI. Introduced from Eurasia. Scattered throughout. *ST*

Descurainia pinnata
BRASSICACEAE

western tansymustard

Prairies, glades, cliffs, shores, roadsides, railroads. Spring, summer.

A glandular annual to 2 ft. tall. Leaves basal and alternate, 2 or 3 times deeply pinnately lobed, to 8 in. long × 4 in. wide, smaller and shorter-petiolate up stem. Flowers regular, to ⅛ in. wide, with 4 spreading, stalked petals, in many-flowered racemes in upper part of plant. Fruit a silique to ½ in. long on spreading-ascending stalk. The introduced flixweed (*D. sophia*) has longer fruit and lacks glandular hairs on stems (though stems do possess stellate hairs). IA, IL, IN, MI, MN, MO, OH, WI. Most abundant in the Tallgrass Prairie and Ozark Highlands, scattered elsewhere. *ST*

Erysimum cheiranthoides
BRASSICACEAE

wormseed mustard, wormseed wallflower

Fields, roadsides, railroads, waste areas. Spring, summer.

An annual to 3 ft. tall. Leaves alternate, simple, to 3 in. long × ½ in. wide, tapering to tips and bases, sparsely toothed, sessile or short-petiolate. Flowers regular, with 4 spreading petals, to ¼ in. wide, in many-flowered racemes in upper part of plant. Fruit an upright silique to 1½ in. long on spreading stalk. One of very few yellow-flowered mustards in the Midwest that both produces siliques and has unlobed leaves. IA, IL, IN, MI, MN, MO, OH, WI. Introduced, likely from Eurasia. Most abundant in the Northern Lakes and northern portion of the Tallgrass Prairie, scattered elsewhere. *KC*

Rorippa palustris
BRASSICACEAE

marsh yellow cress, bog yellow cress

Marshes, wet meadows, mudflats, ditches, cropland. Spring, summer, fall.

An annual to 3 ft. tall. Leaves alternate, to 7 in. long × 2 in. wide, petiolate below, becoming sessile or nearly so and smaller above, variable, deeply pinnately lobed in lower part of leaf but unlobed in upper part, with blunt to sharp teeth on margins. Flowers regular, to ⅛ in. wide, with 4 petals, in many-flowered racemes in upper part of plant, often clustered at tip with fruit developed below. Fruit a knobby silique to ¼ in. long on spreading stalk. Plants with hairy stems and leaf undersides are var. *hispida*. IA, IL, IN, MI, MN, MO, OH, WI. Widespread. *PR*

Rorippa sessiliflora
BRASSICACEAE

sessile-flowered yellow cress

Marshes, floodplains, sloughs, pond margins, cropland, ditches. Spring, summer, fall.

An annual to 3 ft. tall. Leaves alternate, to 5 in. long × 2 in. wide, petiolate below, becoming sessile or nearly so and smaller above, variable, deeply pinnately lobed in lower part of leaf but unlobed in upper part, with blunt teeth on margins. Flowers regular, to ⅛ in. wide, with 4 petals, sessile or nearly so in many-flowered spikes/racemes in upper part of plant, often clustered at tip with fruit developed below. Fruit a silique to ⅓ in. long, sessile or nearly so. IA, IL, IN, MN, MO, OH, WI. Widespread in southern portion of the Tallgrass Prairie, scattered to rare elsewhere. *DT*

Rorippa sinuata
BRASSICACEAE

spreading yellow cress

Open floodplains, moist fields, shores, disturbed sites. Spring, summer.

A creeping, branched perennial to 16 in. tall (often shorter), rooting at nodes. Leaves alternate, to 3 in. long × 1 in. wide, smaller up stem, short-petiolate, deeply pinnately lobed with up to 15 lobes, lobes sometimes with blunt teeth. Flowers regular, to ¼ in. wide, with 4 petals, in many-flowered racemes in upper part of plant, often clustered at tip with fruit developed below. Fruit a curved silique to ½ in. long on spreading-ascending stalk. IA, IL, MN, MO, WI. Mostly along major rivers in western portion of the Midwest. *PD*

Rorippa sylvestris
BRASSICACEAE

creeping yellow cress

Wet meadows, swamps, shores, cropland, ditches, disturbed sites. Spring, summer.

A creeping, colony-forming perennial with upright stems to 2 ft. tall. Basal leaves in rosette, petiolate, to 8 in. long × 2 in. wide, deeply pinnately lobed with up to 13 lobes that are shallowly lobed or bluntly toothed; stem leaves alternate, similar to basal but smaller up stem, short-petiolate. Flowers regular, to ⅓ in. wide, with 4 petals, in many-flowered racemes in upper part of plant, often clustered at tip with fruit developed below. Fruit a straight to slightly curved silique to ½ in. long on spreading-ascending stalk. **IA, IL, IN, MI, MN, MO, OH, WI. Introduced from Eurasia. Scattered.** *ST*

Sisymbrium altissimum
BRASSICACEAE

tall tumblemustard

Fields, pastures, shores, roadsides, railroads, waste areas. Spring, summer.

A glaucous annual to 3½ ft. tall with sparse spreading white hairs. Leaves basal and alternate, petiolate, to 12 in. long × 3 in. wide, deeply pinnately lobed (appearing compound), lobes lobed to toothed, becoming short-petiolate and smaller with thread-like pinnate lobes above. Flowers regular with 4 spreading pale yellow to whitish petals surrounding 6 stamens and single pistil, to ⅓ in. wide, in many-flowered racemes at ends of branches in upper part of plant. Fruit an ascending silique to 4 in. long on ascending ⅓-in. stalk. **IA, IL, IN, MI, MN, MO, OH, WI. Introduced from Eurasia. Scattered to widespread, more frequent north.** *KC*

Sisymbrium officinale
BRASSICACEAE

hedge mustard

Disturbed sites, fallow fields, barnyards, agricultural areas. Spring, fall.

An erect annual, unbranched to branched, stems densely to sparsely hairy, 1–3 ft. tall. Leaves alternate, lower leaves petiolate, lanceolate and pinnately lobed, the lobes entire or toothed; the upper leaves may be smaller and sessile, hairy to glabrous, to 8 in. long × 2 in. wide. Flowers regular, ⅛–¼ in. wide, petals and sepals 4, on erect racemes with narrow fruits tightly appressed to the stem. The latter feature helps to distinguish it from the similar tall hedge mustard (*S. loeselii*), which has fruits widely spreading. IA, IL, IN, MI, MN, MO, OH, WI. Introduced from Eurasia. Widespread throughout. *ST*

Opuntia cespitosa
CACTACEAE

eastern prickly pear

Barrens on sand or rock, dunes, dry cliffs, glades. Spring, summer.

A prostrate and mat-forming cactus, fleshy, mostly glabrous with unlobed rounded segments (pads) dotted with clusters of finely barbed hairs, sometimes with 1 or 2 long spines per cluster, pads 2–5 in. long and about as wide. Flowers regular, about 3 in. wide with numerous tepals, reddish-orange within, 1 to several attached to narrow margin of pad. Mostly western in our region is plains prickly pear (*O. macrorhiza*) with 3 or more spines per cluster, and fragile prickly pear (*O. fragilis*) with small (less than 2 in. long) cylindrical stems. IA, IL, IN, MI, MN, MO, OH, WI. Widely scattered. *SN*

Lonicera flava
CAPRIFOLIACEAE

yellow honeysuckle

Forests, savannas, glade margins, streambanks. Spring, summer.

A slender woody vine to 20 ft. long. Leaves gray-green (at least in center of upper few pairs), opposite, entire, blunt-tipped or pointed, to 3½ in. long × 2½ in. wide, short-petiolate to sessile, the upper 1–3 pairs clasping and fused around stem. Flowers irregular, to 1 ⅖ in. long, yellow to orange, tubular, terminating in 2 lips, upper lip broad with 4 shallow lobes, lower lip narrow, tube without protrusion at base; in tight cluster of up to 12 at ends of branches. Fruit an orange to red berry. IL, MO, OH. **Nearly restricted to the Ozark Highlands, rare in the Tallgrass Prairie and Eastern Forests.** *SN*

Lonicera hirsuta
CAPRIFOLIACEAE

hairy honeysuckle

Forests, thickets, cedar swamps, streambanks. Summer.

A slender glandular-hairy woody vine to 16 ft. long. Leaves densely hairy, deeply veined with wrinkled appearance, opposite, entire, blunt-tipped or pointed, to 5 in. long × 3 in. wide, short-petiolate to sessile, the upper 1 or 2 pairs clasping and fused around stem. Flowers irregular, to 1 in. long, yellow (sometimes orange to red), tubular, terminating in 2 lips, upper lip broad with 4 lobes, lower lip narrow, tube with protrusion at base; numerous in tight cluster at ends of branches. Fruit an oblong orange-red berry. Grape honeysuckle (*L. reticulata*) has glabrous, glaucous leaves, with those beneath flowers nearly round and sometimes notched at tip. MI, MN, WI. **Widespread northward.** *PD*

Crocanthemum bicknellii
CISTACEAE

hoary frostweed

Glades, dunes, rocky barrens. Spring, summer.

An erect perennial, stems single or branched above, with stellate hairs, 8–18 in. tall. Leaves alternate, short-petiolate, entire, mostly elliptic, upper surface bearing stellate hairs, about 1 in. long × ¼ in. wide. Flowers of two types: the petaliferous is regular, about 1 in. wide, 5-petaled, with up to 10 flowers atop stem; the small apetalous ones form dense axillary clusters. Common name alludes to ice ribbons that form at stem base during early frosts. Syn. *Helianthemum bicknellii*. Common frostweed (*C. canadense*) has 1 or 2 petaliferous flowers and upper surface of leaves bearing stellate and simple hairs. IA, IL, IN, MI, MN, MO, OH, WI. **Primarily in the Tallgrass Prairie.** *PD*

Clintonia borealis
CONVALLARIACEAE

bluebead lily, corn lily

Moist forests, swamps, bogs. Spring.

An unbranched, colony-forming perennial with naked stem to 14 in. tall. Leaves 2–4, basal, thick and fleshy, glabrous and glossy, with ciliate margins, yellow-green, to 8 in. long × 3 in. wide, with distinct midvein, parallel veins less deeply impressed. Flowers regular, bell-shaped, to 1 in. long, with 6 flaring and recurved tepals, stamens and pistil exserted; up to 6 stalked, nodding flowers in upper part of plant. Fruit a blue ¼-in. round berry on erect stalk (see inset). Crushed leaves smell like cucumber. IL, IN, MI, MN, OH, WI. **Widespread in much of the Northern Lakes, absent elsewhere.** *SN*

Uvularia grandiflora
CONVALLARIACEAE

large-flowered bellwort

Rich mesic forests. Spring.

An erect perennial, branched, stems mostly glabrous, 1–2 ft. tall. Leaves alternate, perfoliate, broadly elliptic, entire, typically hairy beneath, 2–4 in. long × 1–2 in. wide. Flowers regular, 1–2 in. long, bell-shaped and dangling, with 6 slightly twisted, narrowly oblanceolate and glabrous tepals, drooping from stem tips. Before expanding after flowering the leaves look wilted. Perfoliate bellwort (*U. perfoliata*), known from eastern OH and southern IN, also has perfoliate leaves and drooping flowers but is typically smaller overall with leaves smooth beneath, and the inner surface of its tepals bear yellowish-orange granules (absent in *U. grandiflora*). IA, IL, IN, MI, MN, MO, OH, WI. **Mostly common throughout.** *SN*

Uvularia sessilifolia
CONVALLARIACEAE

sessile-leaved bellwort

Usually acidic soils in floodplain forests, flatwoods. Spring.

An erect perennial, branched, glabrous, stoloniferous and colonial, 5–12 in. tall. Leaves alternate, sessile, elliptic and entire, 1½–2½ in. long × ⅔–1 in. wide. Flowers regular, about 1 in. long, bell-shaped and dangling, with 6 linear, smooth and untwisted tepals with upturned tips, occurring singly from upper leaf axils. This little bellwort differs from others of the genus in lacking twisted tepals and perfoliate leaves. IA, IL, IN, MI, MN, MO, OH, WI. **Widespread but large blocks of the interior and western and northern Midwest are without populations.** *AG*

Sedum acre
CRASSULACEAE

mossy stonecrop

Fields, dunes, rocky and sandy shores, outcrops, cemeteries, roadsides. Summer.

A glabrous, fleshy, mat-forming perennial to 5 in. tall. Leaves alternate, entire, ovate, oval in cross section, sessile, to ⅓ in. long × ⅓ in. wide, ascending and dense along stem, often reddish at tips. Flowers regular, to ¾ in. wide, with 5 spreading petals, 10 ascending stamens, and 5 ascending pistils, subtended by 5 fleshy sepals, in clusters of 3–6 at top of plant. Stringy stonecrop (*S. sarmentosum*) has leaves thickened but flat in cross section and in whorls of 3. IA, IL, IN, MI, MN, MO, OH, WI. Introduced from Europe. Rare to scattered throughout, more frequent in the Northern Lakes. *MJH*

Diervilla lonicera
DIERVILLACEAE

dwarf bush honeysuckle, northern bush honeysuckle

Sandy or rocky savannas and forests, commonly with conifers. Summer.

A branched shrub to 3 ft. tall. Leaves opposite, petiolate, minutely toothed, ovate with abruptly pointed tip, to 5 in. long × 2¾ in. wide. Flowers irregular, to ¾ in. long and wide, yellow to red, funnel-shaped, with 5 spreading to recurved lobes, in clusters of up to 7 at tips of branches. Fruit a hardened, elongated capsule with 5 spreading calyx lobes persistent at tip. Vegetatively similar honeysuckles (*Lonicera*) have entire leaf margins. IA, IL, IN, MI, MN, OH, WI. Mostly in the Northern Lakes with fewer occurrences in adjacent portions of the Tallgrass Prairie and Eastern Forests. *SN*

Euphorbia cyparissias
EUPHORBIACEAE

cypress spurge, graveyard weed

Fields, roadsides, railroads, old homesites, cemeteries. Spring, summer, fall.

A monoecious perennial to 2 ft. tall. Leaves alternate but whorled immediately below inflorescence, blue-green, entire, to 1½ in. long × ⅛ in. wide, containing milky sap. Flowers petal-less, greenish, to ½ in. wide, with 2 kidney-shaped yellowish bracts below 4 half-moon-shaped petal-like yellow glands, in showy, tight, terminal clusters. Fruit a stalked 3-parted capsule. Leafy spurge (*E. virgata*; see inset) has similar inflorescences but is larger with broader leaves and can be quite invasive. IA, IL, IN, MI, MN, MO, OH, WI. Introduced from Europe. Widespread, with fewer occurrences in MO. *MJH/SN*

Astragalus canadensis
FABACEAE

Canada milkvetch, rattleweed

Prairies, savannas, dry cliff ledges, shorelines. Summer.

An erect perennial, branched, short-hairy, some hairs 2-pronged, 1–3 ft. tall, commonly colonial from rhizomes. Leaves alternate, about 5–7 in. long, petiolate, pinnately compound, with 10–15 pairs of oblong leaflets plus a terminal one, ⅓–1½ in. long × ½ in. wide. Flowers irregular with typical pea family shape, ½–¾ in. long, the banner petal often with upcurved tip, in erect spike-like racemes 2–7 in. long. Fruit a legume. Aka rattleweed because of sounds produced from shaking the dried fruits. IA, IL, IN, MI, MN, MO, OH, WI. Mostly in the Tallgrass Prairie and Ozark Highlands. *SN*

Baptisia tinctoria

FABACEAE

yellow false indigo, horseflyweed

Savannas, open woodlands.
Spring, summer.

A bushy perennial to 3 ft. tall. Leaves alternate, short-petiolate, compound with 3 sessile, obovate, blunt-tipped leaflets to 1 in. long × ½ in. wide, glabrous, entire. Flowers irregular with typical pea-family shape, to ½ in. long, on short stalks, numerous in few-flowered clusters at the ends of branches. Fruit a black, oblong, inflated legume. Flowers and leaves are much smaller and more numerous than our other *Baptisia* species, although plants are similar in size. IA, IL, IN, MI, MN, OH, WI. **Infrequent in eastern portion of the Northern Lakes and very rare in the Tallgrass Prairie and Eastern Forests.** *MJH*

Chamaecrista fasciculata

FABACEAE

partridge pea

Prairies, barrens, woodlands, old fields.
Summer.

An erect annual, sometimes branched above, stems sparsely hairy, 1–3 ft. tall. Leaves alternate, petiolate with the petiole bearing a cup-shaped gland, pinnately compound with 10–18 pairs of glabrous, oblong leaflets ⅓–¾ in. long × 1/16–¼ in. wide. Flowers irregular, about 1 in. wide, 5-petaled, unequal in size, yellow or with some reddish at base, 1–5 flowers in short axillary clusters. Fruit a legume. Small-flowered partridge pea (*C. nictitans*) is less than half the height of partridge pea and has flowers about a third the size. IA, IL, IN, MI, MN, MO, OH, WI. **Common except in the Northern Lakes.** *MAH*

Crotalaria sagittalis
FABACEAE

rattlebox

Barrens, sand savannas, glades, dry open woodlands. Summer.

An erect to ascending annual, branched, hairy, 6–12 in. tall. Leaves alternate, petiolate, stipulate, elliptical to lanceolate, 1–3 in. long × ½–1½ in. wide. Flowers irregular with typical pea family shape, ⅓–½ in. long, with an erect, rounded banner petal and 2 lateral wings enclosing the keel, flowers 2–4 from a stalk attached opposite a leaf. Fruit a legume. Rattlebox is so named because its inflated fruits, when dried and shaken, produce a rattling sound, not unlike that of a rattlesnake. IA, IL, IN, MI, MN, MO, OH, WI. **Principally in the region's western half, likely introduced farther east.** *PD*

Lotus corniculatus
FABACEAE

bird's-foot trefoil

Fields, roadsides, pastures. Summer.

An erect to sprawling perennial, branched, mostly glabrous, 6–24 in. tall. Leaves alternate, pinnately compound, with 5 oval-oblong leaflets, 3 at tip separated by a short stalk from 2 stipule-like ones at base, ¼–¾ in. long and about half as wide. Flowers irregular with typical pea family shape, about ½ in. long, occurring 3–10+ in head-like clusters on long stalks. Fruit a legume. This weed is a familiar feature along many highways. IA, IL, IN, MI, MN, MO, OH, WI. **Introduced from Europe. Widespread throughout.** *MAH*

Medicago lupulina
FABACEAE

black medick

Fields, roadsides, waste areas.
Spring, summer.

A sprawling annual or biennial, branched, glabrous to hairy, 8–24 in. long. Leaves alternate,
divided into 3 leaflets, petiolate with lanceolate stipules, leaflets oval to obovate, ¼–¾ in.
long × ⅛–⅓ in. wide. Flowers irregular with
typical pea family shape, about ⅛ in. long,
occurring 10–20+ in dense round heads. Fruit
a black kidney-shaped legume. This invasive
weed may look like low hop clover (*Trifolium
campestre*) or lesser hop clover (*T. dubium*) but
differs in having visible fruit. IA, IL, IN, MI,
MN, MO, OH, WI. Introduced from Eurasia.
Probably in every county. *MAH*

Melilotus officinalis
FABACEAE

yellow sweet clover

Fields, pastures, roadsides, prairies,
waste areas. Summer.

An erect biennial, branched, glabrous, 3–5 ft.
tall. Leaves alternate, petiolate, compound,
divided into 3 oval to oblanceolate leaflets
⅓–1½ in. long × ¼–⅔ in. wide with finely
saw-toothed margins. Flowers irregular with
typical pea family shape, 3/16–¼ in. long,
occurring on erect, narrow and loosely flowered racemes. Fruit is a strongly wrinkled
legume. Although its common name is sweet
clover, it is not in *Trifolium*, the genus of
plants most people recognize as clover. IA,
IL, IN, MI, MN, MO, OH, WI. Introduced
from Eurasia. Abundant, probably in every
county. *SN*

Senna marilandica

FABACEAE

southern wild senna

Dry upland forests, glades, savannas.
Summer.

An erect perennial, stems single or branched
above, glabrous or sparsely so with appressed
hairs, 2–6 ft. tall. Leaves alternate, pinnately
compound, to 8 in. long, with 4–10 pairs
of oblong leaflets 1–2½ in. long × ⅓–1 in.
wide, petioles with 2 stipules and a short,
dome-shaped, black nectary gland near base.
Flowers irregular, ¾ in. wide, with 5 spatulate
petals rounded at tips and conspicuous brown
anthers, in terminal and axillary racemes. Fruit
a segmented legume with rectangular partitions
and appressed hairs. Northern wild senna (*S.
hebecarpa*) has spreading hairs on stem and
fruit, the latter with square partitions. IA, IL,
IN, MO, OH, WI. **Mostly in the lower Mid-**
west. *MAH*

Stylosanthes biflora

FABACEAE

pencil flower

Barrens, dry upland woodlands,
acidic glades, rocky sterile slopes, prairies.
Spring, summer.

A mostly ascending or prostrate perennial, branched, hairy, 4–10+ in. tall. Leaves
alternate, petiolate with stipules fused to
the petiole and surrounding the stem, compound with 3 narrowly lanceolate, entire
and bristle-tipped leaflets, these ½–1½ in.
long × ⅛–¼ in. wide. Flowers irregular with
typical pea family shape, ¼–⅓ in. wide, the
calyx ⅛ to nearly ¼ in. long. Inflorescence of
few-flowered terminal and axillary clusters.
Both the common and genus names refer to
the tubular (pencil-like?) calyx. IL, IN, MO,
OH. **Relatively common in hills in the lower**
Midwest. *MAH*

Trifolium campestre
FABACEAE

low hop clover

Fields, pastures, lawns, roadsides, waste areas. Spring, summer.

A branching, gray-green annual to 1 ft. tall. Leaves alternate, long-petiolate, compound with 3 minutely toothed leaflets to ¾ in. long × ⅓ in. wide, lateral leaflets stalkless or nearly so, terminal leaflet longer-stalked; a pair of leaf-like stipules at base of each leaf. Flowers irregular with typical pea family shape, to ¼ in. wide, numerous in stalked, tight, nearly spherical clusters to ½ in. long. Golden clover (*T. aureum*) has all leaflets stalkless or nearly so and flowering clusters ½ in. long or larger. Lesser hop clover (*T. dubium*) has flowering clusters to ⅓ in. long. IA, IL, IN, MI, MN, MO, OH, WI. Introduced from Eurasia. Scattered to widespread throughout. *MAH*

Hypericum ascyron
HYPERICACEAE

great St. John's-wort

Meadows, wet prairies, fens, streambanks. Summer.

An erect perennial, branched, glabrous, 3–6 ft. tall. Leaves opposite, sessile, entire, lanceolate to oblong, 2–4 in. long × 1–2 in. wide. Flowers regular, 1–2 in. wide, with 5 spreading obovate-oblanceolate petals and 100+ stamens, occurring singly or in small clusters at branch tips. In our region great St. John's-wort is the tallest herbaceous member of the genus with some of the largest flowers and leaves. Syn. *H. pyramidatum*. IA, IL, IN, MI, MN, MO, OH, WI. Occurs primarily in the upper Midwest. *PS*

Hypericum gentianoides
HYPERICACEAE

orange grass, pineweed

Glades, barrens, sand flats, old fields. Summer.

An erect annual, stiff and wiry, glabrous, 4–12 in. tall. Leaves opposite, entire, linear and scale-like, appressed to stem, ⅛ in. long or less. Flowers regular, about ⅓ in. wide, with 5 oblong petals, stamens less than 20, occurring singly in leaf axils or a few at branch tips. This species is wiry like some grasses and reputed to smell like an orange when crushed. Nits and lice (*H. drummondii*) is similar, but its leaves are longer, up to ¾ in. long. **IA, IL, IN, MI, MO, OH, WI. Occurs mostly in a band from southern Wisconsin south to the Ozark Highlands.** *SN*

Hypericum hypericoides
HYPERICACEAE

St. Andrew's cross

Dry upland woodlands, sandstone glades, old fields. Summer.

A profusely branched and sprawling low shrub, glabrous, 2–12 in. tall. Leaves opposite, sessile, entire, elliptic to oblanceolate, ¼–1 in. long × ¼–⅓ in. wide. Flowers regular, about ½ in. wide and long, consisting of 2 opposing pairs of petals forming a cross or X. Inflorescence mostly of single flowers occurring at branch tips. This species differs from other *Hypericum* species treated here in possessing only 4 (not 5) petals. The taller subsp. *hypericoides* exists only in southern MO; subsp. *multicaule* is more widespread. Syn. *Ascyrum hypericoides*. **IL, IN, MO, OH. Found mostly in the Ozark Highlands and unglaciated hills of the Eastern Forests.** *SN*

Hypericum mutilum
HYPERICACEAE

dwarf St. John's-wort, weak St. John's-wort

Marshes, wet meadows, wet spots in glades, swamps, seeps, ditches. Summer, fall.

A glabrous, branched annual or short-lived perennial to 2 ft. tall. Leaves opposite, entire, oval, sessile, to 1½ in. long × ¾ in. wide. Flowers regular, to ¼ in. wide, with 5 petals, up to 15 stamens, and 3-styled pistil; in few-flowered clusters in upper part of plant. Northern St. John's-wort (*H. boreale*) has leafy bracts. Larger St. John's-wort (*H. majus*) and Canadian St. John's-wort (*H. canadense*) have lanceolate leaves; those of latter are narrower and usually with single vein. Small-flowered St. John's-wort (*H. gymnanthum*) has clasping deltoid leaves. IA, IL, IN, MI, MN, MO, OH, WI. Widespread, except in northern and northwestern extent. *ST*

Hypericum perforatum
HYPERICACEAE

common St. John's-wort, Klamath weed

Roadsides, railroads, disturbed fields. Summer.

An erect perennial, branched, ridged below each leaf, glabrous, 1–3 ft. tall. Leaves opposite, sessile, entire, oblong to lanceolate-elliptic, with translucent dots scattered about and a few black dots mostly on the margins, ¼–1½ in. long × ¹⁄₁₆–½ in. wide. Flowers regular, about 1 in. wide, 5-petaled, these black-dotted mostly along margins and tips. Stamens more than 20. Flowers numerous in rounded panicles. This species of St. John's-wort is sold as an herbal supplement for depression and nerve-related maladies. IA, IL, IN, MI, MN, MO, OH, WI. Introduced from Eurasia. Widespread and common in all states. *SN*

Hypericum punctatum
HYPERICACEAE

spotted St. John's-wort

Moist to dry prairies, fields, savannas, open woods, thickets. Summer, fall.

A glabrous perennial to 2½ ft. tall. Leaves opposite, entire, sessile or short-petiolate, to 2½ in. long × 1 in. wide, blunt-tipped, with black spots beneath and along margins above. Flowers regular, to ½ in. wide, with 5 often black-spotted petals (see inset), in tight clusters at top of plant. Fruit a capsule. Flowers do not produce nectar, but pollen attracts pollinators. False spotted St. John's-wort (*H. pseudomaculatum*) has larger flowers and leaves pointed at tip. IA, IL, IN, MI, MN, MO, OH, WI. **Widespread except in northern and northwestern portions of the Midwest.** *DT/ST*

Hypericum sphaerocarpum
HYPERICACEAE

round-fruited St. John's-wort

Prairies, glades, fens, rocky streambanks. Summer.

An erect perennial, glabrous, branching above, base somewhat woody, colonial, 1–2 ft. tall. Leaves opposite, sessile, entire, narrowly elliptic, 1½–3 in. long × ⅛–½ in. wide. Flowers regular, about ½ in. wide, with 5 oblanceolate to elliptic petals, stamens more than 20, in many-flowered somewhat flat-topped panicles. Fruit a rounded capsule. The generic common name is said to honor St. John the Baptist, whose feast day and presumed birthday (24 June) generally align with the blooming of hypericums, at least in Europe. IA, IL, IN, MI, MO, OH, WI. **Principally in the lower Midwest, especially in the Ozark Highlands and Tallgrass Prairie.** *CB*

Hypoxis hirsuta
HYPOXIDACEAE

yellow star grass

Prairies, glades, dry woodlands, fens.
Spring.

An erect, hairy perennial with flower stalk 2–7
in. tall. Leaves basal and bunched, linear, 5–10
in. long × ⅛–⅓ in. wide. Flowers regular, to 1
in. wide, with 6 lanceolate tepals, arranged in
a cluster of 1–5 terminating the stem and over-
topped by its grass-like leaves. Contrary to its
common name, this species is not a true grass.
IA, IL, IN, MI, MN, MO, OH, WI. Most prevalent in the Tallgrass Prairie. *MAH*

Iris pseudacorus
IRIDACEAE

yellow iris, yellowflag

Floodplains, marshes, swamps, seeps,
ponds, lakes, rivers, ditches. Spring,
summer.

A glabrous, gray-green perennial to 4 ft. tall.
Leaves primarily basal and fan-like, some
alternate and smaller along flowering stem,
entire, lanceolate, erect, to 4 ft. long × 1 in.
wide. Flowers regular, to 4 in. wide, with 3
upright petals to 2 in. long and 3 spreading
to reflexed petal-like sepals to 3 in. long with
reddish veins at base, in several-flowered clusters at top of stem. Fruit a nodding, angular
capsule to 3 in. long. Vegetatively very similar
to southern blueflag (*I. virginica*), but larger in
all respects. IA, IL, IN, MI, MN, MO, OH, WI.
Introduced from Europe. Widely scattered,
most frequent near the Great Lakes. *NP*

Agastache nepetoides

LAMIACEAE

yellow giant hyssop

Thickets, moist woodlands, stream corridors. Summer.

A perennial with a sharply 4-angled stem, glabrous to finely hairy, branched above, 3–5 ft. tall. Leaves opposite, petiolate, ovate to lanceolate, coarsely toothed, 3–6 in. long × 2–3 in. wide. Flowers irregular, ⅓–½ in. long, tubular, 5-lobed with 2 upper lobes and 3 lower, the central of which is curved downward, calyx lobes 5, blunt and glabrous, typically less than ⅛ in. long. Inflorescence of dense, terminal spikes 3–8 in. long. IA, IL, IN, MI, MN, MO, OH, WI. **Generally common and widespread in the lower two-thirds of Midwest.** *MAH*

Collinsonia canadensis

LAMIACEAE

horse balm, rich weed

Rich mesic forests, ravines, stream terraces. Summer.

An erect perennial, stem mostly unbranched except above, glabrous, 4-angled, 1–3 ft. tall. Leaves opposite, long-petiolate (petioles commonly purple at base), ovate, serrate, glabrous to short-hairy, 2–6 in. long × 1–4 in. wide. Flowers irregular, about ½ in. long, tubular, 2-lipped, the upper lip with 2 upper and 2 lateral spreading lobes, yellowish and commonly purple-striped, the longer lower lip yellowish and heavily white-fringed, in a terminal branched panicle. Flowers of horse balm are said to be lemon-scented, but usually only when crushed or bruised. IL, IN, MI, MO, OH, WI. **Mostly in Eastern Forests.** *BS*

Utricularia cornuta
LENTIBULARIACEAE

horned bladderwort

Marshes, pannes, swamps, fens, bogs, shores. Summer.

A carnivorous, glabrous, colony-forming perennial to 14 in. tall, usually much shorter. Basal leaves buried in soil, alternate, filiform, bearing bladders that trap and digest tiny animals; stem leaves reduced to few, alternate, minuscule scales. Flowers irregular, exotically shaped and long-spurred, to ¾ in. long, 2-lipped, with lower lip arched and upper lip hood-like, up to 6 short-stalked in terminal raceme, fewer during abnormally wet or dry years. Fruit a round to teardrop-shaped capsule. IL, IN, MI, MN, OH, WI. Frequent in the Northern Lakes, rare in the Eastern Forests. *MJH*

Utricularia intermedia
LENTIBULARIACEAE

flat-leaved bladderwort

Fens, bogs, marshes, lakes, often in alkaline water. Summer.

A perennial aquatic, creeping when stranded, branched, 2–6 in. long, carnivorous. Leaves alternate, finely dissected and fan-shaped (see inset), forking 2 or 3 times, segments flat, ¾–1 in. long and wide. Bladders capable of trapping and digesting tiny aquatic animals, borne on separate leafless stems, are often buried. Flowers irregular, ½ in. long, 2-lipped, the lower lip broad, flat, and much larger than upper, a narrow spur pressed close beneath the lower. Often occurs in small channels in wetlands created by muskrats and other animals. IA, IL, IN, MI, MN, OH, WI. Occurs mostly in the Northern Lakes. *SN*

Utricularia macrorhiza
LENTIBULARIACEAE

common bladderwort

Lakes, ponds, deep marshes. Summer.

A carnivorous perennial aquatic, to 3 ft. long. Leaves alternate, 2 in. long, segments thread-like, forking 5+ times, bearing bladders capable of trapping and digesting tiny aquatic animals (see inset). Flowers irregular, ½–¾ in. long, with 2 lips equal in length, the lower humped in middle with edges down-turned and bearing a forward-projecting spur beneath, in several-flowered racemes on stalks 4–8 in. tall. Syn. *U. vulgaris*. Bog bladderwort (*U. geminiscapa*) and humped bladderwort (*U. gibba*) have smaller flowers. The latter's leaves differ from both, forking only 1 or 2 times. **IA, IL, IN, MI, MN, MO, OH, WI. Common, mostly in the Northern Lakes.** *SN*

Utricularia minor
LENTIBULARIACEAE

lesser bladderwort

Submergent marshes, sedge meadows, fens, swales, pannes, lakes; often in calcareous soil or water. Spring, summer.

A carnivorous, glabrous, colony-forming perennial to 6 in. tall. Basal leaves in water or mud, alternate, fan-like, to ⅔ in. long and wide, palmately segmented several times, segments narrowly linear and flat, bearing bladders that trap and digest tiny animals; flowering stem naked, emerged from water or mud. Flowers irregular, pale to creamy yellow, to ¼ in. long, 2-lipped, with lower horizontal lip twice as long as upper ascending lip, with red streaks internally, spur short and inconspicuous; up to 9 short-stalked in terminal raceme. **IA, IL, IN, MI, MN, MO, OH, WI. Frequent in the Northern Lakes, widely scattered elsewhere.** *SN*

Erythronium americanum
LILIACEAE

yellow trout lily, yellow dogtooth violet

Moist forests, floodplain forests. Spring.

An unbranched colony-forming perennial with naked stem to 6 in. tall. Leaves basal, 2 on flowering plants, 1 on sterile plants, glabrous and waxy, brown-mottled, to 6 in. long × 2 in. wide. Flower regular, solitary, nodding, to 1¼ in. long with 6 recurved tepals, stamens and pistil exserted, anthers reddish-purple or yellow. Fruit a capsule. Yellow fawnlily (*E. rostratum*), rare, is known from southern MO and southern OH and has appendages at base of tepals that are lacking in *E. americanum*. IA, IL, IN, MI, MN, MO, OH, WI. Widespread in the Northern Lakes, Eastern Forests, eastern Tallgrass Prairie, and eastern Ozark Highlands, scattered elsewhere. *MAH*

Linum sulcatum
LINACEAE

grooved yellow flax

Dry prairies, glades, usually calcareous. Summer.

An erect annual, branched above, grooved, glabrous to minutely hairy, to 2½ ft. tall. Leaves alternate, vertically aligned with the stem, linear-lanceolate, with a pair of tiny glandular stipules (appearing as black dots; see inset) at their base, ¼–1 in. long × ⅛ in. wide. Flowers regular, about ½ in. wide, petals and sepals 5 each, in a spreading panicle. The similar stiff yellow flax (*L. medium*) is perennial with slightly smaller flowers and no stipules. IA, IL, IN, MI, MN, MO, OH, WI. Most common in the Midwest's western half, especially in the Ozark Highlands and Tallgrass Prairie. *DT*

Linum virginianum
LINACEAE

woodland flax

Forests, open woods. Summer, fall.

A glabrous perennial to 2½ ft. tall. Leaves numerous, opposite below, alternate above, sessile, entire, to 1 in. long × ¼ in. wide, smaller above. Flowers regular, with 5 spreading petals, to ⅖ in. wide, with 5 styles and 5 stamens with brown anthers ascending from middle, numerous in flattish-topped branched inflorescence at top of plant. Fruit a spherical, hardened capsule to ⅛ in. wide that splits into segments like an orange. Ridged flax (*L. striatum*) is very similar but with less flat-topped and more elongated inflorescence and with decurrent tissue between leaves along stem. IL, IN, MI, MO, OH. Scattered mostly southward and eastward, also around the Great Lakes and in southern MI. *KM*

Abutilon theophrasti
MALVACEAE

velvetleaf

Agricultural fields, construction sites, waste areas. Summer.

A branched annual, stout with velvet-like hair, 1–5 ft. tall. Leaves alternate, on long petioles, cordate with entire margins or shallowly toothed, 4–6 in. long × 4–5 in. wide. Flowers regular, ½–1 in. wide, 5-petaled, from axils of leaves. Fruit a button-shaped capsule, hairy and segmented with distinctive spreading beaks. Velvetleaf is highly invasive in agricultural fields. Also called pie-marker: its seed capsules have been used to imprint patterns on pie crust. IA, IL, IN, MI, MN, MO, OH, WI. Introduced from Asia. Common throughout except for far northern counties of the Northern Lakes. *SN*

Malvastrum angustum
MALVACEAE

hispid false mallow

Shallow soil in prairies and glades. Summer.

An appressed-hairy annual to 2½ ft. tall. Leaves simple, toothed, alternate, to 1 ⅖ in. long × ⅖ in. wide, short-petiolate. Flowers regular, to ⅖ in. wide, inconspicuous with 5 shallowly lobed petals shorter than calyx or barely longer, which obscures them, often closed and appearing pentagonal but spreading when open, with central stigma-tipped column covered with stamens; flowers short-stalked, solitary in leaf axils in upper part of plant. Calyx becomes pinkish-brown at maturity. Syns. *M. hispidum, Sphaeralcea angusta*. IA, IL, MO. Uncommon to rare in the Ozark Highlands and the Tallgrass Prairie. *SH*

Sida spinosa
MALVACEAE

prickly sida, prickly fanpetals

Roadsides, pastures, agricultural fields, waste areas. Summer.

An erect annual, branched, stems hairy, 1–2 ft. tall. Leaves alternate, petiolate with a hardened spine-like projection sometimes present just beneath the petioles, blades ovate to narrowly oblong, toothed, hairy beneath, 1–2½ in. long × ½–1 in. wide. Flowers regular, ⅓–½ in. wide, shaped vaguely like a pinwheel with 5 asymmetrically shaped petals. The similar Coastal Plain sida (*S. elliottii*) occurs in southern MO and southern IL; it has narrower leaves and larger flowers. IA, IL, IN, MI, MO, OH. Introduced from tropical America. Common, mostly in the lower Midwest. *DT*

Lysimachia ciliata
MYRSINACEAE

fringed loosestrife

Lowland forests, swamps, riverbanks, ditches. Spring, summer.

An erect perennial, mostly unbranched, mostly glabrous, colonial, 1–4 ft. tall. Leaves opposite, petioles narrowly winged with long hairs along margins, blades broadly lance-olate, non-dotted, short hairs on margin but otherwise glabrous, 2–5 in. long × 1–2½ in. wide. Flowers regular, about 1 in. wide, very short-tubular, with 5 elliptic to obovate lobes, each usually red/orange at the base and bearing a sharp-pointed tip. Inflorescence of mostly long-stalked flowers nodding from upper leaf axils. The flowers produce oil, fed upon by bees. IA, IL, IN, MI, MN, MO, OH, WI. Common, possibly in every county. *MJH*

Lysimachia lanceolata
MYRSINACEAE

lance-leaved loosestrife

Dry to mesic upland forests, prairies, meadows. Summer.

An erect perennial, unbranched to upwardly branched, mostly glabrous, stems slender and 4-angled, stoloniferous, 8–18 in. tall. Leaves opposite or a few whorled above, principal ones sessile to short-petiolate, lanceolate, pale beneath, may have hairy margins, especially at base, 2–5 in. long × ¼–¾ in. wide. Flowers regular, about ¾ in. wide, with 5 elliptic to obovate lobes having ragged margins and red/orange bases, on 1–4 long stalks from the upper leaf axils. The similar lowland loose-strife (*L. hybrida*) has stout stems with ovate leaves evenly green beneath. IA, IL, IN, MI, MO, OH, WI. Widespread, rare to absent in far northern and western counties. *MJH*

Lysimachia nummularia
MYRSINACEAE

moneywort

Floodplain forests, swamps, seeps, fens, ditches. Summer.

A semi-evergreen perennial, creeping to ascending, mat-forming, branched, glabrous, growing 1–3+ ft. long. Leaves opposite, short-petiolate, mostly round, with reddish-brown dots, ½–1½ in. long and about as wide. Flowers regular, ¾–1 in. wide, with 5 oblong lobes with ragged tips. Flowers from leaf axils. In the Midwest, moneywort spreads quite efficiently by vegetative means only; reportedly, it does not produce seeds here. IA, IL, IN, MI, MN, MO, OH, WI. Introduced from Eurasia. Widely scattered and common except for far northern and western counties. *MAH*

Lysimachia quadriflora
MYRSINACEAE

narrow-leaved loosestrife, prairie loosestrife, four-flowered loosestrife

Fens, moist to wet prairies, seeps. Summer, fall.

A branching perennial to 2½ ft. tall. Leaves numerous, simple, opposite, entire, rolled under along margins, sessile, with prominent midvein, linear, to 3½ in. long × ¼ in. wide. Flowers regular, to 1 in. wide, with 5 spreading-ascending petals tipped by abrupt point; solitary and nodding on stalks to 1½ in. long from axils of middle and upper leaves, often appearing to be in whorls of 4 due to very short stem internodes. Fruit an erect, round capsule. IA, IL, IN, MI, MN, MO, OH, WI. Throughout but spotty. *SN*

Lysimachia quadrifolia
MYRSINACEAE

whorled loosestrife

Dry upland forests, dunes, savannas.
Summer.

An erect perennial, unbranched, stems glabrous to hairy, 1–1½ ft. tall. Leaves whorled,
usually in 4s, sessile or petiolate, lanceolate,
hairy beneath and on margins, with scattered
and fine, dark streaking or dots, 1–4 in. long ×
½–1½ in. wide. Flowers regular, ½–¾ in. wide,
with 5 ovate-lanceolate lobes often with red/
orange bases and smooth margins, 1–5 on long
stalks from axils of upper whorls. This species
favors well-drained soils, often on acidic substrates such as dunes and sandstone-derived
soils. IL, IN, MI, MN, OH, WI. Occurs principally in the eastern Midwest, absent from
most of western half. *PR*

Lysimachia terrestris
MYRSINACEAE

swamp-candles

Marshes, swamps, fens, bogs, ditches.
Summer.

A glabrous perennial to 3 ft. tall. Leaves simple, opposite, entire, dark green, rolled under
along margins, scale-like below, larger above,
sessile or short-petiolate, to 4 in. long × ¾
in. wide, undersides covered with tiny dots,
producing axillary red caterpillar-like bulbils
to ¾ in. in summer (for vegetative reproduction). Flowers regular, to ¾ in. wide, with 5
spreading yellow petals with red spots at base,
numerous, on stalks to ¾ in. in long terminal
raceme. IA, IL, IN, MI, MN, MO, OH, WI.
Widespread in the Northern Lakes and adjacent portions of the Tallgrass Prairie, scattered in remainder of the Tallgrass Prairie
and the Eastern Forests. *SN*

Lysimachia thyrsiflora
MYRSINACEAE

tufted loosestrife

Marshes, swamps, floodplains, fens, bogs, ditches. Spring, summer.

A perennial to 3 ft. tall. Leaves simple, opposite, entire, scale-like below, larger above, sessile or short-petiolate, to 6 in. long × 2 in. wide, undersides covered with tiny dots. Flowers regular, to ⅓ in. wide, with 5–7 spreading-ascending petals surrounding protruding stamens; in solitary, dense, thimble-shaped clusters to 1½ in. long × 1 in. wide on stalks to 2 in. long from axils of mid-stem leaves. IA, IL, IN, MI, MN, MO, OH, WI. Widespread in the Northern Lakes and adjacent portions of the Tallgrass Prairie, scattered in remainder of the Tallgrass Prairie and the Eastern Forests. SN

Nuphar advena
NYMPHAEACEAE

spatterdock

Bogs, marshes, swamps, lakes, ponds, rivers. Spring, summer, fall.

A rhizomatous colonial perennial to 6+ ft. tall, leaves floating or emerged to 2 ft. from water. Leaves upright to horizontal, to 16 in. long × 12 in. wide, cordate with rounded basal lobes, entire, shiny, petiole round in cross section. Flowers regular, to 1¾ in. wide, with 6 conspicuous yellow sepals (sometimes red internally) and numerous inconspicuous yellow petals, stamens, and large disk-like stigma, solitary on stalks emerged a few inches from water. Bullhead pond-lily (*N. variegata*) has floating leaves with nearly winged petioles. Small yellow pond-lily (*N. microphylla*) has 5 sepals and smaller leaves and flowers. IL, IN, MI, MO, OH, WI. Scattered to widespread in southeastern quarter of Midwest. SN

Ludwigia alternifolia
ONAGRACEAE

seedbox

Marshes, wet prairies, river and lake
borders, ditches. Summer.

An erect perennial, branched, glabrous or
hairy, 2–4 ft. tall. Leaves alternate, mostly
sessile, lanceolate, 2–4 in. long × ⅓–¾ in.
wide. Flowers regular, ½–¾ in. wide, usually
with 4 rounded petals and 4 broadly lanceolate sepals, occurring singly in leaf axils. Fruit
a capsule shaped like a square box that when
dried can produce a rattle sound when shaken.
The fruit's distinctive shape and persistence
allows for identification well into winter. IA,
IL, IN, MI, MO, OH, WI. Populations occur
mostly in the lower Midwest. *SN*

Ludwigia peploides
ONAGRACEAE

floating primrose willow

Shallow pond and lake margins,
slow-moving watercourses. Summer.

A branched perennial, creeping and
mat-forming, mostly glabrous, 1–3+ ft. long.
Leaves alternate, long-petiolate, elliptic to
oblanceolate, shiny, 1–3 in. long × ¼–1½ in.
wide. Flowers regular, ¾–1 in. wide, with 5
obovate, rounded petals, occurring singly in
leaf axils. Fruit a narrow cylindrical capsule.
This plant is equally at home on wet soil or
water. In shallow ponds it can form mats that
completely cover the water's surface. Syn.
Jussiaea repens. IA, IL, IN, MO, OH. Occurs
mostly in the lower third of the Midwest but
is spreading northward. *SN*

Oenothera biennis
ONAGRACEAE

common evening primrose

Prairies, fallow fields, roadsides, railroads, disturbed sites. Summer, fall.

An erect biennial, unbranched or branched above, glabrous to hairy, 1–5 ft. tall. Leaves alternate, mostly sessile, lanceolate to lance-oblong, entire or toothed and wavy-margined, 2–7 in. long × ½–1½ in. wide. Flowers regular, 1–2 in. wide, with 4 obovate petals with notched tips, sepals reflexed, borne from spikes atop the stem. Fruit a narrow cylindrical capsule. Similar species include dune evening primrose (*O. oakesiana*), small-flowered evening primrose (*O. parviflora*), and downy evening primrose (*O. villosa*). IA, IL, IN, MI, MN, MO, OH, WI. Probably occurs in every Midwestern county. *MAH*

Oenothera clelandii
ONAGRACEAE

Cleland's evening primrose

Sand prairies, fields, savannas, barrens, dunes, roadsides. Summer, fall.

A hairy biennial to 3 ft. tall (usually shorter). Leaves in basal rosette first year, then alternate, to 3½ in. long × ½ in. wide, sessile to short-petiolate, entire or sparsely toothed. Flowers regular, to 1¼ in. wide, with 4 spreading, pointed petals surrounding 8 long stamens and solitary style of about same length, subtended by 4 reflexed sepals, numerous in leafy terminal spike. Fruit bluntly 4-lobed. Sand evening primrose (*O. rhombipetala*) has larger flowers and styles longer than stamens. IA, IL, IN, MI, MN, MO, OH, WI. Widespread in southwestern portion of the Northern Lakes and northeastern portion of the Tallgrass Prairie, rare elsewhere. *SN*

Oenothera laciniata
ONAGRACEAE

cut-leaved evening primrose

Prairies, fallow fields, roadsides, railroads, disturbed sites. Summer.

An erect to ascending or reclining annual, hairy, 8–18 in. tall. Leaves alternate, sessile, oblong to oblanceolate, irregularly, sharply, and deeply toothed or lobed, 1½–3½ in. long × ½–1 in. wide. Flowers regular, ½–1½ in. wide, somewhat pale, with 4 obovate petals notched at tip, sepals reflexed, borne singly from upper leaf axils. Fruit a narrow cylindrical capsule. This species favors sandy substrates. The wide-ranging large-flowered evening primrose (*O. grandis*) is similar but has larger flowers. IA, IL, IN, MI, MN, MO, OH, WI. Relatively common except in the Midwest's northern third. *SN*

Oenothera linifolia
ONAGRACEAE

threadleaf evening primrose

Prairies and open sites with acidic substrates (sandstone, chert, granite). Spring, summer.

An erect annual, glabrous to hairy, 4–18 in. tall. Leaves alternate, sessile, narrow and thread-like, entire, ⅓–1½ in. long × 1/16 in. wide. Flowers regular, about ½ in. wide, with 4 oblong-obovate petals with notched tips, sepals reflexed, ovary hairy, on terminal compact racemes that elongate at maturity. Fruit a club-shaped capsule. The epithet *linifolia* ("flax-leaved") is a reference to *Linum*, the genus of flax. IL, MO. Mostly common, in the Ozark Highlands and in far southern IL. *SN*

Oenothera macrocarpa
ONAGRACEAE

Missouri evening primrose

Calcareous glades, rocky barrens, various exposed mostly rocky surfaces. Spring, summer.

A sprawling to erect perennial, unbranched with appressed silky hairs, 8–18 in. tall. Leaves alternate, petiolate to sessile, linear-lanceolate, entire, 2–5 in. long × ½–1¼ in. wide. Flowers regular, 3–5 in. wide, with 4 broadly obovate petals with slightly notched tips, sepals reflexed and commonly red-spotted, borne singly from upper leaf axils. Fruit a broad, 4-winged capsule. The flowers of this evening primrose are exceptionally large and showy; like many *Oenothera* species, they open in the evening and fade the following day. **MO. Occurs naturally in the Ozark Highlands. A rare few escaped occurrences exist elsewhere.** *ST*

Oenothera perennis
ONAGRACEAE

small sundrops

Moist prairies, fields, meadows, savannas, open woods, roadsides. Summer.

A perennial to 2 ft. tall (often shorter). Leaves alternate, to 2¼ in. long × ¾ in. wide, sessile, entire to minutely sparsely toothed, with short axillary leafy branches and basal leaves sometimes present. Flowers regular, to ¾ in. wide, with 4 spreading, rounded petals surrounding 8 stamens and solitary style, subtended by 4 green to pink partially reflexed sepals, numerous in leafy spike-like racemes at ends of branches. Fruit a capsule, 4-winged in cross section. **IA, IL, IN, MI, MN, MO, OH, WI. Most frequent in eastern portion of the Eastern Forests and in the Northern Lakes and adjacent Tallgrass Prairie, rare elsewhere.** *PS*

Oenothera pilosella
ONAGRACEAE

prairie sundrops

Prairies, fields, moist meadows, savannas, open woodlands, roadsides, railroads. Summer.

A branched, soft-hairy perennial to 2 ft. tall. Leaves alternate, to 3 in. long × 1 in. wide, sessile, entire to minutely sparsely toothed. Flowers regular, showy, to 2 in. wide, with 4 spreading, shallowly notched petals surrounding 8 stamens and solitary style, subtended by 4 yellow to pinkish partially reflexed hairy but glandless sepals, several in leafy clusters at ends of branches. Fruit a capsule, 4-winged in cross section. Narrow-leaved sundrops (*O. fruticosa*) has glabrous or glandular-hairy sepals. IA, IL, IN, MI, MO, OH, WI. Most frequent in eastern portions of the Tallgrass Prairie and Ozark Highlands, scattered elsewhere. *SN*

Oenothera serrulata
ONAGRACEAE

toothed evening primrose

Dry prairies and fields. Summer.

A hairy perennial to 2 ft. tall (often shorter). Leaves alternate, nearly linear, to 3 in. long × ¼ in. wide, sessile, irregularly sharp-toothed, especially in upper part of leaf (rarely entire), often folded or rolled lengthwise. Flowers regular, to 1 in. wide, with 4 spreading, blunt, ragged-tipped petals surrounding 8 stamens and solitary style, subtended by 4 reflexed green to pink sepals, numerous and sessile in upper leaf axils. Fruit a capsule, bluntly 4-lobed in cross section. Syn. *Calylophus serrulatus*. IA, IL, IN, MI, MN, MO, WI. Primarily in northwestern portion of the Tallgrass Prairie, scattered to rare elsewhere; introduced in MI. *KC*

Cypripedium parviflorum
ORCHIDACEAE

yellow lady's-slipper

Upland forests, dunes, fens, sedge meadows. Spring, summer.

An erect perennial, unbranched, hairy, 5–24 in. tall. Leaves alternate, sheathing, elliptic-ovate, entire, 3–7 in. long × 2–4 in. wide. Flowers irregular, 3-petaled, the lowermost (lip) pouch-shaped and rounded, 1–2¼ in. long, the others yellowish-green or dark purple, hairy, linear-lanceolate and twisted, sepals 3 (lateral sepals fused), yellowish to dark purple/brown. The larger var. *pubescens* (pictured) has yellowish sepals and lateral petals; the smaller vars. *makasin* and *parviflorum* have uniformly dark sepals and lateral petals. The young, uppermost sheathing bract of the latter two is mostly glabrous, or densely silvery-hairy, respectively. IA, IL, IN, MI, MN, MO, OH, WI. **Widespread but very local.** *MAH*

Aureolaria flava
OROBANCHACEAE

smooth false foxglove

Dry upland forests. Summer.

An erect perennial with stems unbranched or branched above, glabrous, purplish and glaucous, 2–4 ft. tall. Leaves opposite, petiolate, ovate-lanceolate, middle and lower deeply pinnately divided, lobes entire or with few teeth, with short hairs on upper surface and glabrous below, 2–5 in. long × ½–2 in. wide. Flowers irregular, up to 2½ in. long, tubular to funnel-shaped with 5 flaring, rounded lobes, glabrous, in leaf axils of upper stem, often in pairs. Hemiparasitic, mostly on oak roots. Downy false foxglove (*A. virginica*) is similar, but its stems possess fine hairs. IL, IN, MI, MO, OH. **Occurs mostly from northern MI southward to eastern OH, west to southeastern MO.** *SN*

Aureolaria grandiflora
OROBANCHACEAE

large-flowered false foxglove

Glades, savannas, open woods. Summer, fall.

A hairy perennial to 4½ ft. tall. Leaves opposite, twice pinnately lobed below, to 7 in. long × 2½ in. wide, smaller and becoming unlobed above, sessile or nearly so. Flowers irregular, to 2¼ in. long, tubular with 5 spreading lobes, solitary on stalks from bracts in upper part of plant. Hemiparasitic. Combleaf yellow false foxglove (*A. pectinata*) is densely glandular-hairy with deeply twice pinnately lobed leaves. Clammy false foxglove (*A. pedicularia*) has sticky leaves and stalked glands in inflorescence. **IA, IL, IN, MN, MO, WI. Widespread in much of the Ozark Highlands, eastern portion of the Tallgrass Prairie, and adjacent Northern Lakes.** *CR*

Dasistoma macrophylla
OROBANCHACEAE

mullein foxglove

Rocky forested slopes, glades, thickets, streambanks. Summer.

An annual or perennial, erect, branched, hairy, 3–7 ft. tall. Leaves opposite, petiolate, ovate-lanceolate, the lower ones deeply pinnately lobed, toothed, and up to 1 ft. long × 4 in. wide, the upper smaller and commonly unlobed. Flowers irregular, ½–⅔ in. long, funnel-shaped and woolly within (see inset), with 5 rounded flaring lobes, mostly sessile and paired in axils of leafy terminal spikes. Mullein foxglove looks much like members of the genus *Aureolaria* but differs in having smaller flowers that are woolly within. Hemiparasitic. **IA, IL, IN, MI, MO, OH, WI. Mostly in the lower Midwest.** *DT*

Pedicularis canadensis
OROBANCHACEAE

wood betony, Canadian lousewort

Dry to dry-mesic woods, barrens, savannas, prairies, usually in acid soils. Spring.

An erect perennial, unbranched, hairy, 6–15 in. tall, colonial. Leaves basal and alternate on flowering stalk, petiolate, lanceolate, pinnately lobed and toothed, fern-like, 2–6 in. long × 1–2 in. wide. Flowers irregular, ¾–1 in. long, tubular and 2-lipped, the upper curved and hood-like, the lower 3-lobed, in a dense spike, ranging in color from burnt orange to reddish-purple to yellow. Wood betony is a hemiparasite on roots of other plants. IA, IL, IN, MI, MN, MO, OH, WI. Occasional to common in most of the region. *MAH*

Oxalis grandis
OXALIDACEAE

large yellow wood sorrel

Mesic to dry-mesic mostly upland forests. Spring.

An erect perennial, sparsely hairy, 10–18 in. tall. Leaves alternate, petiolate, compound, clover-like, bearing 3 obcordate leaflets with thin purple margins, leaflets ¾–1½ in. long and slightly wider. Flowers regular, ½–1 in. wide with 5 oblong petals usually marked with red at bases, in a cluster of 2 or 3 usually extending above the leaves. Illinois wood sorrel (*O. illinoensis*) is very similar but lacks purple leaf margins, and its roots bear small tubers (absent in *O. grandis*); it occurs in southern IL and IN. IN, OH. Fairly widespread and common in OH and southern IN. *PR*

Oxalis stricta
OXALIDACEAE

common yellow wood sorrel

Mesic and dry-mesic forests, fields,
disturbed sites. Spring, summer, fall.

An annual or short-lived perennial, erect to
ascending, branched, with mostly spreading
soft hairs possessing partitions (view with
hand lens), 5–12 in. tall. Leaves alternate,
petiolate, compound, clover-like, bearing 3
obcordate leaflets, ¼–¾ in. long and slightly
wider. Flowers regular, ¼–½ in. wide, with 5
oblong petals, in clusters of 1–3 usually at or
extending above the leaves. Fruit a cylindrical
capsule on upright stalk. Slender yellow wood
sorrel (*O. dillenii*) is very similar, but stalks
bearing capsules are reflexed and stem hairs
are mostly appressed and lack partitions. **IA,
IL, IN, MI, MN, MO, OH, WI. Abundant
throughout the Midwest.** *MAH*

Corydalis flavula
PAPAVERACEAE

pale corydalis

Rich mesic forests, ravines, slopes. Spring.

An erect or reclining annual, branching, glabrous, 4–10 in. tall. Leaves alternate, lower
ones long-petiolate, upper ones to nearly
sessile, pinnately divided and finely dissected,
leaflets about 1 in. long × ¾ in. wide. Flowers
irregular, about ¼ in. long, tubular, 4-petaled,
inner 2 joined, upper and lower petals 3- or
4-toothed at flared apex, upper with incurved
basal spur. Inflorescence of terminal racemes.
Golden corydalis (*C. aurea*) and slender
corydalis (*C. micrantha*) are restricted to more
western parts of the Midwest; their flowers are
similar but slightly larger, with straight spurs.
**IA, IL, IN, MI, MO, OH. Occurs primarily in
the Midwest's southern third.** *SN*

Stylophorum diphyllum
PAPAVERACEAE

wood-poppy, celandine-poppy

Rich forests, ravines. Spring.

A perennial to 1½ ft. tall. Leaves basal with opposite pair on stem, twice pinnately lobed with rounded tips, to 6 in. long × 4 in. wide, petiolate, containing yellow sap. Flowers regular, to 2¼ in. wide, with 4 spreading-ascending rounded petals surrounding numerous stamens and single pistil, few on short stalks at top of plant. Fruit a nodding, prickly-hairy capsule to 1 in. wide. Greater celandine (*Chelidonium majus*; see inset), introduced from Eurasia, has ¾-in. flowers that bloom slightly later, alternate stem leaves, and narrow, erect fruit. **IL, IN, MI, MO, OH. Scattered, most frequent in the Eastern Forests and along the eastern and southern sides of Lake Michigan.** *SN*

Passiflora lutea
PASSIFLORACEAE

yellow passionflower

Mesic and dry-mesic upland and lowland forests, thickets. Spring, summer.

A perennial vine climbing by tendrils, branched, glabrous to hairy, 5–10 ft. long. Leaves alternate, petiolate, palmately and shallowly 3-lobed, entire, 1–4 in. long and slightly wider. Flowers regular, about 1 in. wide, with a 3-styled pistil and 5 stamens encircled by 5 oblong sepals and 5 similar-looking petals and a series of thread-like filaments. Flowers solitary in leaf axils. Fruit a ½-in. round berry, purple to black. **IL, IN, MO, OH. Occurs relatively commonly in the southern third of the Midwest.** *SN*

Mimulus glabratus
PHRYMACEAE

round-leaf monkey-flower

Fens, cedar swamps, springs, calcareous shores. Summer, fall.

A weak, mat-forming perennial, sometimes with upright stems to 1½ ft. tall. Leaves opposite, short-petiolate below, clasping above, roundish, to 1 in. wide, shallowly toothed. Flowers irregular with 2 upper and 3 lower lobes, tubular, yellow with a few red spots in throat, to ½ in. wide, solitary and short-stalked from upper leaf axils. Syn. *Erythranthe glabrata*. Michigan monkey-flower (*M. michiganensis*), a Midwest endemic known only from Straits of Mackinac and Grand Traverse regions of MI, is more upright with more conspicuously toothed leaves and longer style (over ¼ in.). IA, IL, MI, MN, MO, WI. Scattered, primarily in the Northern Lakes and northern portion of the Tallgrass Prairie. *CR*

Linaria vulgaris
PLANTAGINACEAE

butter-and-eggs, yellow toadflax

Fields, gravel pits, vacant lots, roadsides, railroads, disturbed sites. Summer, fall.

A mostly unbranched, glabrous, gray-green perennial to 2 ft. tall. Leaves alternate, dense along lower stem, simple, entire, linear, sessile, to 2½ in. long × ⅛ in. wide. Flowers irregular, to 1 in. long, oriented upward, tubular with 2 upper and 3 lower spreading pale yellow lobes (the "butter") and a darker orange-yellow protrusion in middle (the "eggs"), with long spur protruding downward, numerous in terminal raceme to 6 in. long at top of plant. Fruit a capsule. IA, IL, IN, MI, MN, MO, OH, WI. Introduced from Eurasia. Widespread, with fewer occurrences in MO. *SN*

Heteranthera dubia
PONTEDERIACEAE

water stargrass

Marshes, sloughs, ponds, lakes, rivers, streams, springs. Summer, fall.

A sprawling emergent or submerged aquatic perennial with elongated stems. Leaves alternate, sessile, entire, to 5½ in. long × ¼ in. wide, with parallel veins but lacking distinct midvein. Flowers regular, to ¾ in. wide, with 6 spreading strap-like tepals, 3 upright inflated stamens, and inflated pistil, solitary at end of tubular stalk to 4 in. long; most frequent on stranded plants, but those on submerged plants lack tepals. Linguini-like leaves and flattish stem similar to those of flatstem pondweed (*Potamogeton zosteriformis*); leaves of latter have distinct midvein. IA, IL, IN, MI, MN, MO, OH, WI. Throughout, with most occurrences in the Northern Lakes and the Ozark Highlands. *SN*

Portulaca oleracea
PORTULACACEAE

common purslane

Fallow fields, gardens, pavement cracks, waste areas. Summer.

A prostrate and mat-forming branching annual with reddish stems, glabrous and succulent, 2–5 in. tall. Leaves alternate or subopposite, sessile, obovate with rounded tips, ¼–1½ in. long × ⅛–¾ in. wide. Flowers regular, about ¼ in. wide, typically 5-petaled, occurring 1 to few at stem tips. Common purslane occurs worldwide; it was thought to be introduced to the Americas, but archeological evidence indicates it was present during pre-Columbian times. It is edible, with high nutritional value. IA, IL, IN, MI, MN, MO, OH, WI. Abundant, probably in every county. *MAH*

Caltha palustris
RANUNCULACEAE

marsh marigold

Fens, spring runs, seepage swamps, marshes. Spring.

An erect to sprawling perennial, branched above, glabrous, 1–2 ft. tall. Leaves alternate, petiolate, broadly rounded to kidney-shaped with a cordate base, shallowly toothed, 2–6 in. long and about as wide. Flowers regular, 1–2 in. wide, with 5 or 6 petal-like sepals in branched clusters atop the stem. Marsh marigold is not to be confused with marigolds of the aster family. It is more closely related to buttercups (*Ranunculus*). **IA, IL, IN, MI, MN, MO, OH, WI. Occurs primarily within the Northern Lakes but extends farther south, mostly into territory that has been glaciated.** *SN*

Ficaria verna
RANUNCULACEAE

lesser celandine

Moist upland and lowland forests. Spring.

An erect to reclining perennial, creeping, glabrous, 4–12 in. long, colonial. Leaves opposite (lower plant) or alternate (upper), petiolate, cordate to semicircular with wavy-toothed margins, some with axillary bulblets, 1–2 in. long and about as wide. Flowers regular, about 1 in. wide, with 8–12 shiny, narrowly elliptical petals, occurring singly from stem tips or leaf axils. This highly invasive spring ephemeral spreads extensively, especially along watercourses, its seeds, bulblets, and tubers dispersing to form new colonies. It is not related to greater celandine (*Chelidonium majus*), another invasive non-native species. **IL, IN, MI, MO, OH, WI. Introduced from Eurasia. Occurrences widely scattered but rapidly spreading.** *MAH*

Ranunculus abortivus

RANUNCULACEAE

small-flowered buttercup

Mesic forests, agricultural fields, disturbed sites. Spring.

An erect annual or biennial, simple or branched, mostly glabrous, 8–18 in. tall. Basal leaves petiolate, kidney-shaped to round, some divided, with scalloped margins, ½–2 in. long and about as wide, stem leaves smaller, sessile, deeply divided into 3 or more linear-lanceolate segments. Flowers regular, about ¼ in. wide, with 5 triangular petals, sepals about length of petals and glabrous, occurring singly at branch tips. Fruit a shiny achene. The lower-Midwest-ranging rock buttercup (*R. micranthus*) and the more eastern Alleghany buttercup (*R. allegheniensis*) possess hairy stems, and hairy sepals, respectively. IA, IL, IN, MI, MN, MO, OH, WI. **Abundant**, likely in every county. *SN*

Ranunculus acris

RANUNCULACEAE

tall buttercup, common buttercup

Moist fields, clearings, thickets, streambanks, roadsides. Spring, summer, fall.

A perennial to 3½ ft. tall. Basal leaves to 5 in. long and wide, with deeply impressed veins, long-petiolate, deeply palmately 3- to 5-lobed, lobes lobed and coarsely toothed; stem leaves few, alternate, becoming smaller, narrower-lobed, and sessile above. Flowers regular, to 1¼ in. wide, with 5 spreading petals and numerous stamens surrounding numerous green pistils, 1 to several on long stalks in branched inflorescence. Fruit an achene with spreading beak, in dense, nearly spherical ⅓-in. cluster. IA, IL, IN, MI, MN, MO, OH, WI. **Introduced from Europe. Widespread in the Northern Lakes and eastern portion of the Eastern Forests, scattered elsewhere.** *SN*

Ranunculus fascicularis
RANUNCULACEAE

early buttercup

Dry prairies, glades, savannas, open woods, rock outcrops. Spring.

A perennial to 6 in. tall. Basal leaves long-petiolate, to 3 in. long × 2½ in. wide, pinnately compound with 3–5 leaflets, each leaflet with narrow lobes and occasionally few teeth; stem leaves few to absent, alternate, smaller, undivided or 3-parted, sessile or nearly so. Flowers regular, to 1 in. wide, with 5 spreading petals, 5 sepals, and numerous stamens surrounding numerous pistils, usually solitary (rarely few) at top of plant. Fruit an achene with ascending beak, in dense, nearly spherical ⅓ in. cluster. IA, IL, IN, MI, MN, MO, OH, WI. **Widespread in the Ozark Highlands and eastern portion of the Tallgrass Prairie, scattered elsewhere.** *SN*

Ranunculus flabellaris
RANUNCULACEAE

yellow water buttercup, yellow water crowfoot

Marshes, swamps, vernal pools, ponds, ditches. Spring, summer.

A glabrous, creeping, branching perennial to 3 ft. long, submerged part of year, stems emergent to 1 ft. above water. Leaves sessile to long-petiolate, alternate, variable, to 3 in. long × 3½ in. wide, deeply 3- to 5-lobed, submerged leaves with lobes with filiform to narrow segments, lobes fewer and narrower above. Flowers regular, to 1 in. wide, with 5 spreading-ascending petals and numerous stamens surrounding numerous pistils, solitary at ends of upper branches. Small yellow water crowfoot (*R. gmelinii*) has flowers to ½ in. wide and smaller leaves. IA, IL, IN, MI, MN, MO, OH, WI. **Scattered, with fewer occurrences in the Ozark Highlands.** *CR*

Ranunculus harveyi

RANUNCULACEAE

Harvey's buttercup

Acid rocky woods. Spring.

An erect perennial, branched, glabrous to sparsely hairy, 8–18 in. tall. Leaves alternate, basal, petiolate, kidney-shaped to round, some divided, ½–1¼ in. long and about as wide, margins scalloped, stem leaves sessile, usually deeply divided into 3 or more linear-lanceolate segments. Flowers regular, ¼–½ in. wide, petals 5, elliptic, longer than sepals, surrounding a head of several achenes with short slender beaks, occurring singly atop stems. Small-flowered buttercup (*R. abortivus*) is somewhat similar, but its flowers are clearly smaller with petals not longer than sepals. IL, IN, MO. Most occurrences are in the Ozark Highlands, quite rare in IL and IN. *ST*

Ranunculus hispidus

RANUNCULACEAE

hispid buttercup

Upland forests (var. *hispidus*); floodplain forests, open wetlands (vars. *caricetorum* and *nitidus*). Spring.

A hairy perennial, erect to tufted (var. *hispidus*) to stoloniferous (vars. *caricetorum* and *nitidus*), 6–30 in. tall. Leaves basal, alternate, petiolate, with 3 lobed or toothed leaflets, ovate in outline, 1–5 in. long and wide. Flowers regular, ¾ in. wide (vars. *hispidus* and *nitidus*) to 1¼ in. wide (var. *caricetorum*), styles elongated and slender, 5 petals, 5 sepals, the latter spreading (vars. *hispidus* and *caricetorum*) or reflexed (var. *nitidus*). The latter (syn. *R. septentrionalis*) is also often less hairy (pictured). The non-native creeping buttercup (*R. repens*) has short, triangular styles. IA, IL, IN, MI, MN, MO, OH, WI. Throughout. *MAH*

Ranunculus pensylvanicus
RANUNCULACEAE

Pennsylvania buttercup, bristly crowfoot

Marshes, wet fields, swamps, thickets, shores, ditches. Summer.

A bristly-hairy annual to 3 ft. tall. Basal leaves long-petiolate, to 6 in. long and wide, compound with 3 stalked leaflets (terminal stalk longest), each leaflet 3-lobed, these lobed or coarsely toothed; stem leaves alternate, smaller, short-petiolate. Flowers regular, to ¼ in. wide, with 5 spreading-ascending petals, 5 longer sepals, and numerous stamens and pistils, solitary on branches in upper part of plant. Fruit an achene with spreading beak. Macoun's buttercup (*R. macounii*) has larger flowers with short sepals. IA, IL, IN, MI, MN, OH, WI. Widespread in the Northern Lakes, scattered in northern portion of the Tallgrass Prairie and the Eastern Forests. *SN*

Ranunculus recurvatus
RANUNCULACEAE

hooked buttercup

Moist to wet forests, swamps, seeps, streambanks. Spring.

An erect perennial, hairy, 8–18 in. tall. Leaves basal and alternate on stem, all petiolate to becoming nearly sessile on upper stem, ovate in overall outline, with 3–5 deeply divided and toothed lobes, 1–3 in. long × 1–4 in. wide. Flowers regular, ¼–½ in. wide, with 5 small, pale, oblong petals slightly shorter than or equal in length to sepals. Fruit an achene with a recurved or hooked beak (style), visible even when immature. These and its small flowers are distinctive. IA, IL, IN, MI, MN, MO, OH, WI. Widespread and common except in the western Tallgrass Prairie. *MAH*

Ranunculus rhomboideus
RANUNCULACEAE

prairie buttercup

Prairies, open woods, rocky ridges, sandy banks. Spring.

A hairy perennial to 10 in. tall. Basal leaves long-petiolate, to 1½ in. long × ¾ in. wide, unlobed to 3-lobed with blunt teeth in upper half; stem leaves few, alternate, sessile, deeply lobed and appearing whorled with 3–5 entire linear segments. Flowers regular, to ½ in. wide, with 5–8 spreading petals and numerous stamens surrounding numerous pistils, solitary at ends of branches in upper part of plant. Fruit an achene with very short, spreading beak, in dense, nearly spherical ⅜-in.-wide cluster. IA, IL, MI, MN, WI. Widespread in transition zone between the Northern Lakes and the Tallgrass Prairie, scattered elsewhere in these regions. *KC*

Ranunculus sardous
RANUNCULACEAE

hairy buttercup

Pastures, meadows, roadsides. Spring, summer.

An erect to spreading annual, branched, hairy, 4–24 in. tall. Leaves alternate, petiolate, mostly basal, compound with 3 deeply lobed leaflets that are triangular-obovate, the terminal one stalked, the lower 2 mostly sessile, ½–1½ in. long and wide. Flowers regular, ½–¾ in. wide, petals 5, obovate and glossy, solitary on axillary branches. Fruit an achene, surface smooth or often with small, scattered bumps. This species is rapidly spreading, especially in pastures. Bulbous buttercup (*R. bulbosus*), likewise non-native, has a bulbous base and slightly larger flowers. IL, IN, MI, MO, OH. Introduced from Europe. Most Midwestern occurrences are in the southernmost counties. *MAH*

Ranunculus sceleratus
RANUNCULACEAE

cursed crowfoot

Marshes, seeps, shallow pools, mudflats, ditches. Spring, summer.

An erect annual or perennial, branched, stems somewhat fleshy and hollow, glabrous, 8–24 in. tall. Leaves basal and alternate on stems, petiolate, those of lower stem kidney-shaped and deeply 3-lobed, ½–2 in. long × 1–3 in. wide, the upper ones smaller, usually with 3 linear-oblong segments. Flowers regular, ¼–⅓ in. wide with 5 oblong-elliptic petals and 5 sepals ringing the base of an elongated, cylindrical fruiting head up to ½ in. tall. *Sceleratus* means "cursed," which according to *Gray's Manual of Botany* suggests a plant that grows "in vile places." IA, IL, IN, MI, MN, MO, OH, WI. **Widespread and mostly common, less so in our southern counties.** *SN*

Agrimonia gryposepala
ROSACEAE

common agrimony

Woodlands, thickets. Summer.

A mostly unbranched perennial with scattered glandular and non-glandular long hairs throughout, up to 4 ft. tall. Leaves alternate, petiolate with deeply toothed stipules, pinnately compound with 4–8 opposing pairs of lanceolate to ovate, toothed leaflets and a single terminal leaflet, the larger ones 2–4 in. long × 1–2 in. wide, alternating with smaller ones. Flowers regular, about ¼ in. wide, 5-petaled. Inflorescence with widely spreading branches beset with glands. Fruit ¼–⅓ in. long, with hooked bristles. In much the same range is roadside agrimony (*A. striata*); its stipules are entire. IA, IL, IN, MI, MN, MO, OH, WI. **Occurs mostly in the region's northern two-thirds.** *KC*

Agrimonia parviflora
ROSACEAE

swamp agrimony

Wet or moist thickets, fields, woodlands, clearings. Summer.

A mostly erect perennial with stems covered in dense glandular hairs, 2–4 ft. tall. Leaves alternate, petiolate with toothed stipules, pinnately compound with 10–20 opposing pairs of narrowly lanceolate, sharply toothed leaflets plus a terminal one, the larger ones 2–3 in. long × ½–1 in. wide and alternating with much smaller ones; undersurfaces dotted with glands. Flowers regular, ⅛–¼ in. wide, 5-petaled, on narrow, widely spreading branches. Fruit up to ¼ in. long, bristly. This species has a greater number of pairs of leaflets per leaf than our other agrimonies. IA, IL, IN, MI, MO, OH, WI. Mostly in the Midwest's southern half. Common. *MJH*

Agrimonia pubescens
ROSACEAE

downy agrimony

Moist to dry woodlands, edges, thickets. Summer.

A mostly unbranched perennial, densely short-hairy, 1–3 ft. tall. Leaves alternate, petiolate with toothed stipules, pinnately compound with 4–8 opposing pairs of lanceolate or ovate, coarsely toothed leaflets plus a single terminal one, the larger ones alternating with much smaller ones, the larger 2–4 in. long × ½–2 in. wide, undersurface hairy. Flowers regular, ⅛–¼ in. wide, regular, 5-petaled, on hairy but eglandular spreading branches. Fruits bristly. The main stalk of the inflorescence in the similar woodland agrimony (*A. rostellata*) is glandular with few if any hairs. IA, IL, IN, MI, MN, MO, OH, WI. Common except in the far northern counties. *DT*

Dasiphora fruticosa
ROSACEAE

shrubby cinquefoil

Fens, calcareous interdunal wetlands, marl flats, rocky shores. Summer, fall.

A branched shrub to 3 ft. tall with shredding bark. Leaves alternate, petiolate, pinnately compound with crowded oblong leaflets, to 2 in. long and wide. Flowers regular, to 1¼ in. wide, with 5 rounded spreading petals, numerous, at ends of branches. Produces flowers all summer long. Usually in intact natural areas, but numerous cultivars, including white- or pink-flowered forms, are often used in landscaping. Syns. *D. floribunda*, *Potentilla fruticosa*. IA, IL, IN, MI, MN, OH, WI. Mostly in the Northern Lakes with fewer occurrences in adjacent portions of the Tallgrass Prairie and Eastern Forests. *SN*

Geum aleppicum
ROSACEAE

yellow avens

Wet meadows, marshes, swamps, thickets, roadsides. Summer.

A glistening hairy perennial to 3½ ft. tall. Leaves pinnately compound with coarsely toothed leaflets of variable size and large leafy stipules at connection to stem, basal leaves long-petiolate, stem leaves alternate, short-petiolate to sessile, terminal leaflet abruptly largest to 4½ in. long × 3½ in. wide. Flowers regular with 5 petals, to ½ in. wide, few at ends of branches at top of plant. Fruiting heads obovate, to ¾ in. long; beaked fruit have hooked tips that stick in fur and clothing. Large-leaved avens (*G. macrophyllum*) is very similar with smaller, round fruiting heads. IA, IL, IN, MI, MN, MO, OH, WI. Widespread in the Northern Lakes, scattered in adjacent regions. *MJH*

Geum vernum

ROSACEAE

spring avens

Moist forests, seeps, woodland edges, old fields, trails. Spring.

A perennial to 2 ft. tall. Basal leaves simple and roundish or pinnately compound, long-petiolate, terminal leaflet largest to 4 in. long × 2 in. wide; stem leaves pinnately compound, with coarsely toothed leaflets of variable size and large leafy stipules at connection to stem, alternate, short-petiolate or sessile. Flowers regular with 5 petals, to ¼ in. wide, several in upper part of plant. Fruiting heads spherical, elevated on stalk above receptacle (see inset), to ⅓ in. wide; beaked fruit have hooked tips. IA, IL, IN, MI, MO, OH, WI. Widespread in the Ozark Highlands and Eastern Forests; common in southern portion of the Tallgrass Prairie and adjacent Northern Lakes. *SN*

Potentilla anserina

ROSACEAE

silverweed

Sand and gravel lake and pond shores, interdunal swales, wet sand prairies, roadsides. Summer, fall.

A hairy perennial to 8 in. tall, spreading by red runners. Leaves in basal rosettes, to 12 in. long × 4 in. wide, widest above middle, silvery-hairy beneath, pinnately compound with up to 11 coarsely toothed primary leaflets, much smaller secondary leaflets between. Flowers regular, with 5 petals, to ¾ in. wide, solitary on naked stems from nodes along creeping stems. Syn. *Argentina anserina*. Pennsylvania cinquefoil (*P. pensylvanica*) has alternate stem leaves and flowers in clusters. Bushy cinquefoil (*P. supina*) has green leaf undersides and lacks creeping stems. IA, IL, IN, MI, MN, OH, WI. Mostly in the Northern Lakes, scattered in the Tallgrass Prairie. *SN*

Potentilla argentea
ROSACEAE

silvery cinquefoil

Prairies, savannas, fields, trails, lawns, roadsides, railroads, disturbed sites. Spring, summer.

A hairy, somewhat sprawling perennial to 20 in. tall, usually much shorter. Leaves basal and alternate, to 1 in. long and wide, palmately lobed with 5 stalkless coarsely toothed leaflets, densely whitened with felt-like hair beneath, on petioles to 3 in. long, smaller and becoming short-petiolate up stem. Flowers regular, with 5 petals, to ⅓ in. wide, 1 to few blooming at a time in clusters at ends of branches in upper part of plant. IA, IL, IN, MI, MN, MO, OH, WI. **Introduced from Europe. Widespread in the Northern Lakes, scattered in the Tallgrass Prairie and Eastern Forests.** *SN*

Potentilla indica
ROSACEAE

mock strawberry, Indian strawberry

Fields, forest openings, gardens, lawns. Spring, summer.

A creeping, mat-forming perennial to 6 in. tall. Leaves basal, to 2½ in. long × 3 in. wide, composed of 3 short-stalked coarsely toothed leaflets with deeply impressed veins, on hairy petioles to 2½ in. long. Flowers regular, 5-petaled, to ¾ in. wide, solitary on long stalks from leaf axils and creeping stem. Fruit an achene, numerous on a nearly spherical, warty, red receptacle, to ½ in. wide (see inset). Syn. *Duchesnea indica*. Foliage and fruit look similar to strawberry (*Fragaria*), but these "strawberries" are dry and lacking in taste. IA, IL, IN, MI, MO, OH, WI. **Introduced from Asia. Scattered throughout, most frequent in the Eastern Forests.** *MAH*

Potentilla norvegica

ROSACEAE

ough cinquefoil, Norwegian cinquefoil

rairies, fields, pastures, shores, gardens, acant lots, roadsides, railroads. Spring, ummer, fall.

A hairy, upright annual or perennial to 2½ t. tall. Leaves basal and alternate, to 6 in. ong × 4 in. wide, pinnately compound with coarsely toothed leaflets (basal sometimes vith 5 leaflets), on petioles to 4 in. long, with onspicuous pair of sharply toothed leaf-like tipules at base, smaller and becoming hort-petiolate to sessile and sometimes sim- le above. Flowers regular with 5 petals, to ½ n. wide, in leaf axils and branched clusters t top of plant. Brook cinquefoil (*P. rivalis*) as flowers to ¼ in. wide and nearly entire tipules. IA, IL, IN, MI, MN, MO, OH, WI. Videspread. *MJH*

Potentilla recta

ROSACEAE

sulphur cinquefoil

Fallow fields, pastures, roadsides, railroads, waste areas. Spring, summer.

An erect perennial, branched above, hairy, 1–2½ ft. tall. Leaves alternate, sessile to pet-iolate, palmately compound, generally with 5–7 narrowly oblanceolate, coarsely toothed leaflets 1–4 in. long × ¼–1½ in. wide. Flowers regular, to ¾ in. wide with 5 heart-shaped petals, occurring several in a widely spreading, somewhat flat-topped inflorescence. The root word of *Potentilla*, *potens* ("powerful"), is a reference to the reported medicinal properties of the genus. IA, IL, IN, MI, MN, MO, OH, WI. Introduced from Europe. Abundant and widespread, probably in every county. *SN*

Potentilla simplex
ROSACEAE

common cinquefoil

Prairies, barrens, woodlands, old fields.
Spring, summer.

A perennial with stems erect early, then
arching, reclining, with stolons forking and
rooting at the tips, glabrous to hairy, 1–3 ft.
long. Leaves alternate, petiolate, palmately
divided into 5 narrowly oblanceolate to elliptic
and toothed leaflets ½–2 in. long × ¼–1 in.
wide. Flowers regular, about ½ in. wide, with 5
heart-shaped petals, occurring singly on long
stalks from leaf axils, lowest flower arising in
the axil of the second developed leaf on the
stolon. In dwarf cinquefoil (*P. canadensis*), the
lowest flower arises from the first developed
leaf. IA, IL, IN, MI, MN, MO, OH, WI. Common and widespread except in the northwestern counties. *CR*

Waldsteinia fragarioides
ROSACEAE

barren strawberry

Rocky ridgetops, steep north-facing slopes,
mesic forests, thickets. Spring.

A low-growing, mat-forming perennial, hairy,
4–6 in. tall. Leaves all basal, petiolate, with
3 semi-evergreen, wedge- or fan-shaped and
coarsely toothed leaflets 1–2 in. long and
about as wide. Flowers regular, ½–¾ in. wide,
with 5 obovate-elliptic petals and 5 lanceolate
sepals, occurring few on upright stalks 4–6
in. tall, arising separately from leaf clusters.
Although with foliage somewhat like that of
strawberry (*Fragaria*), it lacks the latter's juicy
red "berries." Syn. *Geum fragarioides*. IL, IN,
MI, MN, MO, OH, WI. Occurs mostly from
the upper Northern Lakes southeastward
into the Eastern Forests. *PS*

Cruciata pedemontana
RUBIACEAE

Piedmont bedstraw, foothill bedstraw

Lawns, pastures, fields, roadsides, railroads, waste areas. Spring, summer.

An ascending, weak, densely short-hairy annual with 4-angled stems to 1½ ft. tall. Leaves numerous, spreading, in whorls of 4, entire, to ½ in. long × ¼ in. wide, sessile. Flowers regular, to ¹⁄₁₆ in. wide with 4 spreading lobes, solitary or in short-stalked clusters from leaf axils. Syn. *Galium pedemontanum*. May be confused with bedstraws in the genus *Galium*, which have white, greenish-white, or purplish flowers. IL, IN, MI, MO, OH. Introduced from Eurasia and northern Africa. Primarily in the Ozark Highlands and southern portion of the Tallgrass Prairie, with fewer occurrences in the Eastern Forests and even fewer in the Northern Lakes. *SN*

Verbascum thapsus
SCROPHULARIACEAE

common mullein

Roadsides, railroads, waste areas. Summer.

An erect biennial, usually unbranched, 3–6 ft. tall. Leaves a basal rosette only during first year followed by a leafy flowering stem in year 2, petiolate to sessile or clasping, oblong to oblanceolate, entire or with shallow teeth, densely hairy and flannel-like, 4–12+ in. long × 1–5 in. wide. Flowers nearly regular, ¾–1¼ in. wide, short-tubular, 5-lobed, the lower 3 slightly longer, numerous in a spike-like terminal inflorescence 6–24 in. long. Individual flowers seemingly bloom randomly within the inflorescence. The standing dead stalks often persist throughout the winter. IA, IL, IN, MI, MN, MO, OH, WI. Introduced from Europe. Abundant and likely in every county. *MAH*

Physalis heterophylla
SOLANACEAE

clammy ground-cherry

Prairies, fields, open rocky woods, thickets, gravel bars, roadsides, waste areas. Spring, summer.

A soft, somewhat sticky-hairy, branching perennial to 2½ ft. tall. Leaves simple, alternate, wavy-margined and irregularly toothed, to 4 in. long × 3 in. wide, petiolate. Flowers regular, bell-shaped, dangling on short stalks from branch axils, corolla very shallowly 5-lobed, to ¾ in. wide, pale yellow with purplish-brown star-shaped center. Fruit a roundish ½-in.-wide yellow berry, enclosed in a 1½-in.-long papery sac formed by mature calyx. Downy ground-cherry (*P. pubescens*), annual, has flowers less than ½ in. wide and smaller papery sac surrounding berry. IA, IL, IN, MI, MN, MO, OH, WI. Widespread, scattered to absent in northernmost portion of Midwest. *MAH*

Physalis longifolia
SOLANACEAE

longleaf ground cherry

Disturbed moist woodlands, waste areas, roadsides, gardens. Summer.

An erect perennial, branched, glabrous or with short hairs, 1–2½ ft. tall. Leaves alternate, petiolate, lance-elliptic, mostly entire, 2–5 in. long × 1½–3 in. wide. Flowers regular, ½–¾ in. wide, funnel-shaped with 5 shallow lobes and a purplish center, occurring singly and dangling from upper leaf axils. Fruit a berry loosely enclosed by a papery calyx. Virginia ground cherry (*P. virginiana*) is similar, but its stems are uniformly hairy. The tomatillo used in salsa is another *Physalis* species. IA, IL, IN, MI, MN, MO, OH, WI. Widespread and common except in the upper Northern Lakes. *MAH*

Solanum rostratum
SOLANACEAE

buffalo bur

Pastures, railroads, waste areas. Summer.

An erect annual, stems branched, hairy and densely armed with yellow spines, 8–24 in. tall. Leaves alternate, petiolate, blades ovate-oblong, somewhat irregularly pinnately to bi-pinnately lobed, the margins wavy, superficially like a watermelon leaf but with scattered spines along veins, 1–4 in. long × ¾–3 in. wide. Flowers regular, about 1 in. wide, 5-petaled, united into a star-shaped disk, in terminal racemes. Fruit a berry enclosed by the spiny calyx tube. This species must be approached carefully due to its vicious spines. IA, IL, IN, MI, MN, MO, OH, WI. Occurs mostly in our western counties and possibly native there. *MJH*

Viola pubescens
VIOLACEAE

downy yellow violet

Forests, forested floodplains. Spring.

A hairy (glabrous or nearly so in var. *scabriuscula*; syn. *V. eriocarpa*) perennial to 14 in. tall. Leaves few, basal and alternate (3 or more stem leaves and 1 or more basal leaves in var. *scabriuscula*, fewer in typical variety), cordate, to 4 in. long × 3½ in. wide, petiolate, toothed, with pair of entire to shallowly toothed stipules to ¾ in. where petioles attach to stem. Flowers irregular, to ¾ in. wide, yellow with purplish veins, with 5 spreading to recurved petals, 1 short-spurred, lateral petals bearded at base, solitary on stalks from leaf axils. IA, IL, IN, MI, MN, MO, OH, WI. Widespread and common. *MAH*

Xyris torta
XYRIDACEAE

twisted yellow-eyed grass

Shores, bogs, swales, wet sand. Summer.

A glabrous perennial to 2½ ft. tall, base bulbous, stem twisted. Leaves basal, to 20 in. long × ¼ in. wide, ascending. Flowers regular, numerous in cylindrical terminal head to 1 in. long, 1 to few blooming at a time, to ¼ in. wide, with 3 spreading petals (see inset), sepals with short hairs length of midvein are hidden behind brown, hardened, scale-like bracts. Bog yellow-eyed grass (X. *difformis*) and northern yellow-eyed grass (X. *montana*) lack bulbous base; former, sepal midveins hairy in upper portion with bracts with green central band; latter, sepals usually glabrous and bracts lacking green central band. IA, IL, IN, MI, MN, MO, OH, WI. Scattered, most frequent near Great Lakes. *PD*

Tribulus terrestris
ZYGOPHYLLACEAE

puncture vine

Sandy cultivated and fallow fields, pastures, waste areas, unpaved roadbeds. Summer.

A prostrate annual, branched, hairy, to 3+ ft. long, mat-forming. Leaves opposite, one often larger than opposing one, petiolate, pinnately compound with 12–16 narrowly oblong leaflets ¼–½ in. long × ⅛–⅓ in. wide. Flowers regular, ½–¾ in. wide, with 5 oblong petals, solitary in leaf axils. Fruit a hard, spiny capsule ⅓–½ in. thick. Fruits of this plant are vicious, reported to be capable of puncturing bicycle tires and piercing shoe soles. IA, IL, IN, MI, MN, MO, OH, WI. Introduced from the Mediterranean region. Occurs throughout but sparsely, especially north and eastward. *MJH*

GREEN
FLOWERS

Manfreda virginica

AGAVACEAE

false aloe

Dry rocky woodlands, glades, barrens. Summer.

An erect perennial, unbranched, glabrous, the flowering stem 3–6 ft. tall. Leaves in a basal rosette, lanceolate to oblanceolate, succulent and thick, often spotted with purple, entire or with fine prickles, 3–12 in. long × 1–3 in. wide. Flowers regular, about 1 in. long, tubular with 6 short, greenish lobes from which protrudes a single white 3-lobed stigma and 6 large whitish anthers (see inset), on a bracteate spike of 25+ flowers. Very fragrant at night. Syn. *Agave virginica*. IL, IN, MO, OH. Found principally in the southern Ozark Highlands and Eastern Forests. *SN*

Amaranthus retroflexus

AMARANTHACEAE

redroot amaranth

Agricultural fields, farmyards, roadsides, pastures, gardens, waste areas. Summer.

An erect, unbranched annual, monoecious, stems hairy, 1–4 ft. tall. Leaves alternate, petiolate, entire with somewhat wavy margins, hairy beneath at least along veins, lanceolate to ovate, 1–5 in. long × ½–3 in. wide. Flowers lacking petals, about ⅛ in. long, with 5 greenish sepals, these blunt and outwardly curved. Inflorescence in compact terminal and axillary spikes. This and several other amaranths are agricultural pests in North America. IA, IL, IN, MI, MN, MO, OH, WI. Introduced from tropical America. Widespread and common throughout. *MJH*

Chenopodium album
AMARANTHACEAE

lamb's-quarters

Fields, open woods, gardens, roadsides, disturbed sites. Summer, fall.

A gray-green annual to 6 ft. tall. Leaves alternate, long-petiolate, irregularly coarsely toothed to shallowly lobed, to 5 in. long × 3 in. wide, mealy-whitened on undersides (and above on younger leaves). Flowers regular, small, lacking petals, with 5-parted calyx, mealy on outer surface, tightly packed in glomerules in dense, branching arrays. Slimleaf goosefoot (*C. pallescens*) and desert goosefoot (*C. pratericola*) are similar with narrower leaves. Summer-cypress (*Bassia scoparia*) has flowers solitary or few in clusters in axils of linear bracts and lacks mealy coating. IA, IL, IN, MI, MN, MO, OH, WI. Possibly introduced from Eurasia. Widespread. *DT*

Chenopodium capitatum
AMARANTHACEAE

strawberry blite

Shores, limestone ledges, dumps,construction sites, roadsides; in sand or gravel. Summer.

A glabrous annual to 3½ ft. tall. Leaves alternate, thin-textured, long-petiolate, triangular, irregularly coarsely toothed, to 4 in. long × 3½ in. wide, becoming short-petiolate, less toothed, and narrower above. Flowers regular, tiny, lacking petals, with 3-parted calyx, numerous and tightly packed in round clusters to ⅜ in. in upper leaf axils and spike-like array at top of plant. Ovary becoming brown and calyx expanding and becoming bright red in fruit, resulting in clusters reminiscent of strawberries. Syn. *Blitum capitatum*. IA, IL, IN, MI, MN, MO, OH, WI. Frequent in northern portion of the Northern Lakes, scattered to rare south and east. *PD*

Chenopodium simplex
AMARANTHACEAE

maple-leaved goosefoot

Woodlands, ravines, rocky slopes, savannas, cliff overhangs. Summer.

An erect or ascending annual, branched, glabrous, 1–4 ft. tall. Leaves alternate, short-petiolate, triangular-ovate with widely spaced broad teeth, thin, 1½–6 in. long × 1–3½ in. wide. Flowers about ⅛ in. wide, petals lacking, sepals 5, greenish-white, in dense clusters at stem tips. Seeds black, relatively large, positioned horizontally within fruit. The seeds are thought to have been utilized as food by Native American peoples; in some of the large rock overhangs they formerly occupied, living plants can still be found. IA, IL, IN, MI, MN, MO, OH, WI. **Most common in the upper Midwest.** *PR*

Cycloloma atriplicifolium
AMARANTHACEAE

winged pigweed, tumbleweed

Dunes, sandy river bars and barrens, railroads. Summer.

An erect annual, rounded and multi-branched, hairy when young becoming nearly glabrous at maturity, 4–24 in. tall. Leaves alternate, sessile to short-petiolate, lanceolate to oblong, sharply toothed with wavy margins, 1–3 in. long × ¼–1 in. wide. Flowers tiny, petals lacking, sepals 5, greenish. Fruits encircled by a wavy-margined, membranous wing. In fall the entire aboveground plant breaks away and becomes a purplish-red tumbleweed. However, the species most referred to as tumbleweed is *Kali tragus*, another member of the family. IA, IL, IN, MI, MN, MO, OH, WI. **For us probably native only in our western counties.** *PR*

Kali tragus
AMARANTHACEAE
Russian-thistle
Sandy fields, dunes, roadsides, railroads. Summer, fall.

A branched, tumbleweed-like annual to 3 ft. tall. Stems and leaves often red, especially late in season. Leaves alternate, needle-like, ending in spine tip, to 2½ in. long below, smaller above, to ⅔ in. into inflorescence. Flowers regular, to ⅓ in. wide, lacking petals, with 5-parted whitish wing-like papery calyx, solitary in leaf axils, subtended by 2 leaf-like bracts, numerous on upper stems and branches. Syn. *Salsola tragus*. American bugseed (*Corispermum americanum*), hairy bugseed (*C. villosum*), and Pallas' bugseed (*C. pallasii*) lack spine tips to leaves and have awned, ovate bracts. IA, IL, IN, MI, MN, MO, OH, WI. **Introduced from Eurasia. Scattered, most frequent north.** *MJH*

Asclepias viridiflora
APOCYNACEAE
green milkweed
Barrens, glades, dry prairies. Spring, summer.

An erect or ascending, unbranched perennial, glabrous to sparsely hairy, 8–24 in. tall, with milky sap. Leaves opposite, short-petiolate, entire, linear-lanceolate, elliptic to oblong, 1–5 in. long × ¼–2½ in. wide. Flowers regular, about ¼ in. wide with 5 reflexed, green corolla lobes and 5 light green flattened and appressed hoods, horns absent, in short-stalked axillary and compact umbels of up to 80 flowers. Fruit a smooth follicle with flat seeds each bearing a tuft of long hairs. IA, IL, IN, MI, MN, MO, OH, WI. **Most common in the Tallgrass Prairie and Ozark Highlands.** *ST*

Asclepias viridis
APOCYNACEAE

green-flowered milkweed, green antelope-horn

Barrens, glades, prairies. Spring, summer.

A mostly glabrous perennial, erect but often sprawling with multiple stems, 1–2 ft. tall, with milky sap. Leaves alternate, sessile, entire, oblong, lanceolate to elliptic-lanceolate, somewhat fleshy, 2–5 in. long × ½–2½ in. wide. Flowers regular, up to 1 in. wide with corolla lobes upright, the hoods purple, club-shaped, with white tips, horns greatly reduced and appearing absent, up to 20 in terminal and axillary umbels. Fruit a thick, ridged follicle with flat seeds each bearing a tuft of long hairs. This milkweed is atypical as its corolla lobes are fully upright. IL, IN, MO, OH. Mostly found in MO and southwestern IL. *SN*

Arisaema dracontium
ARACEAE

green dragon

Floodplain forests, mesic forests. Spring.

An erect, unbranched perennial, glabrous, monoecious, 1–3 ft. tall. Leaf usually 1, from base of plant on a long petiole up to 3 ft. long, palmately divided into 7–13 lobes, the latter elliptic to oblanceolate, 4–8 in. long × 1–4 in. wide, positioned somewhat horizontally in a semicircle. Flowers tiny, lacking petals, located at base of long tail-like spadix that protrudes 2–6 in. beyond mouth of tubular spathe atop stalk from plant base. Fruit a brilliant red (and poisonous) berry, many in a cluster (see inset). Both common name and specific epithet allude to a fanciful resemblance of the spathe and spadix to a mythical dragon. IA, IL, IN, MI, MN, MO, OH, WI. Widespread and common, rare northward. *MJH/MAH*

Arisaema triphyllum
ARACEAE

Jack-in-the-pulpit

Rich mesic forests, thickets. Spring.

An erect, unbranched perennial, glabrous, 1–2+ ft. tall. Leaves 1–2, from base of plant on long petiole to 2+ ft. tall, divided into 3 elliptic-ovate leaflets 3–8 in. long × 1–3 in. wide. Flowers tiny and rarely seen, petal-less, either male or female (or both) depending on year, borne at base of a 2- to 3-in.-long club-shaped spadix located within a vase-shaped hooded spathe occurring singly atop a stalk from plant base. Fruit a brilliant red berry, many in a cluster. Although eaten by some wildlife, the fruit and other parts of plant are not considered safe for human consumption. IA, IL, IN, MI, MN, MO, OH, WI. **Common throughout, perhaps in every county.** *SN*

Peltandra virginica
ARACEAE

arrow arum

Swamps, shaded pond, lake, and river margins. Spring, summer.

An erect perennial, unbranched, glabrous, 1–3 ft. tall. Leaves basal on long stalks, broadly sagittate, pinnately veined, blade 8–20 in. long × 2–7 in. wide. Flowers tiny, separate male and female located at base of long spadix partially enclosed in a narrow tubular spathe, the latter occurring singly from plant base. Arrow arum's tropical-looking leaves resemble those of certain *Sagittaria* species, but the latter are palmately veined. IA, IL, IN, MI, MN, MO, OH, WI. **Widely scattered but largely absent in our region to the north and west of IL.** *NP*

Hydrocotyle americana
ARALIACEAE

American water-pennywort

Marshes, springs, swamps, forest streambanks, floodplains. Spring, summer.

A glabrous, creeping, colonial perennial, to 10 in. tall, rooting at nodes. Leaves alternate and basal, to 2 in. long and wide, simple, scallop-margined, kidney-shaped, with petiole attaching at base in notch. Flowers regular, to ⅛ in. wide, with 5 spreading petals, in tight, nearly stalkless clusters in leaf axils. Manyflower marsh-pennywort (*H. umbellata*), mainly in southern MI and northern IN, has peltate leaves. Floating pennywort (*H. ranunculoides*), with long-stalked umbels from leaf axils, is very scattered in the Midwest but seems to be increasing. **IN, MI, MN, OH, WI. Mostly in the eastern portion of the Northern Lakes, with few occurrences elsewhere.** *JT*

Panax quinquefolius
ARALIACEAE

American ginseng

Rich, moist upland forests, ravines. Summer.

An erect perennial, glabrous, 4–18 in. tall. Leaves whorled atop stem, usually 3 or 4, palmately compound with 5 stalked leaflets 2–5 in. long × 1–2½ in. wide, these obovate with toothed margins,. Flowers regular, about ⅛ in. wide, 5-petaled, in a small umbel of 5–15+ flowers centered atop stem (see inset). Fruit a red berry. Ginseng is a well-known herbal plant, both cultivated and, increasingly, wild-collected. Most of the roots are shipped to China, where it is used as a supposed cure-all for various ailments. **IA, IL, IN, MI, MN, MO, OH, WI. Throughout the Midwest except in far northern counties.** *MAH/CR*

Asparagus officinalis
ASPARAGACEAE

wild asparagus, garden asparagus

Fields, pastures, fencerows, thickets, roadsides, railroads, waste areas. Spring.

An abundantly branched, glabrous, gray-green perennial to 7 ft. tall. True leaves alternate, scale-like, triangular, yellow to purple, appressed to stem. Soft, needle-like cladophylls (leaf-like branches) to 1 in. long abundant on branches in whorls of 4 or 5, giving feathery appearance. Flowers regular, greenish-white to greenish-yellow, bell-shaped with 6 straight to recurved tepals, to ⅓ in. long, stalked in abundant clusters of 1–3 in upper part of plant. Fruit a red berry. Young shoots are edible. IA, IL, IN, MI, MN, MO, OH, WI. Introduced from the Old World. Widespread. *MAH*

Ambrosia artemisiifolia
ASTERACEAE

common ragweed

Disturbed fields, agricultural land, waste areas, roadsides. Summer.

A variously hairy, branching annual, monoecious, 1–3 ft. tall. Lower leaves opposite, upper ones commonly alternate, petiolate, deeply dissected into narrow lobes, 2–4 in. long × 1–2 in. wide. Flowers in composite heads to ⅛ in. wide, the male flower heads numerous in spikes 1–6 in. long with a few separate female flowers at the base. Its seeds are beneficial to wildlife as a food source. Pollen of all ragweed species is highly allergenic and a major cause of hay fever. IA, IL, IN, MI, MN, MO, OH, WI. Common throughout. *MAH*

Ambrosia bidentata
ASTERACEAE

lanceleaf ragweed

Prairies, fields, roadsides, waste areas. Summer, fall.

A bristly-hairy annual to 3½ ft. tall, monoecious. Leaves opposite below, alternate above, lanceolate, unlobed or with a pair (rarely 3 or 4) of small lobes at base, sessile, to 2¾ in. long × ½ in. wide. Flowers tiny, in yellow to green composite heads, male heads in single dense terminal spike, female heads in small axillary spikes or solitary. Distinguished from other ragweed species by primarily alternate and unlobed to few lobed leaves, and by single dense terminal spike of male flower heads. IA, IL, IN, MN, MO, OH. **Common in the Ozark Highlands and southern portion of the Tallgrass Prairie, uncommon in the Eastern Forests.** *PR*

Ambrosia trifida
ASTERACEAE

giant ragweed, horseweed

Bottomland forests, fields, ditches, disturbed sites. Summer, fall.

An annual with hairy stems, commonly branched above, monoecious, 4–8 ft. tall. Leaves opposite, petiolate, typically divided into 3 large lobes, some with an additional pair, those of smaller shaded plants sometimes unlobed, rough to the touch due to short stiff hairs, up to 12 in. long × 4–8 in. wide. Flowers in composite heads ⅛–¼ in. wide, male flowers numerous on 3- to 6-in.-long terminal spikes, female flowers in leaf axils. In our southern region its roots may be parasitized by river broomrape (*Orobanche riparia*), an attractive though rare wildflower. IA, IL, IN, MI, MN, MO, OH, WI. **Abundant throughout.** *SN*

Artemisia campestris
ASTERACEAE

wild wormwood, field sagewort

Prairies, glades, fields, savannas, dunes, gravel bars, railroads. Summer, fall.

A biennial to 3½ ft. tall. Basal leaves to 4 in. long × 3 in. wide, petiolate, deeply divided 1–3 times into linear lobes. Stem leaves mostly in lower half, alternate, smaller and sometimes unlobed above. Flowers in composite heads; no ray flowers, disk flowers inconspicuous, in numerous small drooping heads. Introduced sweet wormwood (*A. annua*) and biennial wormwood (*A. biennis*) have toothed leaf segments; the latter has sessile, upright flower heads. Mugwort (*A. vulgaris*), introduced, has dark green leaves whitened beneath, with broader segments. IA, IL, IN, MI, MN, MO, OH, WI. Frequent in the Northern Lakes and adjacent Tallgrass Prairie, scattered southward. *SN*

Cyclachaena xanthiifolia
ASTERACEAE

big marsh-elder, giant sumpweed

Moist fields, shores, vacant lots, barnyards, roadsides, disturbed sites. Summer, fall.

A colonial annual to 6 ft. tall. Leaves opposite below, alternate in inflorescence, rough-hairy, long-petiolate, to 7½ in. long × 5½ in. wide, 3-lobed and coarsely toothed, becoming smaller, narrower, and unlobed above. Flowers in composite heads; ray flowers absent, disk flowers tiny, in greenish-white downward-facing heads to ⅛ in., heads numerous and dense in sticky-hairy branching arrays at top of plant and from upper leaf axils. Behavior similar to giant ragweed (*Ambrosia trifida*); leaves similar to cocklebur (*Xanthium strumarium*). Syn. *Iva xanthiifolia*. IA, IL, IN, MI, MN, MO, OH, WI. Scattered in our northwest, becoming rare east and south. *PD*

Erechtites hieraciifolius
ASTERACEAE

fireweed, pilewort

Moist to dry woodlands, barrens, wetland edges, ditches, disturbed sites. Summer, fall.

An erect annual, unbranched, glabrous to hairy, 2–7 ft. tall. Leaves alternate, mostly sessile, elliptic to lanceolate with conspicuous marginal teeth, 3–7 in. long × 2–3 in. wide. Flowers in composite heads about 1 in. long × ¼ in. wide, urn-shaped, ray flowers absent, disk flowers barely evident except for the protruding styles, in a somewhat flat-topped terminal inflorescence. This plant shows a positive response following fire, hence one of its common names. Syn. *Senecio hieraciifolius*. IA, IL, IN, MI, MN, MO, OH, WI. Mostly throughout and common. *MAH*

Iva annua
ASTERACEAE

rough marsh elder, sumpweed

Agricultural fields, creek banks, sloughs, waste areas. Summer, fall.

An erect annual, branching, stems rough-hairy, 1–3 ft. tall. Leaves opposite (some upper ones alternate), lanceolate to ovate, petiolate, glabrous to somewhat roughened with toothed margins, 1–6 in. long × 1–4 in. wide. Flowers in composite heads about ¼ in. wide, ray flowers lacking, disk flowers 10–20, inner male, outer female, in drooping heads arranged on spike-like racemes in leaf axils and atop stem. Archeological evidence has shown that it was domesticated here over 4000 years ago; such cultivated forms no longer exist. IA, IL, IN, MI, MO, OH, WI. Most common in MO, southern IL, and southwestern IN. *MAH*

Xanthium strumarium
ASTERACEAE

cocklebur

Sandbars, mudflats, cultivated fields, waste areas. Summer.

An erect, branched or unbranched monoecious annual with short white hairs, 2–5 ft. tall. Leaves alternate, long-petiolate, broadly ovate, kidney-shaped or triangular with flat to cordate base and 3–5 lobes, 2–6 in. long × 1–5 in. wide. Flowers in composite heads about ½ in. wide, ray flowers lacking, separate male and female heads in terminal and axillary clusters, the male heads uppermost with several minute flowers, female heads 2–4 per spike. Fruit an ellipsoid bur about 1 in. long and bearing hooked prickles. IA, IL, IN, MI, MN, MO, OH, WI. Common, mostly throughout. *MJH*

Cannabis sativa
CANNABACEAE

hemp, marijuana

Fields, vacant lots, ditches, roadsides. Summer, fall.

An annual to 9 ft. tall, dioecious (male pictured). Leaves opposite and long-petiolate below, alternate and nearly sessile above, to 10 in. long and wide, palmately compound with 3–9 coarsely toothed and deeply veined narrowly elliptic leaflets longer in middle and progressively shorter radially. Flowers regular, tiny, petal-less, in axillary and terminal panicles to 1 ft. long on male plants and glandular axillary spikes to 1 in. on female plants. Rough cinquefoil (*Potentilla recta*) has similar foliage, but leaflets are broader and less pointed. IA, IL, IN, MI, MN, MO, OH, WI. Introduced from Asia. Widespread throughout much of the Tallgrass Prairie, scattered in the Northern Lakes and Eastern Forests. *MJH*

Humulus lupulus
CANNABACEAE

common hops

Forest edges, fencerows, roadsides. Summer.

A perennial climbing vine, 30+ ft. long, dioecious, with rough, prickly hairs. Leaves opposite, petiole shorter than blade, unlobed to broadly ovate and mostly 3-lobed, cordate, toothed, surfaces rough, 1–6 in. long and about as wide. Flowers about ¼ in. wide, petal-less, female flowers (see inset) becoming enclosed by scales bearing yellow aromatic glands, forming a "cone" 1–1½ in. long. The 3 varieties known in the Midwest (1 introduced from Europe) are separated by leaf characteristics. The introduced Japanese hops (*H. japonicus*) has 5- to 9-lobed leaf blades with petioles longer than blades. IA, IL, IN, MI, MN, MO, OH, WI. **Scattered throughout and locally common.** *SN/MAH*

Paronychia canadensis
CARYOPHYLLACEAE

smooth forked nailwort, forked chickweed

Dry upland forests, bluffs, sand savannas, glades. Summer, fall.

An erect, highly branched annual with forking stems, glabrous, 1–12 in. tall. Leaves opposite, entire, sessile, elliptic to obovate, with scattered glandular dots, ¼–1¼ in. long × ⅛–⅓ in. wide. Flowers regular, minute and lacking petals, sepals 5, 3-nerved, solitary in axils of leaves. Fruit is a capsule. Being greenish and tiny, the flowers are easily overlooked. Most passersby that see this delicate little plant think it is either yet to bloom or past blooming. IA, IL, IN, MI, MN, MO, OH, WI. **Widespread but local to absent in the upper Midwest.** *KB*

Paronychia fastigiata
CARYOPHYLLACEAE

hairy nailwort, forked chickweed

Dry open woods, rocky areas, waste areas. Spring, summer, fall.

An easily overlooked branching annual to 1 ft. tall with finely hairy stem. Leaves opposite, entire, simple, sessile, oval, to 1 in. long × ¼ in. wide on main stem (smaller on branches), with pair of pointed, papery stipules to ⅓ in. long where leaves connect to stem. Flowers regular, numerous, inconspicuous, minuscule (shorter than stipules), lacking petals, sessile, solitary in branch axils. Grows in acidic substrates. IA, IL, IN, MI, MN, MO, OH, WI. **Widespread in the Ozark Highlands and surrounding areas, scattered elsewhere, primarily south.** *JB*

Scleranthus annuus
CARYOPHYLLACEAE

knawel, German knotweed

Glades, sandy fields, gardens, roadsides, railroads, waste areas. Spring, summer, fall.

An easily overlooked branching, ascending annual to 5 in. tall. Leaves opposite, entire, simple, sessile, to 1 in. long, needle-like, sometimes crowded and appearing whorled. Flowers regular, inconspicuous, to ⅛ in. wide, with 5 ascending (erect in fruit) persistent sepals that are green with whitish margins, lacking petals, very numerous and sessile in dense clusters at ends of branches, or solitary in leaf axils, surprisingly attractive when viewed with a hand lens. IL, IN, MI, MN, MO, OH, WI. **Introduced from Eurasia. Primarily in the Northern Lakes and Ozark Highlands but scattered throughout.** *SN*

Euonymus obovatus
CELASTRACEAE

running strawberry bush

Moist forests, rocky woodlands, swamps.
Spring, summer.

A perennial, sprawling shrub to 1½ ft. tall.
Leaves opposite, finely toothed, to 2½ in.
long × 1¾ in. wide, broadest above middle,
short-petiolate. Flowers regular, inconspicuous, greenish-yellow to purplish, with 5
spreading petals, to ⅓ in. wide, up to 4 in
stalked clusters from leaf axils. Fruit an
orange to pink warty capsule to ¾ in. wide that
opens to expose fleshy red arils (see inset).
Common name comes from slight resemblance of fruit to a strawberry. IL, IN, MI, MO,
OH, WI. Widespread in the Eastern Forests,
southeastern portion of the Northern Lakes,
and extreme eastern Tallgrass Prairie, common in the Ozark Highlands. *MAH/SN*

Medeola virginiana
CONVALLARIACEAE

Indian cucumber-root

Mesic, usually acid forests. Spring, summer.

An erect perennial, unbranched, wiry, stems
glabrous but woolly when young, colonial,
8–24 in. tall. Leaves in a single whorl (or 2
including whorl of leaf-like, ovate, involucral
bracts found in flowering/fruiting individuals), 5–9, sessile, oblanceolate to oblong, 3–6
in. long × ½–2 in. wide. Flowers regular, ½–¾
in. wide, with 6 lanceolate, recurved tepals, 6
stamens, and 3 widely spreading styles. Inflorescence of 3–5 flowers that droop from upper
whorl. Sterile plants look much like large
whorled pogonia (*Isotria verticillata*), which
has hollow stems. IL, IN, MI, MO, OH, WI.
Occurs mostly in Midwest's eastern half. *SN*

Polygonatum biflorum
CONVALLARIACEAE

smooth Solomon's seal

Mesic and dry-mesic forests, thickets, ravines. Spring, summer.

An erect to arching perennial, unbranched, glabrous and often glaucous, 1–2 ft. tall but some up to 5 ft. Leaves alternate, somewhat 2-ranked, sessile to partially clasping, oval, 2–6 in. long × 1–3 in. wide. Flowers regular, ½–¾ in. long, tubular, with 6 short lobes barely flaring, occurring mostly in pairs but up to 5, dangling from leaf axils. Fruit a berry. This species is highly variable in size and flower number. Hairy Solomon's seal (*P. pubescens*), common in the Northern Lakes and Eastern Forests, is similar, but its leaves bear hairy veins beneath. IA, IL, IN, MI, MN, MO, OH, WI. **Common throughout except in upper Northern Lakes.** *MJH*

Dioscorea villosa
DIOSCOREACEAE

wild yam

Upland forests, lowlands, thickets. Spring, summer.

A perennial twining and climbing vine, dioecious (male plant pictured), glabrous or sparsely hairy, up to 20 ft. long. Leaves alternate or lowest ones whorled, petiolate, cordate with conspicuous curving parallel veins, 2–5 in. long and almost as wide. Flowers regular, about ⅛ in. wide, with 6 petal-like segments, occurring in dangling axillary clusters. Fruit a rounded, 3-winged capsule about 1 in. wide. The invasive Chinese yam (*D. polystachya*), mostly in our southern counties, has miniature potato-like "tubers" in the leaf axils (see inset). IA, IL, IN, MI, MN, MO, OH, WI. **Common except in the far north and west.** *SN*

Acalypha gracilens
EUPHORBIACEAE

slender three-seeded mercury

Woodlands, trails, roadsides, disturbed sites. Summer, fall.

A simple or branched monoecious annual, mostly hairy, to 15 in. tall. Leaves narrow, entire or with shallow rounded teeth, 1–2½ in. long on petioles about ¼ length of the leaves. Flowers minute, petals lacking, located in leaf axils, the associated bracts possess 9–11 broadly triangular teeth with stalked glands. Fruit a 3-seeded capsule. One-seeded mercury (*A. monococca*) is similar but possesses a one-seeded capsule; it occurs primarily in the Ozark Highlands. IA, IL, IN, MI, MO, OH, WI. Scattered, mostly in the lower Midwest. *DT*

Acalypha ostryifolia
EUPHORBIACEAE

rough-podded copperleaf, hornbeam three-seeded mercury

Fields, roadsides, agricultural land, waste areas. Summer.

An erect simple or branched monoecious annual with finely hairy stems, up to 20 in. tall. Leaves alternate, toothed, broadly oval to cordate, sparsely hairy, 2–4 in. long × 1–2 in. wide, on petioles ½ or more the length of the leaves. Flowers minute, lacking petals, emerging from bracts with numerous deeply cut lobes located in terminal spikes. Fruit a 3-seeded prickly capsule. This species is quite weedy. IA, IL, IN, MO, OH. Common, principally in southern third of the Midwest. *MAH*

Acalypha rhomboidea
EUPHORBIACEAE

common three-seeded mercury

Moist woodlands, streambanks, trails, disturbed sites. Summer.

An erect, branched monoecious annual, mostly smooth or with scattered spreading hairs, up to 20 in. tall. Leaves mostly lanceolate, or ovate to somewhat diamond-shaped, 2–4 in. long × 1½ in. wide, on petioles ½ or more length of the blades. Flowers minute, petals lacking, located in leaf axils, the associated bracts possess 7–9 triangular teeth. Fruit a 3-seeded capsule (see inset). Syn. *A. virginica* var. *rhomboidea*. The similar Deam's mercury (*A. deamii*) has 2-seeded capsules. IA, IL, IN, MI, MN, MO, OH, WI. **Throughout the Midwest except for the far northern counties.** *MAH*

Euphorbia commutata
EUPHORBIACEAE

wood spurge

Rocky woodlands, slopes, gravel stream terraces. Spring, summer.

A biennial or perennial, monoecious, erect, branched, glabrous, with milky sap, 4–12 in. tall. Leaves alternate, short-petiolate or sessile, oblanceolate, entire, ¼–1½ in. long × ⅛–⅓ in. wide. Leaves of inflorescence opposite, sessile, ovate, some fused at base. Flowers minuscule, without sepals and petals, the cup-like cyathium containing several male flowers (each consisting of a single stamen) and a single separate pistil (female flower), these surrounded by 4 yellowish, 2-pronged glands. Fruit a smooth 3-lobed capsule. The similar blunt-leaved spurge (*E. spathulata*) has toothed leaves and a warty capsule. IA, IL, IN, MI, MO, OH, WI. **Mostly in the Eastern Forests and Ozark Highlands.** *SN*

Euphorbia cyathophora
EUPHORBIACEAE

paintedleaf, wild poinsettia, fire-on-the-mountain

Prairies, glades, open rocky woods, gravel bars, roadsides, railroads, waste areas. Summer, fall.

A monoecious annual to 3 ft. tall. Leaves alternate below, opposite above and on flowering stems, variable, unlobed or lobed and irregularly toothed, short-petiolate, to 4 in. long × 1½ in. wide, containing milky sap; green bracts with red bases. Flowers inconspicuous, greenish, lacking petals and petal-like appendages, in clusters at top of plant. Fruit a stalked 3-parted capsule. The colorful bracts provide most of the visual interest. Related to the ornamental poinsettia (*E. pulcherrima*). IA, IL, IN, MN, MO, OH, WI. Widespread in the Ozark Highlands and frequent in the Tallgrass Prairie, with few occurrences elsewhere. *PD*

Euphorbia davidii
EUPHORBIACEAE

David's spurge

Creek banks, fallow fields, disturbed sites. Summer, fall.

An erect annual, mostly unbranched, hairy, monoecious, 6–24 in. tall. Leaves opposite, petiolate, elliptic to lanceolate, toothed, bearing stiff hairs with widened bases beneath, ½–4 in. long × ¼–1½ in. wide. Flowers minuscule, lacking sepals and petals, the cup-like cyathium containing several male flowers (each consisting of a single stamen) and a single separate pistil (female flower) surrounded by fringed lobes and a 2-lipped gland. Bracts of inflorescence commonly pale at base. Toothed spurge (*E. dentata*) almost identical but leaf undersurfaces bear only a few slender hairs, these without widened bases. IA, IL, IN, MI, MN, MO, OH, WI. Widely scattered, likely more common than reported. *MAH*

Phyllanthus caroliniensis
EUPHORBIACEAE

Carolina leaf-flower

Floodplain forests, flatwoods, mudflats, moist agricultural fields. Summer, fall.

An erect to ascending monoecious annual, horizontally branched, mostly glabrous, 3–10 in. tall. Leaves alternate, short-petiolate, entire, elliptical to oblong, 2-ranked, ¼–1 in. long × ⅛–½ in. wide. Flowers regular and minute, about ¹⁄₁₆ in. wide, petals absent, sepals 6, 1–3 in leaf axils. Carolina leaf-flower is a wide-ranging species, occurring as far south as northern South America. Its attractive foliage is usually all that is noticed, its flowers being tiny and somewhat hidden. IL, IN, MO, OH. **Occurs principally in the Midwest's southern third.** *MAH*

Frasera caroliniensis
GENTIANACEAE

American columbo

Dry rocky woodlands, glades, sand and gravel bluffs. Spring, summer.

A relatively short-lived monocarpic perennial, erect, unbranched, glabrous, 5–7 ft. tall. Leaves in a basal rosette and in whorls of 4 or 5 along stem, elliptic-oblanceolate, 8–15 in. long × 3–4 in. wide. Flowers regular, about 1 in. wide, short-tubular with 4 lobes, each purplish-speckled and adorned with a fringed green bump (gland) toward its base (see inset). Inflorescence a panicle atop stem and from upper nodes. Not unlike century plants (*Agave*) of the desert, a given plant may grow for several years before it flowers and subsequently dies. IL, IN, MI, MO, OH. **Occurs primarily in the Eastern Forests.** *ST/SN*

Halenia deflexa
GENTIANACEAE

spurred gentian, green gentian

Swamps, forests, thickets, bogs, fens. Summer.

An inconspicuous, square-stemmed annual to 2½ ft. tall. Leaves opposite, entire, glabrous, to 1½ in. long × ½ in. wide, sessile, with impressed parallel veins. Flowers regular, tubular, closed, yellow-green to pale pink or purple, to ½ in. long, tapering from base to tip and therefore somewhat triangular, with 4 fused petals, each petal with a ¼-in.-long spur at base, stalked in clusters at top of plant. Fruit a conical capsule that sticks out from mature flower. MI, MN, WI. **Restricted to northern portion of the Northern Lakes, where widespread.** *MJH*

Proserpinaca palustris
HALORAGACEAE

common mermaidweed

Ponds, streams, interdunal wetlands, marshes, wet shores, ditches, mudflats. Summer.

A glabrous, creeping to ascending perennial to 3 ft., often emergent, to 1 ft. tall. Leaves simple and sharp-toothed (above) to deeply lobed or compound with numerous pinnately arranged filiform segments (below, often submerged in water), alternate, sessile to short-petiolate, to 2 in. long × 1½ in. wide (narrower when simple). Flowers regular, inconspicuous, to ¼ in. wide, lacking petals, 3-sided and somewhat conical in fruit (see inset), sessile, 1–5 in axils of simple leaves. IA, IL, IN, MI, MO, OH, WI. **Scattered throughout, with greatest frequency in areas immediately adjacent to the Great Lakes.** *MJH/SN*

Lemna turionifera
LEMNACEAE

red duckweed

Wet places. Spring, summer, fall.

A glabrous, free-floating, colonial aquatic perennial (pictured, most of plants). Leaves to ⅙ in. long, nearly as wide, broadly oval, flat, usually in clusters of 3 or 4, with single root from underside of each blade, usually with at least some reddish-purple coloration on underside. Flowers regular, minuscule, petal-less, solitary in cup-like structure at base of blade. Numerous *Lemna* species are very similar. *Spirodela polyrhiza* (also pictured) and *Landoltia punctata* have larger leaves with red undersides and multiple roots per blade; leaves round in former, elongated in latter. *Wolffiella gladiata* is rootless with sword-shaped leaves. IA, IL, IN, MI, MN, MO, OH, WI. Widespread in the Northern Lakes, scattered elsewhere. *NP*

Ludwigia palustris
ONAGRACEAE

water-purslane, marsh-purslane

Marshes, swamps, seeps, lakes, rivers, ditches. Summer, fall.

A colony-forming glabrous and somewhat succulent perennial to 12 in. tall, sprawling when out of water but upright when submerged. Often reddish- or bronze-tinged. Leaves simple, opposite, short-petiolate, spatulate, entire, to 1½ in. long × ¾ in. wide. Flowers regular, inconspicuous, to ¼ in. long × ⅛ in. wide, lacking petals, with pointed 4-lobed calyx surrounding 4 short stamens and single style, solitary and sessile in leaf axils. Fruit a capsule. Easily overlooked. IA, IL, IN, MI, MN, MO, OH, WI. Widespread and common except in western IA, most of MN, and most of northern MO, where absent. *PR*

Ludwigia polycarpa
ONAGRACEAE

false loosestrife, water-purslane

Marshes, swamps, sloughs, wet prairies, ponds, rivers, ditches. Summer, fall.

A glabrous perennial to 3 ft. tall. Often reddish-tinged. Leaves abundant, simple, alternate, short-petiolate or sessile, broadest at middle and tapering to either end, entire (but rough-margined), to 4 in. long × ½ in. wide. Flowers regular, inconspicuous, to ¼ in. long and wide, lacking petals, with pointed 4-lobed green to pink or reddish-brown calyx surrounding 4 short stamens and single style, solitary and sessile or nearly so in mid and upper leaf axils. Fruit a capsule. Easily overlooked. IA, IL, IN, MI, MN, MO, OH, WI. Scattered throughout but mostly absent from the Ozark Highlands and northern portions of the Northern Lakes. *ST*

Coeloglossum viride
ORCHIDACEAE

long-bracted orchid

Mesic upland forests, forested dunes. Spring, summer.

An erect perennial, unbranched, glabrous, 8–18 in. tall. Leaves alternate, sessile, elliptic to oblanceolate, entire, 2–5 in. long × ¾–2½ in. wide. Flowers irregular, ⅓–½ in. long, with 3 petals, the lip petal oblong with 2 lateral lobes at tip, and 2 smaller petals as well as 3 petal-like sepals forming a hood above (see inset). A pouch-shaped spur occurs beneath. Inflorescence a tight, spike-like raceme with a leafy bract extending beyond each flower. This orchid is easily overlooked due to its green coloration. Syn. *Dactylorhiza viridis*. IA, IL, IN, MI, MN, MO, OH, WI. Occurs mostly in the upper Midwest. Uncommon. *CR/KC*

Corallorhiza trifida
ORCHIDACEAE

early coralroot, yellow coralroot

Moist forests, swamps, bogs. Spring.

A perennial with green to yellow stem to 1 ft. tall (usually much shorter). Leaves inconspicuous and scale-like, reduced to yellowish, alternate scales at base of plant. Flowers irregular, to ½ in. wide, with 3 yellow to green sepals and 2 similar-looking lateral petals (petals and sepals sometimes with purplish tips), and a white (sometimes purple-spotted at base) lip petal; flowers up to 20 in spike-like inflorescence. Fruit a capsule. Partially mycoheterotrophic, but has chlorophyll and obtains some nutrients through photosynthesis. IL, IN, MI, MN, OH, WI. **Almost entirely restricted to the Northern Lakes.** *PS*

Epipactis helleborine
ORCHIDACEAE

broad-leaved helleborine

Upland forests, dunes, thickets, gardens. Summer.

An erect perennial, unbranched, hairy, 1–1½ ft. tall. Leaves alternate, sheathing, elliptic-lanceolate, entire, 2–6 in. long × 1–3 in. wide. Flowers irregular, up to ¾ in. wide, with 3 petals, lowermost (the lip) cup-shaped at base and divided by a prominent constriction from a triangular-ovate apex, purplish to green, as are ovate-lanceolate petals and 3 ovate sepals (see inset). Inflorescence a spike-like raceme bearing leafy bracts. The only non-native orchid naturalized in the Midwest. IA, IL, IN, MI, MN, MO, OH, WI. **Introduced intentionally from Eurasia in the late 1800s. Concentrated mostly in MI and Lake Michigan border counties.** *MJH/PG*

Liparis loeselii
ORCHIDACEAE

Loesel's twayblade

Fens, sedge meadows, lake borders.
Spring, summer.

An erect perennial, unbranched, glabrous,
4–7 in. tall. Leaves 2, basal, elliptic-lanceolate,
shiny, 2–5 in. long × 1–2 in. wide. Flowers
irregular, about ½ in. long, the lip petal obovate to oblong, yellowish-green, translucent,
arched near middle, with 2 thread-like petals
and 3 linear-lanceolate sepals, 10–15 in a terminal raceme. Fruit a capsule longer than its
stalk. The Greek *liparos* ("fat," "oily") alludes
to the shiny, almost greasy look of the leaves.
IA, IL, IN, MI, MN, MO, OH, WI. Most common in the Northern Lakes but expanding
southward. *KB*

Malaxis unifolia
ORCHIDACEAE

green adder's mouth

Mesic to dry-mesic acidic upland forests,
sphagnum bogs. Spring, summer.

An erect perennial, unbranched, glabrous,
4–8 in. tall. Leaf 1, sheathing, ovate to
ovate-lanceolate, shiny, 1½–2½ in. long × 1–1½
in. wide. Flowers irregular, minuscule, about
1/16 in. long, lip petal cordate-ovate with 3 teeth
at apex, the lateral teeth longer, lateral petals
linear, sepals elliptic-lanceolate, 25–50 flowers
per inflorescence (see inset). The genus *Malaxis* has the smallest flowers of our Midwestern orchids. The lip apex of the similar white
adder's mouth (*M. monophyllos*) comes to a single, long-tapering point. IA, IL, IN, MI, MN,
MO, OH, WI. Mostly in the Northern Lakes,
local and rare southward. *PG/CR*

Platanthera aquilonis
ORCHIDACEAE

northern green orchid

Wet meadows, fens, calcareous interdunal wetlands, swamps, seeps, thickets, streambanks. Summer.

A glabrous perennial to 2 ft. tall. Leaves few, alternate, entire, sheathing stem, to 9 in. long × 1½ in. wide, smaller up stem and becoming bracts in inflorescence. Flowers irregular, to ⅜ in. wide, spurred, with upper sepal and lateral petals forming hood and lateral sepals and lip petal spreading, short-stalked, up to 40 in loose raceme. Tall northern bog orchid (*P. huronensis*) has fragrant whitish flowers and is taller with more flowers in raceme. IA, IL, IN, MI, MN, OH, WI. Scattered, mostly northern, in the Northern Lakes and the Tallgrass Prairie; extirpated from the Eastern Forests. *PD*

Platanthera flava
ORCHIDACEAE

tubercled orchid

Floodplain forests, swamps, vernal pools, sand prairies. Summer.

An erect perennial, unbranched, glabrous, 1–1½ ft. tall. Leaves alternate, lanceolate to elliptic, sheathing, 3–6 in. long × 1–1½ in. wide. Flowers irregular, about ¼ in. long, with 3 petals and 3 ovate sepals, lip petal lowermost, oblong or semi-orbicular to square with a tooth on each side at base and a small bump (tubercle) in middle, basal spur about ¼ in. long, other petals ovate. Inflorescence 15–40 flowers in tight spike-like raceme. Floral bracts longer than most flowers in var. *herbiola*, whereas only the lowest flowers are exceeded by bracts in the rare southern var. *flava*. IA, IL, IN, MI, MN, MO, OH, WI. Widely scattered, uncommon. *PR*

Platanthera hookeri
ORCHIDACEAE

Hooker's orchid

Sandy soil in forests and thickets, usually on dunes; swamps. Summer.

An easily overlooked glabrous perennial to 16 in. tall. Leaves 2, basal, entire, lying flat on ground or nearly so, broadly oval, to 5½ in. long × 4 in. wide. Flowers irregular, yellowish or green, to 1 in. long, long-spurred, with upper sepal and lateral petals forming hood, lateral sepals recurved back, and lip petal spreading horizontally and often curling upward, short-stalked, up to 25 in loose terminal raceme. IA, IL, IN, MI, MN, OH, WI. Scattered in the Northern Lakes and far north-central Tallgrass Prairie; extirpated from the Eastern Forests. *PS*

Platanthera lacera
ORCHIDACEAE

ragged fringed orchid, green fringed orchid

Sphagnum bogs, fens, sand prairies, flatwoods, old fields. Summer.

An erect perennial, unbranched, glabrous, 8–24 in. tall. Leaves alternate, oblong-lanceolate to elliptic, sheathing, 4–6 in. long × 1–3 in. wide. Flowers irregular, about ¾ in. long, with 3 petals and 3 ovate sepals, lip petal lowermost, with 3 deeply fringed, thread-like lobes and a basal spur about ½ in. long, upper petals linear-oblong and entire. Inflorescence to 8 in. long with 15–45 flowers in a terminal raceme. IA, IL, IN, MI, MN, MO, OH, WI. Widely scattered but absent in much of the western Tallgrass Prairie. *PR*

Platanthera obtusata
ORCHIDACEAE

blunt-leaved orchid

Conifer swamps, forests, bogs. Summer.

A glabrous perennial to 1 ft. tall. Leaf solitary (rarely 2), basal, entire, ascending, to 6 in. long × 2 in. wide, spatulate, broadest above middle and blunt-tipped. Flowers irregular, to ½ in. long × ¼ in. wide, long-spurred, with upper sepal and lateral petals loosely forming hood, lateral sepals recurved back, and long and narrow lip petal spreading downward, stalked, up to 20 in loose terminal raceme. Fruit an erect capsule. Pollinated by certain moths and mosquitoes. **MI, MN, WI. Restricted to northern portion of the Northern Lakes.** *PD*

Penthorum sedoides
PENTHORACEAE

ditch stonecrop

Floodplain forests, swamps, marshes, mudflats, ditches. Summer.

An erect perennial, branched above, mostly glabrous, 1–2 ft. tall. Leaves alternate, sessile or short-petiolate, lanceolate-elliptical, with small sharp teeth along margins, 2–4 in. long × ½–1 in. wide. Flowers regular, about ¼ in. wide, petals commonly lacking, sepals 5 and persistent, several arranged 1-sided along horizontal, curving branches. Fruit a 5-horned capsule that commonly turns pink to bright red on plants growing in full sun (see inset). **IA, IL, IN, MI, MN, MO, OH, WI. Common, possibly in every county.** *MAH/SN*

Besseya bullii
PLANTAGINACEAE

kittentails

Gravelly and sandy prairies, savannas, open woodlands, bluffs. Spring, summer.

A gray-hairy perennial to 16 in. tall. Basal rosette leaves simple, petiolate, cordate, to 4 in. long × 3 in. wide, with whitish palmate veins, margins with small, blunt teeth. Stem leaves alternate, sessile, clasping, smaller, crowded along stem, becoming bracts in inflorescence. Flowers inconspicuous, to ¼ in. long, irregular, 2-lipped, with 1 or 2 lobes on top lip and 1–3 lobes on bottom lip, nearly covered by 4-parted calyx, in dense, many-flowered, solitary spike to 6 in. long. IA, IL, IN, MI, MN, OH, WI. A Midwest endemic. Very uncommon in southern portion of the Northern Lakes and northern portion of the Tallgrass Prairie. *CN*

Callitriche heterophylla
PLANTAGINACEAE

greater water-starwort, two-headed water-starwort

Marshes, mudflats, ponds, rivers, swamps, streams. Summer.

A glabrous, colonial, aquatic annual to 10 in. Leaves opposite, entire; submerged leaves linear, to ½ in. long, less than ⅛ in. wide, connected across stem; floating leaves clustered, shorter, to ¼ in. wide, broadest above middle, short-petiolate. Flowers regular, minuscule, inconspicuous, petal-less, solitary in submerged leaf axils. Fruit a minute capsule as long as wide, unwinged at tip. Vernal water-starwort (*C. palustris*), more northern, has fruit slightly longer than wide and winged at tip; it can be slightly more terrestrial. IA, IL, IN, MI, MN, MO, OH, WI. Widespread in the Ozark Highlands and eastern portion of the Eastern Forests, scattered to rare elsewhere. *EH*

Callitriche terrestris
PLANTAGINACEAE

terrestrial starwort

Exposed soil and rock. Spring, summer.

A prostrate annual, branched and
mat-forming, glabrous, 1–2 in. long. Leaves
opposite, short-petiolate, elliptic-oblong,
⅛–¼ in. long × ¹⁄₁₆–⅛ in. wide. Flowers minuscule, petals and sepals lacking, monoecious,
consisting of either a single stamen or pistil,
occurring in leaf axils. Fruit deeply lobed
on a short stalk. This can be separated from
other starwort species in the region by its
exclusively non-aquatic habitat and flowers/
fruits on stalks (sessile in others). It may have
the smallest flowers of all non-aquatic plants
in the Midwest. Look for it especially on compacted soils of trails and dirt roadbeds. IL,
**IN, MI, MO, OH, WI. Predominantly in the
lower Midwest.** *MAH*

Rumex altissimus
POLYGONACEAE

pale dock

Prairies, marshes, floodplains, ditches.
Spring, summer.

A glabrous, colony-forming perennial to 4
ft. tall. Leaves alternate, entire, to 10 in. long
× 3 in. wide, petiolate, with disintegrating
ocreae (nodal sheaths). Flowers regular, small,
inconspicuous, with 6 tepals, on stalks to ⅖ in.,
numerous in dangling whorls in tightly packed
erect, branched, terminal cluster to 12 in. long.
Usually 1 (rarely 2) of 3 inner tepals develops a
tubercle. Mexican dock (*R. triangulivalvis*) and
swamp dock (*R. verticillatus*) have narrower
leaves and tubercles on all 3 inner tepals; latter
with longer, narrower leaves and longer-stalked
flowers. IA, IL, IN, MI, MN, MO, OH, WI.
**Common and widespread in southern por-
tion of the Tallgrass Prairie, scattered elsewhere.** *PR*

Rumex britannica
POLYGONACEAE

great water dock

Marshes, fens, seeps, ditches. Summer.

A glabrous, erect perennial to 6 ft. tall.
Leaves alternate, simple, bluntly toothed and
wavy-margined, to 24 in. long × 6 in. wide,
petiolate, smaller and short-petiolate above,
with disintegrating ocreae (nodal sheaths).
Flowers regular, to ¼ in. long with 6 tepals, on
stalks to ⅖ in., numerous in dangling whorls
in tightly packed erect, branched clusters to 16
in. long at ends of branches at top of plant. All
3 inner tepals develop a tubercle at maturity.
Syn. *R. orbiculatus*. **IA, IL, IN, MI, MN, OH,
WI. Widespread in the Northern Lakes and
northern portion of the Tallgrass Prairie,
scattered in the Eastern Forests.** *PS*

Rumex crispus
POLYGONACEAE

curly dock

Cultivated fields, roadsides, waste areas.
Spring, summer.

An erect perennial, branched above, glabrous,
1–4 ft. tall. Leaves alternate, petiolate, ocreae
(nodal sheaths) thin and often disintegrating
with age, blades oblong-lanceolate, margins
entire to scalloped, wavy or crinkled, 6–12 in.
long × 1–3 in. wide. Flowers regular, about ¼
in. long, with 6 tepals, the inner 3 ovate and
enlarged in fruit, usually only 1 with a tubercle
that is at least half the length of fruiting tepal,
each on a jointed stalk, arranged in terminal
clusters. In patience dock (*R. patientia*) the
tubercle is less than half the length of fruiting tepal. **IA, IL, IN, MI, MN, MO, OH, WI.
Introduced from Eurasia. Likely in every
county.** *MAH*

Rumex fueginus
POLYGONACEAE

golden dock

Marshes, shores, floodplains, ditches, roadsides, disturbed sites. Summer.

A sprawling annual to 2 ft. tall. Leaves alternate, simple, entire, slightly wavy-margined, to 10 in. long × 1 in. wide, petiolate, smaller and short-petiolate above, with disintegrating ocreae (nodal sheaths). Flowers regular, less than ⅛ in. long, with 6 tepals, inner 3 tepals with distinct bristle-like margins, short-stalked, numerous in dangling whorls in tightly packed, leafy, spike-like clusters in upper part of plant. All 3 inner tepals develop a tubercle at maturity. Syn. *R. maritimus.* IA, IL, IN, MI, MN, MO, OH, WI. Widespread in northwestern portion of the Midwest, scattered to rare elsewhere. *PD*

Rumex obtusifolius
POLYGONACEAE

bitter dock

Fields, forest edges, floodplains, gravel bars, roadsides, waste areas. Spring, summer.

A glabrous perennial to 4 ft. tall. Leaves basal and alternate, bluntly toothed and wavy-margined, to 12 in. long × 6 in. wide, cordate at base, deeply veined, petiolate, smaller and short-petiolate above, with disintegrating ocreae (nodal sheaths). Flowers regular, to ¼ in. long, with 6 tepals, inner 3 tepals with sharply toothed margins (see inset), on stalks to ½ in., numerous in dangling whorls in tightly packed erect, branched clusters to 12 in. long. One to 3 inner tepals develop a tubercle. IA, IL, IN, MI, MN, MO, OH, WI. Introduced from Europe. Scattered to widespread, more frequent in eastern portion of the Midwest. *SN*

Potamogeton nodosus
POTAMOGETONACEAE

longleaf pondweed

Deep marshes, mudflats, lakes, ponds, rivers, streams, ditches. Spring, summer.

A glabrous, branching, aquatic perennial to 8 ft. long. Leaves alternate, entire, parallel-veined, with erect stipule to 3½ in. long where petiole meets stem. Submerged leaves to 5 in. long × 1¼ in. wide, limp, petiolate; floating leaves shiny, to 6 in. long × 1¾ in. wide, tapering to tip and base, long-petiolate. Flowers regular, individually inconspicuous, to ⅛ in. wide in dense spikes to 2 in. long on stout reddish stalks from upper leaf axils, emergent. Other species in the genus are distinguished by leaf width and shape, presence or absence of floating leaves, and other technical characters. IA, IL, IN, MI, MN, MO, OH, WI. Frequent. *PR*

Chrysosplenium americanum
SAXIFRAGACEAE

American golden-saxifrage

Shaded seeps, along streams. Spring.

A matted, creeping, colony-forming perennial to 4 in. tall. Leaves opposite, sessile or short-petiolate, glabrous, roundish with scalloped margins, to ½ in. long × ¾ in. wide. Flowers regular, inconspicuous, to ⅛ in. wide, lacking petals, with 8 short reddish to orange anthers, solitary at ends of branched stems. Fruit a capsule. Easily overlooked in habitat similar to watercress (*Nasturtium officinale*), which it superficially resembles vegetatively at quick glance. IN, MI, MN, OH, WI. Primarily in the Northern Lakes and OH portion of the Eastern Forests, with a few occurrences in WI portion of the Tallgrass Prairie. *SN*

Heuchera richardsonii
SAXIFRAGACEAE

prairie alumroot

Prairies, glades, savannas, woodlands, usually in rocky or poor soils. Spring, summer.

A hairy perennial to 3 ft. tall. Leaves in basal rosette, shallowly palmately 3- to 5-lobed and cordate at base, with coarsely toothed margins, to 3 in. long and wide, long-petiolate. Flowers irregular, inconspicuous, tubular, to ¼ in. long, upper side longer than lower side, greenish, with 5 barely exserted orange anthers, numerous in narrow, branched cluster at top of naked stem. The more eastern common alumroot (*H. americana*) has more regular flowers with anthers longer-exserted. *Heuchera* cultivars are common in landscaping. IA, IL, IN, MI, MN, MO, WI. Widespread in the Ozark Highlands, the Tallgrass Prairie, and western portion of the Northern Lakes. *PR*

Mitella nuda
SAXIFRAGACEAE

naked miterwort, naked bishop's cap

Moist mossy forests, cedar swamps, thickets, bogs. Spring.

A hairy, often glandular, colony-forming perennial to 8 in. tall. Leaves basal, roundish with 3–5 shallow, palmate lobes and cordate base, with scalloped margins, to 1½ in. long and wide, long-petiolate. Flowers regular, yellowish-green to whitish, to ¼ in. wide, with 5 spreading feather-like petals between 5 spreading triangular sepals (see inset), short-stalked in terminal raceme on naked, solitary stem. Fruit a capsule that splits into 2 valves to expose 2 clusters of round, shiny black seeds. IA, MI, MN, WI. Widespread in the Northern Lakes, scattered in adjacent portions of the Tallgrass Prairie. *MJH/PD*

Scheuchzeria palustris
SCHEUCHZERIACEAE

podgrass, rannoch-rush, bog arrowgrass

Bogs, fens. Spring, summer.

A perennial to 16 in. tall. Leaves alternate, entire, linear, sessile and sheathing, to 16 in. long × ⅛ in. wide, shorter above, erect or nearly so, with distinctive small open pore at tip. Flowers easily overlooked, regular, greenish to purplish, to ¼ in. wide, with 6 inconspicuous tepals surrounding 6⅜-in. stamens and 3 erect pistils, 3–6 alternate along stem in terminal raceme. Fruit (pictured) a cluster of 3 spreading, sessile, teardrop-shaped, yellowish capsules, more conspicuous than flowers. **IA, IL, IN, MI, MN, OH, WI. Uncommon to rare, most frequent in the Northern Lakes.** *SN*

Smilax ecirrhata
SMILACACEAE

upright carrion flower

Rich mesic forests, stream terraces, ravines. Spring.

An erect, dioecious and herbaceous perennial, unbranched, mostly glabrous, usually lacking tendrils and prickles, 1–2 ft. tall. Leaves alternate, petiolate, often crowded on upper stem like a whorl, usually under 10 in number, broadly ovate with cordate bases, hairy below, entire, 3–5 in. long × 2–3½ in. wide. Flowers regular, about ¼ in. wide, with 6 greenish tepals. Inflorescence of compact, rounded umbels with up to 25 malodorous flowers on a peduncle below the leaves. Illinois carrion flower (*S. illinoensis*) is commonly taller, bearing some upper axillary tendrils and more than 25 flowers per umbel. **IA, IL, IN, MI, MN, MO, OH, WI. Common in most of the Midwest.** *CR*

Smilax lasioneura
SMILACACEAE
Blue Ridge carrionflower
Forests, thickets, floodplains. Spring.

A sprawling to erect herbaceous perennial vine to 8 ft. Leaves alternate, entire, ovate to cordate, 3 in. long × 2½ in. wide, covered in short hairs beneath, short-petiolate, with pair of tendrils where petiole attaches to stem. Flowers regular, ¼ in. wide, with 6 spreading tepals, stalked, radiating and forming tight, spherical, long-stalked clusters to 2 in. wide. Fruit blue, spherical, fleshy. Downy carrionflower (*S. pulverulenta*) has black fruit. Smooth carrionflower (*S. herbacea*) has leaves glabrous beneath. IA, IL, IN, MI, MN, MO, OH, WI. Scattered to widespread in the Ozark Highlands, Tallgrass Prairie, and much of the Northern Lakes, less frequent eastward; scattered in the Eastern Forests, especially westward. *ST*

Trillium viride
TRILLIACEAE
green trillium
Rich mesic woodlands, rocky limestone slopes, prairies. Spring.

An erect perennial, unbranched, glabrous to sparsely hairy, 8–18 in. tall. Leaf-like bracts 3 atop stalk, whorled, sessile, elliptic, somewhat to not mottled, dotted with tiny whitish stomates on upper surface (best seen with hand lens), 3–8 in. long × 2–3 in. wide. Flowers regular, 1½–2 in. long, with 3 narrowly oblanceolate, green to light purple, erect petals narrowed at base and 3 spreading lanceolate sepals with a sessile base, solitary and sessile atop leaf-like bracts. The flowers have an odor of rotten apples or other decaying fruit. A Midwest endemic. IL, MO. Found mostly in southwestern IL and eastern MO. Not common. *SN*

Sparganium americanum
TYPHACEAE

American bur-reed

Marshes, bogs, sloughs, lakeshores, streams, ditches. Summer.

A monoecious colonial perennial, to 40 in. tall. Leaves alternate, appearing basal, entire, to 40 in. long × ½ in. wide, strongly keeled along midrib. Flowers regular, individually inconspicuous, male and female sessile in separate dense spherical clusters along zigzag, spike-like terminal inflorescence (often branched at base); male above, female with single style and stigma (less than ⅛ in. long), maturing into bur-like heads to 1 in. wide in bract axils with fruit to ⅛ in. wide. Similar species differ in fruiting head size, stigma length, and fruiting head position relative to bracts. IA, IL, IN, MI, MN, MO, OH, WI. Scattered, most frequent in the Ozark Highlands and the Northern Lakes. *RS*

Sparganium eurycarpum
TYPHACEAE

giant bur-reed

Marshes, bogs, shallow lakes, river borders. Summer.

An erect, monoecious perennial, branched in inflorescence, glabrous, rhizomatous and colonial, 2–5 ft. tall. Leaves alternate, sheathing, linear and parallel-veined, keeled, V-shaped in cross section, spongy, to 3+ ft. long × ¼–½ in. wide. Flowers minute, tepals scale-like, female flowers with 2 stigmas, numerous in 1–5 rounded heads positioned lowermost on inflorescence, male flowering heads 2–10+ and uppermost on inflorescence. Individual fruits broadly angular above with a flattened summit (except for beak), each about ¼ in. wide (see inset). IA, IL, IN, MI, MN, MO, OH, WI. Common, more so northward. *SN*

Boehmeria cylindrica
URTICACEAE

false nettle

Floodplains, swamps, marshes, trails. Summer, fall.

A perennial to 3 ft. tall, monoecious or dioecious. Leaves opposite, ovate, coarsely toothed, petiolate, to 6 in. long × 3 in. wide, with 3 deeply impressed parallel veins connected by many lateral veins, appearing wrinkled. Flowers regular, inconspicuous, lacking petals, crowded to separated in spikes to 4 in. long from axils of upper leaves. Lacks stinging hairs common on other members of family. Leaves drooping and folded along midvein, on shorter, pink petioles in var. *drummondiana*. Clearweeds (*Pilea*) also lack stinging hairs but are shorter with translucent stems. IA, IL, IN, MI, MN, MO, OH, WI. **Common throughout, except in northern parts of the Northern Lakes and northwestern parts of the Tallgrass Prairie.** *MAH*

Laportea canadensis
URTICACEAE

wood nettle, stinging nettle

Floodplain forests, stream terraces, rich mesic upland forests. Summer.

An erect, monoecious perennial, mostly unbranched, beset with stinging hairs, 1–3+ ft. tall, colonial. Leaves alternate, petiolate, serrate, broadly ovate to oval, 2–8 in. long × 1½–5 in. wide. Flowers about ⅛ in. wide, petals lacking, sepals 5, male flower clusters widely spreading from lower leaf axils, female from upper leaf axils. Skin contact with the nettle's stinging hairs is not soon forgotten; sap from jewelweed (either *Impatiens capensis* or *I. pallida*) is said to bring relief. IA, IL, IN, MI, MN, MO, OH, WI. **Widespread and common throughout the Midwest.** *SN*

Parietaria pensylvanica
URTICACEAE

Pennsylvania pellitory

Cliff bases, house and building walls,
tree tip-up mounds, streambanks.
Spring, summer.

An ascending to erect annual, monoecious
(or some flowers perfect), mostly unbranched,
hairy (but no stinging hairs), 4–12 in. tall.
Leaves alternate, petiolate, lanceolate, entire,
surface veins distinctly impressed, 1–3 in. long
× ¼–1¼ in. wide. Flowers minute, petals lacking, surrounded by leafy, finger-like bracts.
Inflorescence of sessile clusters in leaf axils.
Although weedy around human habitation,
pellitory does occur in natural settings, such as
at bases of rock cliffs. IA, IL, IN, MI, MN, MO,
OH, WI. Occurs in most of the region but
absent in much of the Northern Lakes. SN

Pilea pumila
URTICACEAE

clearweed

Rich moist forests, floodplains, gravel bars,
ditches, pond and swamp borders. Summer.

An erect monoecious annual, mostly
unbranched, glabrous, stems somewhat translucent, 6–18 in. tall. Leaves opposite, petiolate,
ovate to elliptic, toothed, 2–4 in. long × 1–2 in.
wide. Flowers minuscule, petals lacking, calyx
of male flowers 4-parted, females 3-parted,
occurring on small clusters in branches from
upper leaf axils, nutlets mostly smooth, green
speckled with black. The more northern bog
clearweed (*P. fontana*) is similar but has warty,
mostly all black nutlets. IA, IL, IN, MI, MN,
MO, OH, WI. Widespread and common
throughout. *MAH*

Urtica dioica
URTICACEAE

stinging nettle

Streambanks, floodplains, ditches, roadsides, disturbed wetlands. Summer.

An erect monoecious perennial, mostly unbranched, stems ridged and glabrous or sparsely covered with stinging hairs, 3–6 ft. tall. Leaves opposite, petiolate, elliptic-ovate to lanceolate-ovate, sharply toothed, often drooping, bristly stinging hairs if present confined mostly to undersurface, 2–6 in. long × 1–2½ in. wide. Flowers minuscule, petals lacking, in branching axillary clusters 1–3 in. long. Our most common subspecies is the native subsp. *gracilis* (pictured). Very rare here is the introduced Eurasian subsp. *dioica*; typically it is bristlier and dioecious. IA, IL, IN, MI, MN, MO, OH, WI. Common and widespread except in the very southernmost counties. *SN*

Valeriana edulis
VALERIANACEAE

common valerian, hairy valerian

Wet prairies, fens. Spring.

A perennial to 4 ft. tall. Leaves with dense, white, short pubescence along margins; basal leaves to 12 in. long × 1 in. wide, unlobed or some 3-lobed; stem leaves few, opposite, shorter, unlobed or with 3–9 pinnate lobes. Flowers regular, to ⅛ in. wide, perfect, male, or female, shallowly tubular with 5 greenish-yellow to white spreading to recurved lobes, numerous in dense to loose panicle terminating stem, forming narrowly domed inflorescence. Midwestern plants are var. *ciliata*. IA, IL, IN, MI, MN, OH, WI. Uncommon in transition zone between the Northern Lakes and Tallgrass Prairie; also rare in southeastern MI and the Eastern Forests. *SN*

Hybanthus concolor
VIOLACEAE

green violet

Rich and usually calcareous, mesic
to dry-mesic forests. Spring.

An erect perennial, unbranched, hairy, 1–2½
ft. tall. Leaves alternate, short-petiolate,
elliptic-lanceolate and long-pointed, entire,
hairy to glabrous, 3–6 in. long × 1–2 in. wide.
Flowers irregular, about ¼ in. long × ⅛ in.
wide, with 5 oblong petals, tips upcurved
except for the larger lower petal, it notched at
the tip and slightly swollen at the base, sepals
5, linear, 1–3 flowers per leaf axil. Syn. *Cubel-
ium concolor*. IA, IL, IN, MI, MN, MO, OH,
WI. Common in the lower Midwest, rare to
absent northward. *SN*

BLUE TO
VIOLET
FLOWERS

Ruellia humilis
ACANTHACEAE

hairy wild petunia

Prairies, glades, open rocky woods, roadsides. Spring, summer, fall.

A densely spreading-hairy perennial to 1½ ft. tall. Leaves opposite, entire, short-petiolate to sessile, to 2½ in. long × 1 in. wide. Flowers irregular (nearly regular), lavender to pale blue with darker veins, funnel-shaped with 5 spreading and overlapping lobes, to 2 in. wide with tube to 2½ in. long, each lasting a single day, stalkless in upper leaf axils. Stalked wild petunia (*R. pedunculata*) has stalked flowers (often subtended by leafy bracts). Carolina wild petunia (*R. caroliniensis*) has clearly petiolate leaves. IA, IL, IN, MI, MN, MO, OH, WI. Widespread in the Ozark Highlands and southern portion of the Tallgrass Prairie, rare to scattered elsewhere. *SN*

Ruellia strepens
ACANTHACEAE

wild petunia

Floodplain forests, streambanks, ravines, rocky slopes. Spring, summer.

An erect perennial, mostly unbranched, glabrous to minutely hairy, stems somewhat 4-angled, 1–2½ ft. tall. Leaves opposite, petiolate, ovate-lanceolate, 2–6 in. long × 1–3 in. wide. Flowers irregular (nearly regular), about 2 in. wide, funnel-shaped with 5 rounded lobes and 5 lanceolate calyx lobes up to ⅛ in. wide. Inflorescence a 1- or 2-flower cluster per upper leaf axil. Flowers resemble the unrelated garden petunia. IA, IL, IN, MI, MO, OH. **Found primarily in the southern half of the region.** *MAH*

Amsonia tabernaemontana
APOCYNACEAE

eastern blue star

Wet woods, swamps, streambanks, ditches. Spring.

A perennial to 3 ft. tall, clump-forming and branched with glabrous stems. Leaves alternate, simple, entire, lanceolate to elliptic in shape, glabrous or finely hairy beneath, dull on upper surface, contains milky sap, 4–6 in. long × 2 in. wide. Flowers regular, ½–¾ in. wide, tubular and star-shaped with 5 narrow lobes, numerous in clusters at stem tips. Fruit an erect follicle. The similar-looking Ozark blue star (*A. illustris*) has shiny leaves and drooping fruit. **IL, IN, MO. Locally common in the far southwestern portion of the Eastern Forests.** *SN*

Vinca minor
APOCYNACEAE

lesser periwinkle, myrtle

Forests, thickets, old homesites, cemeteries, roadsides. Spring.

A mat-forming perennial shrub, to 3 ft. long × 8 in. tall, rooting at nodes. Leaves opposite, entire, short-petiolate, glossy, thick-textured, evergreen, to 2 in. long × 1 in. wide, broadest near middle, with light midvein and pale underside. Flowers regular, blue to lavender, to 1 in. wide, tubular with 5 pinwheel-like spreading lobes, solitary on stalks from upper leaf axils. Rarely develops fruit and thus spreads primarily vegetatively. Greater periwinkle (*V. major*) has leaves broadest near base with ciliate margins; variegated forms often encountered in cultivation. **IA, IL, IN, MI, MN, MO, OH, WI. Introduced from Europe. Scattered throughout, most frequent in the Eastern Forests.** *SN*

Arctium minus
ASTERACEAE

lesser burdock, common burdock

Streambanks, fields, thickets, successional woods, roadsides, waste areas. Summer, fall.

A biennial to 6 ft. tall. Leaves gray-green above, white-hairy beneath; basal leaves scalloped, wavy-margined, to 2 ft. long × 1½ ft. wide, long-petiolate; stem leaves alternate, smaller, short-petiolate. Flowers in composite heads; numerous pink to purple (rarely white) disk flowers in several spherical 1-in. heads, sessile to short-stalked in axillary and terminal arrays longer than wide, involucres covered with hooked bristles that attach to animal fur and human clothing. Great burdock (*A. lappa*) has larger, obtuse-tipped leaves and stalked flower heads in arrays wider than long. IA, IL, IN, MI, MN, MO, OH, WI. Introduced from Eurasia. Throughout, less frequent south. *MAH*

Centaurea cyanus
ASTERACEAE

cornflower, bachelor's button

Fields, roadsides. Spring, summer.

A weak perennial to 3 ft. tall. Leaves alternate, to 4 in. long, entire, lower sometimes with a few lobes, covered in white cobwebby hairs. Flowers in composite heads; the small purple-blue disk is surrounded by numerous 5-lobed blue ray flowers, in terminal heads to 1¼ in. wide at ends of branches. Commonly grown as an ornamental, often in wildflower seed mixes; cultivated forms sometimes produce flower heads ranging from pink to purple to white. Flowers were worn by young men when in love (hence bachelor's button). IA, IL, IN, MI, MN, MO, OH, WI. Introduced from the Mediterranean. Widely scattered throughout. *KY*

Cichorium intybus
ASTERACEAE

chicory

Roadsides, fencerows, fields, other disturbed areas. Spring, summer, fall.

An erect or ascending perennial, commonly branched, stems mostly glabrous, with milky sap, 1–3 ft. tall. Leaves alternate and basal, basal short-petiolate, stem leaves sessile and clasping, oblanceolate, pinnately lobed and toothed to entire, glabrous to hairy, 2–12 in. long and up to 3 in. wide. Flowers in composite heads to 1½ in. wide, ray flowers 10–30, disk flowers lacking, arranged in spreading, spike-like branches. Chicory is a relative of endive (*C. endivia*), a leafy vegetable used in salads. IA, IL, IN, MI, MN, MO, OH, WI. **Introduced from Europe. Widespread and common throughout. *SN***

Conoclinium coelestinum
ASTERACEAE

mist flower

Meadows, moist prairies, open woodlands, roadsides. Summer, fall.

An erect to ascending perennial, branched, short-hairy, 1–2 ft. tall, colonial. Leaves opposite, petiolate, triangular to ovate with scalloped margins, undersurface glandular, 1–4 in. long × ½–2 in. wide. Flowers in composite heads about ¼ in. wide, ray flowers lacking, disk flowers 5-lobed with long-exserted styles, in arrays 2–4 in. wide atop stem branches. Mist flower resembles floss flower (*Ageratum houstonianum*), a popular garden annual native to Mexico and Central America. IL, IN, MI, MO, OH. **Occurs primarily in the Midwest's southern third. Populations farther north are likely introductions. *MAH***

Ionactis linariifolius
ASTERACEAE

flax-leaved aster, stiff aster

Sand prairies, glades, savannas, dry woods. Summer, fall.

A glabrous perennial to 1½ ft. tall. Leaves alternate, rigid, linear, sessile or nearly so, densely spaced (appearing nearly whorled), entire, glossy, with strong midvein, to 1½ in. long × ⅛ in. wide. Flowers in composite heads; a yellow disk is surrounded by up to 20 pale blue to lavender ray flowers in heads to 1¼ in. wide, solitary or in few-headed clusters at top of plant. Syn. *Aster linariifolius*. IA, IL, IN, MO, OH, WI. Widespread in the Ozark Highlands, spotty in the Eastern Forests and eastern portion of the Tallgrass Prairie and adjacent Northern Lakes. *SN*

Lactuca floridana
ASTERACEAE

woodland lettuce

Moist forests, mesic savannas. Summer, fall.

An annual or biennial, unbranched, mostly glabrous, 2–8 ft. tall, with milky sap. Lower leaves with winged petioles, narrowly ovate or obovate and usually deeply pinnately lobed (see inset); upper leaves lanceolate, deeply lobed or entire, 6–12 in. long × 1–5 in. wide. Flowers in composite heads approximately ½ in. wide with 10–15 usually blue ray flowers, disk flowers lacking, in a widely branched terminal array. Fruit an achene with white pappus. Tall blue lettuce (*L. biennis*) and western lettuce (*L. ludoviciana*) differ by having a brown pappus, and usually yellow flowers, respectively. IA, IL, IN, MI, MN, MO, OH, WI. Wide ranging, mostly in lower Midwest. *MAH/SN*

Lactuca tatarica
ASTERACEAE

showy blue lettuce

Prairies, rocky openings, disturbed sites.
Summer, fall.

A gray-green perennial to 3 ft. tall. Leaves
alternate, mostly entire and unlobed, sessile
to clasping (basal leaves petiolate), to 6 in.
long × 1½ in. wide, containing milky sap.
Flowers in composite heads; numerous ray
flowers in heads to 1 in. wide, disk flowers
absent, phyllaries appressed in several series
within involucre, heads numerous in open
panicle-like array. Fruit with white pappus on
beaked achenes, in globose heads. Syn. *Mulge-
dium pulchellum*. IA, IL, IN, MI, MN, MO, OH,
WI. Primarily in northwestern portion of
the Tallgrass Prairie and western portion of
the Northern Lakes, absent from the Eastern
Forests and Ozark Highlands. SN

Lygodesmia juncea
ASTERACEAE

rush skeletonplant, skeletonweed

Prairies, fields, roadsides, railroads.
Summer.

A glabrous, gray-green, spindly, branched
perennial to 18 in. tall. Leaves entire, alternate,
stiff, tapering from base to tip, to 2 in. long ×
⅛ in. wide, smaller and scale-like above, sessile, containing cream-colored sap. Flowers in
composite heads; 4–6 spreading ray flowers in
¾-in.-wide heads, disk flowers absent, heads
subtended by ⅗-in.-long involucre with 4–6
long inner phyllaries and small outer phyllaries at base; heads solitary at tips of branches.
Fruit with silky pappus. *Chondrilla juncea* (aka
skeletonweed) has yellow ray flowers. IA, IN,
MN, MO. Native mostly in western portion of
the Tallgrass Prairie, with a few occurrences
introduced elsewhere. MJH

Symphyotrichum ciliolatum
ASTERACEAE

northern heart-leaved aster, Lindley's aster

Fields, forests and forest clearings, roadsides. Summer, fall.

A colony-forming perennial to 3 ft. tall. Leaves basal and alternate, simple, cordate, regularly toothed, to 5 in. long × 2 in. wide, glabrous and shiny on top surface, with broadly winged petiole that somewhat sheaths stem and has ciliate margins, becoming short-petiolate to sessile above. Flowers in composite heads; up to 25 pale blue to lavender ray flowers surround yellow disk (turning reddish) in heads to 1½ in. wide, with appressed phyllaries, heads numerous in open, branched panicle-like array at top of plant, branches with numerous short bracts. **IL, MI, MN, WI. Widespread in the Northern Lakes, particularly the northern portion.** *SN*

Symphyotrichum cordifolium
ASTERACEAE

heartleaf aster, blue wood aster

Moist forests, floodplains. Summer, fall.

A perennial to 3 ft. tall. Leaves basal and alternate, simple, cordate, coarsely and sharply toothed, to 5 in. long × 3 in. wide, with narrow, unwinged or barely winged petiole to 3½ in. long, becoming smaller, oval, and short-petiolate to sessile above. Flowers in composite heads; up to 15 pale blue-lavender to nearly white ray flowers surround cream-colored to yellow disk (turning purplish-red) in heads to ½ in. wide, with appressed phyllaries, heads numerous in open, branched, panicle-like array at top of plant. **IA, IL, IN, MI, MN, MO, OH, WI. Scattered to widespread.** *MAH*

Symphyotrichum drummondii
ASTERACEAE

Drummond's aster

Prairies, savannas, forests. Summer, fall.

A perennial to 4 ft. tall. Stems densely gray-hairy. Leaves basal and alternate, cordate, shallowly toothed, to 4 in. long × 3 in. wide, with winged petiole, becoming smaller, narrow, and short-petiolate to sessile above. Flowers in composite heads; up to 15 pale blue-lavender ray flowers surround cream-colored to yellow disk (turning purplish-red) with appressed to slightly spreading phyllaries, ½-in. heads numerous in branched panicle-like array. Many-rayed aster (*S. anomalum*) has flower heads over ¾ in. wide subtended by recurved phyllaries. Wavyleaf aster (*S. undulatum*) has wavy leaf margins and lower leaves with clasping petioles. IA, IL, IN, MI, MN, MO, OH, WI. **Widespread southwestward, scattered elsewhere.** *PD*

Symphyotrichum firmum
ASTERACEAE

shining aster, smooth swamp aster

Sedge meadows, fens, moist fields, swamp openings, shores, ditches. Summer, fall.

An erect, colony-forming perennial to 8 ft. tall with stem glabrous or hairy in lines, often purple at nodes. Leaves alternate, simple, sparsely shallowly toothed, to 4 in. long × ¾ in. wide, numerous and crowded, becoming smaller into inflorescence, clasping with lobes wrapped around stem. Flowers in composite heads; up to 40 white to pale blue ray flowers surround yellow disk (turning reddish-orange) in heads to 1¼ in. wide, with appressed to somewhat spreading phyllaries, at ends of branches forming dome-shaped cluster at top of plant. IA, IL, IN, MI, MN, MO, OH, WI. **Widespread in the Northern Lakes, scattered elsewhere.** *SN*

Symphyotrichum laeve
ASTERACEAE

smooth aster

Prairies, glades, barrens, open woodlands. Summer, fall.

An erect perennial, unbranched, mostly glabrous throughout, 1½–3 ft. tall. Leaves alternate, sessile and mostly entire, clasping, lanceolate to oblanceolate, commonly glaucous, to 4 in. long × 1¼ in. wide. Flowers in composite heads ¾–1 in. wide, ray flowers 10–20, slightly more disk flowers, phyllaries firm and appressed, in a spreading array atop the plant. This species is less common than range maps indicate; it is mostly confined to remnant native grasslands. Features can be quite variable, especially in leaf length and width. IA, IL, IN, MI, MN, MO, OH, WI. Widespread, more common in our western counties. *KC*

Symphyotrichum novae-angliae
ASTERACEAE

New England aster

Prairies, meadows, old fields, roadsides, open woodlands. Summer, fall.

An erect perennial, branched above and gray-hairy, 2–5 ft. tall. Leaves alternate, entire, narrowly oblanceolate, clasping at base, 1–4 in. long × ¼–¾ in. wide. Flowers in composite heads up to 1½ in. wide, ray flowers purple to pink, 50–100 and a similar number of disk flowers, phyllaries long-tapering, spreading to reflexed and glandular, in a spreading array atop the plant. This aster is easy to recognize, with its strongly clasping and hairy leaves and large heads of deep purple or bright pink ray flowers. IA, IL, IN, MI, MN, MO, OH, WI. Mostly widespread and common. *SN*

Symphyotrichum oblongifolium
ASTERACEAE

aromatic aster

Alkaline glades, cliff ledges, cliff tops, open gravelly slopes, prairies. Summer, fall.

An erect perennial with branching stems from creeping rhizomes, with mix of glandular and eglandular hairs, 1–2½ ft. tall. Leaves alternate, entire, mostly sessile, oblong, firm, 1–3 in. long and up to ¾ in. wide with short spreading hairs, some glandular. Flowers in composite heads about 1 in. across, ray flowers 15–40 and about 50 disk flowers, phyllaries spreading to recurved and glandular, in a rather compact array. Leaves of silky aster (*S. sericeum*) are densely hairy with appressed, silky hairs giving a silvery look. IA, IL, IN, MN, MO, OH, WI. Mostly in the Ozark Highlands and Tallgrass Prairie. *KC*

Symphyotrichum oolentangiense
ASTERACEAE

sky blue aster

Prairies, glades, savannas, cliff edges, rocky woods. Summer, fall.

An erect perennial, stems unbranched, glabrous or somewhat roughened, 1–3 ft. tall. Leaves alternate, entire or with shallow teeth, firm, basal and lower leaves narrowly ovate-cordate on distinct, sometimes winged petioles, blades 2–6 in. long × ½–2 in. wide, upper leaves linear and mostly sessile. Flowers in composite heads ¾–1 in. wide, ray flowers 10–25, disk flowers 15–30, phyllaries mostly glabrous and appressed, in a widely spreading terminal array. Named after the Olentangy River in OH. The common name alludes to the flower color. IA, IL, IN, MI, MN, MO, OH, WI. Occurs primarily in the Tallgrass Prairie and Ozark Highlands. *MJH*

Symphyotrichum patens
ASTERACEAE

late purple aster

Glades, barrens, dry open woodlands. Summer, fall.

An erect perennial, branched, hairy, 1–3 ft. tall. Leaves alternate, entire, lanceolate to oblanceolate, sessile, the base strongly clasping the stem, 1–2 in. long × 1 in. wide. Flowers in composite heads ¾–1 in. wide, ray flowers 10–25, disk flowers up to 50, phyllaries mostly appressed, inflorescence widely spreading. Phyllaries of the widespread var. *patens* are mostly glandular, whereas those of the more western var. *patentissimum* are generally not. Thin-leaf late purple aster (*S. phlogifolium*), known mostly in OH, has larger, thinner, smooth leaves. IL, IN, MO, OH. Occurs principally in Ozark Highlands and southern Eastern Forests. *EH*

Symphyotrichum praealtum
ASTERACEAE

willow aster

Prairies, wet meadows, fens. Summer, fall.

An erect perennial, branching above, stem glabrous to moderately hairy, colonial from creeping rhizomes, 2–5 ft. tall. Leaves alternate, entire with margins slightly rolled under, narrowly elliptic to lanceolate to linear, somewhat roughened above to glabrous below with a noticeable network of veins, 1–5 in. long × ¼–¾ in. wide, getting progressively smaller upward. Flowers in composite heads about ¾ in. wide, ray flowers 15–35, disk flowers 20–35+, phyllaries appressed, on widely spreading branches. The several named varieties display too much intergradation to be listed here. IA, IL, IN, MI, MN, MO, OH, WI. Principally in the Tallgrass Prairie. *MAH*

Symphyotrichum prenanthoides
ASTERACEAE

crooked aster, zigzag aster

Rocky streambanks, floodplain forests, seeps. Summer, fall.

A weakly erect zigzag-stemmed perennial, mostly glabrous, colonial from creeping rhizomes, 1–2½ ft. tall. Leaves alternate, lanceolate to ovate-lanceolate, strongly toothed, glabrous or somewhat rough above, abruptly narrowing into a winged petiole that forms a broad clasping base, 3–5 in. long × 1–2 in. wide. Flowers in composite heads about 1 in. across, ray flowers 15–30, disk flowers 40–65, phyllaries with spreading or recurved pointed tips, arranged in a spreading array atop plant. IA, IL, IN, MN, OH, WI. Occurs mostly in the Eastern Forests and states of the upper Mississippi River Valley. *KC*

Symphyotrichum puniceum
ASTERACEAE

bristly aster, swamp aster, purplestem aster

Marshes, sedge meadows, swamps, fens, shores, ditches. Summer, fall.

An ascending to erect, solitary perennial to 6 ft. tall with bristly-hairy purple stem. Leaves alternate, simple, sparsely shallowly toothed, to 8 in. long × 1¼ in. wide, widely spaced, becoming smaller into inflorescence, clasping with lobes wrapped around and sheathing stem. Flowers in composite heads; up to 60 pale blue-lavender ray flowers surround yellow disk (turning reddish-orange) in heads to 1½ in. wide, with appressed to spreading phyllaries, at ends of branches forming dome-shaped array at top of plant. IA, IL, IN, MI, MN, MO, OH, WI. Widespread in the Northern Lakes, scattered elsewhere. *SN*

Symphyotrichum sericeum
ASTERACEAE

silky aster, western silver aster

Prairies, glades, fields, savannas, open woods. Summer, fall.

A sprawling to erect, wiry-stemmed, silvery-silky perennial to 2 ft. tall. Leaves alternate, simple, entire, to 1¼ in. long × ⅓ in. wide, clasping stem. Flowers in composite heads; up to 25 lavender to pale purple ray flowers surround white to cream-colored disk (turning reddish-brown) in heads to 1¼ in. wide, subtended by broad, silvery-silky, spreading phyllaries; heads stalked at ends of branches in upper part of plant. IA, IL, IN, MI, MN, MO, WI. Widespread in the Ozark Highlands, scattered in northern portion of the Tallgrass Prairie and adjacent portions of the Northern Lakes. *BS*

Symphyotrichum shortii
ASTERACEAE

Short's aster

Moist to dry upland forests, rocky woodlands. Summer, fall.

An erect perennial, mostly unbranched, lower stem glabrous, becoming hairy upward, 1½–4 ft. tall. Leaves alternate, entire, mostly glabrous (undersurface hairy), ovate-lanceolate with slightly cordate base, petiolate (petiole un-winged), 2–6 in. long × 1–2½ in. wide. Flowers in composite heads about 1 in. wide, ray flowers 12–20, disk flowers 15–25, phyllaries appressed and finely hairy, in a widely spreading terminal array. Short's aster is named for its discoverer, famed 19th-century botanist Charles W. Short. IA, IL, IN, MI, MN, OH, WI. Common in IL, IN, OH, and portions of eastern IA and southern WI. *MAH*

Symphyotrichum turbinellum
ASTERACEAE

turbinate aster

Acidic dry upland forests, blufftops, rocky glades, barrens. Summer, fall.

An erect perennial, glabrous to sparsely hairy in upper plant, 2–3 ft. tall. Leaves alternate, entire, dark green, linear to lance-elliptic, sessile, 2–4 in. long × ⅓–¾ in. wide. Flowers in composite heads about 1 in. wide, ray flowers 15–25, disk flowers 15–20+, phyllaries appressed, in a widely spreading terminal array. The epithet *turbinellum* refers to the involucre expanding upward in this species, shaped somewhat like a snow cone cup, or a toy top (at least when pressed; not as evident in live specimens). IA, IL, MO. **Occurs in most of MO, southern half of IL, and IA.** *ST*

Vernonia baldwinii
ASTERACEAE

western ironweed, Baldwin's ironweed

Prairies, fields, glades, open woods, thickets, roadsides. Summer, fall.

A hairy perennial to 5 ft. tall. Leaves alternate, simple, sharply toothed, sessile to short-petiolate, to 5 in. long × 2 in. wide, deeply veined. Flowers in composite heads; up to 30 disk flowers in ¾-in. heads, ray flowers absent, each head subtended by involucre of several series of spreading-tipped phyllaries, heads numerous, stalked, forming flat-topped array at top of plant. Curlytop ironweed (*V. arkansana*) has glabrous, glaucous stems, leaves to 1 in. wide, 50+ disk flowers per head, and spreading but incurved filiform phyllaries. IA, IL, IN, MN, MO, WI. **Widespread in the Ozark Highlands and southwestern portion of the Tallgrass Prairie, rare elsewhere.** *SN*

Vernonia fasciculata
ASTERACEAE

common ironweed, prairie ironweed, smooth ironweed

Wet prairies, marshes, shores, ditches. Summer, fall.

A glabrous perennial to 5 ft. tall with glaucous stem. Leaves alternate, simple, sharply toothed with whitish-tipped teeth and whitened mid-vein, sessile to short-petiolate, to 6 in. long × 1 in. wide, undersides pitted. Flowers in composite heads; up to 30 disk flowers in heads to ¾ in. across, ray flowers absent, each head subtended by involucre of several series of appressed phyllaries, heads numerous, stalked, forming flat-topped array at top of plant. Fruiting heads with cinnamon-brown pappus. IA, IL, IN, MI, MN, MO, OH, WI. Widespread in the Tallgrass Prairie and adjacent Northern Lakes, scattered to rare elsewhere. *BS*

Vernonia gigantea
ASTERACEAE

tall ironweed

Moist prairies, meadows, old fields, pastures, wet thickets, swamp borders. Summer, fall.

An erect perennial, minutely hairy, 4–8 ft. tall. Leaves alternate, short-petiolate, lanceolate, margins sharply toothed, leaf surfaces somewhat sparsely hairy to roughened with short, stiff hairs, undersurface with few or no resin dots, 4–10 in. long × 1–3 in. wide. Flowers in composite heads ½–¾ in. wide, ray flowers absent, disk flowers 10–30, phyllaries rounded or broadly angled with a short tip. Array of flat-topped clusters atop plant. IA, IL, IN, MI, MO, OH. Occurs mostly in the lower Midwest. *DT*

Vernonia missurica

ASTERACEAE

Missouri ironweed

Prairies, sedge meadows, fields, floodplains, streambanks, thickets, roadsides. Summer, fall.

A perennial to 6 ft. tall with densely hairy stems and leaf undersides. Leaves alternate, simple, shallowly toothed, sessile to short-petiolate, to 7 in. long × 2 in. wide. Flowers in composite heads; 30+ disk flowers in heads to ¾ in. across, ray flowers absent, each head subtended by urn-shaped involucre of several series of appressed phyllaries, heads numerous, stalked, forming flat-topped array at top of plant. Fruiting heads with cinnamon-brown pappus. Hybridizes with tall ironweed (*V. gigantea*). IA, IL, IN, MI, MO, OH. **Widespread in southern portion of the Tallgrass Prairie and southeastern portion of the Northern Lakes, scattered elsewhere.** *SN*

Caulophyllum giganteum

BERBERIDACEAE

giant blue cohosh

Rich mesic forests, rocky slopes, ravines. Spring.

An erect perennial, glabrous, 1–2 ft. tall. Leaves 2 per plant when flowering, upper smaller than lower, the latter mostly 2 or 3 times ternately compound, somewhat glaucous, obovate in outline as are leaflets, the latter commonly divided with rounded teeth, 2–4 in. long and about as wide, usually purplish and not fully expanded at start of flowering. Flowers regular, about ⅔ in. wide, petal-like sepals 6, mostly dark purple, true petals tiny, the style about ¹⁄₁₆ in. long. Inflorescence consisting of 5–15 flowers in open clusters. Berry-like seeds are fleshy and poisonous. The styles are about twice the length of those in *C. thalictroides*. IN, MI, OH. **Primarily occurs in eastern OH and MI.** *LC*

Cynoglossum virginianum
BORAGINACEAE

wild comfrey, hound's-tongue

Well-drained forested slopes, often rocky or sandy woods. Spring.

A perennial, 2–3 ft. tall, hairy. Leaves mostly basal, 7–10 in. long × 2–4 in. wide, oval and coarsely hairy (rough to the touch), clasping and progressively smaller above. Flowers regular, pale blue to white, ¼–½ in. wide with 5 rounded lobes somewhat overlapping, arranged singly on a few widely spreading terminal branches. Fruit a nutlet covered with tiny prickles allowing easy attachment to clothing and fur. In the upper Midwest is the similar northern wild comfrey (*C. boreale*); it has smaller flowers with non-overlapping petals. IL, IN, MO, OH. Common, mostly in the lower Midwest. *SN*

Echium vulgare
BORAGINACEAE

viper's bugloss, blueweed

Fields, vacant lots, roadsides, railroads, gravelly or sandy waste areas. Summer, fall.

A bristly-hairy, somewhat spiny biennial to 3 ft. tall. Stem purple-spotted. Leaves basal and alternate, sessile, to 9 in. long × 1¾ in. wide, becoming smaller along stem, with entire, often wavy margins. Flowers regular, blue (sometimes white or pinkish), funnel-shaped with 5 spreading lobes, fused at base, to 1 in. long × ¾ in. wide, with long-exserted stamens and pistil, numerous on 1-sided downcurved branches from main stem throughout much of plant. Increasing in ruderal areas with coarse soils. IA, IL, IN, MI, MN, MO, OH, WI. Introduced from Eurasia. Scattered throughout the Midwest, but more common near Great Lakes. *SN*

Heliotropium indicum
BORAGINACEAE

Indian heliotrope

Lowland forests, floodplains, ditches. Summer.

An erect annual, branched, sparsely hairy, 1–2 ft. tall. Leaves alternate, petiolate, ovate with rough and noticeably wrinkled upper surface with wavy margins, 1½–5 in. long × 1–3 in. wide. Flowers regular, ⅛ in. wide, funnel-shaped with 5 rounded lobes, numerous in compact rows along the upper side of an arching and coiled (scorpioid) terminal spike. This species is found in warmer regions throughout the world, including North America. It has been established in our region for over 100 years. IL, IN, MO, OH. Introduced from Asia. Relatively common in far southern counties. *ST*

Hydrophyllum appendiculatum
BORAGINACEAE

appendaged waterleaf

Rich mesic forests, stream terraces, ravines. Spring.

An erect biennial, branched above, hairy, 1–2 ft. tall. Leaves in early spring are basal, pinnately lobed and splotched with gray, later flowering stem leaves are alternate, palmately 5-lobed, 2½–6 in. long and wide. Flowers regular, ½–¾ in. wide, bell-shaped and 5-lobed, these oval and slightly wavy-margined, with 5 calyx lobes that have reflexed, narrowly triangular appendages alternating between them. Inflorescence of loosely flowered clusters standing above the leaves. IA, IL, IN, MI, MN, MO, OH, WI. Common in the southern two-thirds of the region, rare to absent in upper third. *SN*

Lappula squarrosa
BORAGINACEAE

European stickseed

Fields, roadsides, railroads, disturbed sites.
Spring, summer.

A hairy annual to 30 in. tall. Stem leaves to
4 in. long × ⅜ in. wide, sessile or nearly so,
smaller above; basal leaves longer, absent
when flowering. Flowers regular, with 5
spreading corolla lobes fused at base into
yellow throat, to ⅛ in. wide, subtended by 5
sepals and leaf-like bract, in racemes at ends
of branches in upper part of plant. Fruit consists of 4 nutlets, each with 2 rows of prickles.
Western stickseed (*L. redowskii*), introduced
from western North America, has nutlets
with a single row of prickles. Mistaken for
forget-me-nots (*Myosotis*). IA, IL, IN, MI, MN,
MO, OH, WI. Introduced from Eurasia. Scattered, more frequent northward. *KB*

Mertensia paniculata
BORAGINACEAE

northern bluebells, tall lungwort

Swamps, forests, woodland edges,
streambanks, roadsides. Spring, summer.

A bristly-hairy perennial to 4 ft. tall. Stem
leaves alternate, short-petiolate, to 5 in. long ×
1½ in. wide, deeply veined; basal leaves larger
and long-petiolate. Flowers regular, glabrous,
to ½ in. long, tubular and bell-shaped with
tube shorter than inflated portion, with 5
short flaring lobes, dangling from slender
stalks in several racemes from upper leaf
axils and upper part of plant. Fruit 1- to
4-parted, each part containing a nutlet. IA,
MI, MN, WI. Primarily in northern portion
of the Northern Lakes, rare in the Tallgrass
Prairie. *SN*

Mertensia virginica
BORAGINACEAE

Virginia bluebells

Rich bottomland forests, mesic ravines, stream terraces. Spring.

An erect perennial, glabrous, colonial, 1–2 ft. tall. Leaves alternate, sessile to short-petiolate, entire, elliptic to oblanceolate, rounded at tip, 2–8 in. long × 1–4 in. wide. Flowers regular, 1 in. long, tubular with 5 shallow, wavy-margined lobes forming a cup-shaped mouth, in nodding branches of multiple flowers. Plant dies down shortly after blooming and fruiting. Common comfrey (*Symphytum officinale*), a Eurasian native escaped from cultivation, is mostly summer-blooming with drooping bell-like blue flowers and hairy stem and inflorescence (see inset). IA, IL, IN, MI, MN, MO, OH, WI. **Common except in the Northern Lakes.** *MAH/SN*

Myosotis laxa
BORAGINACEAE

small forget-me-not

Sedge meadows, swamps, streambanks. Spring, summer, fall.

A hairy, colony-forming, short-lived perennial to 16 in. tall with weak, tangled stems. Leaves alternate, entire, short-petiolate, to 2½ in. long × ⅔ in. wide below, becoming smaller and sessile above. Flowers regular, with 5 spreading pale blue lobes fused at base into yellow throat, to ¼ in. wide, subtended by 5-lobed tubular calyx covered with appressed hairs, in racemes at ends of branches in upper part of plant, tips of flowering branches tightly curled and unfurling as flowers open. IN, MI, MN, OH, WI. **Scattered in the Northern Lakes and few areas in the adjacent Tallgrass Prairie, also in extreme eastern portion of the Eastern Forests.** *SN*

Myosotis scorpioides
BORAGINACEAE

true forget-me-not

Swamps, vernal pools, seeps, ponds, rivers, streams, ditches. Spring, summer, fall.

A hairy, colony-forming perennial to 20 in. tall with creeping to erect stems, rooting at nodes. Leaves alternate, entire, short-petiolate, to 3 in. long × ¾ in. wide below, smaller and sessile above. Flowers regular, with 5 spreading blue lobes fused at base into yellow throat, to ⅓ in. wide, subtended by 5-lobed tubular calyx covered with appressed hairs, in racemes in upper part of plant, tips of flowering branches tightly curled and unfurling as flowers open. Garden forget-me-not (*M. sylvatica*) has hook-tipped spreading hairs on calyx. IA, IL, IN, MI, MN, MO, OH, WI. Introduced from Eurasia. Widespread and common in the Northern Lakes, scattered elsewhere. *SN*

Phacelia bipinnatifida
BORAGINACEAE

fernleaf phacelia

Rich mesic forests, rocky slopes, ravines. Spring.

An erect biennial, branched, covered with sticky glandular hairs, 1–1½ ft. tall. Leaves alternate, petiolate, blotched with gray mottling, broadly triangular in outline, pinnately divided into 3–5 toothed leaflets 2–3 in. long × 1½–2 in. wide. Flowers regular, about ½ in. wide, pale lavender to dark purple, somewhat bell-shaped with 5 rounded lobes, in spreading branches atop the stems. The plant is quite sticky and emits a distinctive odor when handled. IL, IN, MO, OH. Occurs roughly within an area of lines connecting Cincinnati to Chicago to southwestern MO. Common. *MAH*

Phacelia hirsuta
BORAGINACEAE

fuzzy phacelia

Rocky prairies, glades, moist ravines, open woods, roadsides. Spring.

A hairy annual to 12 in. tall. Leaves alternate, to 2½ in. long × 1 ⅓ in. wide, deeply pinnately lobed with 5–9 lobes, petiolate below, sessile above. Flowers regular, to ½ in. wide, with 5 purplish-blue (rarely white) ascending to spreading corolla lobes, each with 2 deep purple spots at base above whitish short-tubular throat, with 5 exserted stamens topped by purplish anthers, short-stalked and numerous in curling cluster at top of stem. **MO. Almost entirely restricted to the Ozark Highlands.** *SN*

Phacelia purshii
BORAGINACEAE

Miami mist

Moist woods and thickets, floodplains, gravel bars. Spring.

An annual to 16 in. tall. Leaves alternate, to 2½ in. long × ¾ in. wide, deeply pinnately lobed with 5–9 lobes, petiolate below, sessile above. Flowers regular, to ¾ in. wide, glandular-hairy, with 5 delicately fringed lavender-blue ascending to spreading lobes that are white at base, with 5 exserted stamens topped by lavender anthers, stalked and numerous in curling clusters in upper part of plant. Brand's phacelia (*P. gilioides*), nearly restricted to the Ozark Highlands, has hairy (not glandular) flowers with shorter-fringed floral lobes with purple spots at base. **IL, IN, MI, MO, OH. Scattered in the Eastern Forests and eastern portion of the Ozark Highlands, rare elsewhere.** *SN*

Cardamine douglassii
BRASSICACEAE

purple cress

Rich, moist upland forests, stream terraces, flatwoods. Spring.

An erect perennial, unbranched, short-hairy, 4–10 in. tall. Leaves alternate, basal leaves long-petiolate, rounded to ovate, commonly withered by flowering time, stem leaves becoming sessile upward, oblong to lanceolate, entire or wavy-margined or with a few teeth, 1–2 in. long × ¼–1¼ in. wide. Flowers regular, about ½ in. wide, 4-petaled, lavender to white, in mostly terminal racemes that elongate when in fruit. Fruit an ascending silique up to 1½ in. long. Purple cress is an early spring ephemeral wildflower. IA, IL, IN, MI, MO, OH, WI. Most occurrences are in the eastern half of the Midwest. *SN*

Hesperis matronalis
BRASSICACEAE

dame's rocket

Mesic forests, stream terraces, roadsides. Spring, summer.

A biennial or short-lived perennial, erect, branched, hairy, 2–4 ft. tall. Leaves alternate, petiolate to sessile upward, lanceolate or ovate-lanceolate, toothed, with branched hairs on the undersurface, 2–7 in. long × ½–2½ in. wide. Flowers regular, about 1 in. wide, petals 4, mostly purple but also white and other colors, overlapping at their lower margins, in a branching terminal panicle. Fruit a silique 2–4 in. long. Often mistaken for a phlox. Dame's rocket resembles another garden plant, silver dollar plant (*Lunaria annua*), but the latter's fruit is broadly circular in shape. IA, IL, IN, MI, MN, MO, OH, WI. Introduced from Europe. Widespread and common. *SN*

Iodanthus pinnatifidus
BRASSICACEAE

purple rocket

Floodplain forests, ravine bottoms, stream terraces, seeps. Spring, summer.

An erect perennial, mostly unbranched, glabrous to hairy, 1–3 ft. tall. Leaves alternate, lanceolate to elliptic, irregularly toothed, petiolate or sessile upward, often purple when young, the lower leaves usually pinnately divided, 2–6 in. long × ½–2½ in. wide. Flowers regular, about ¼–⅓ in. wide, with 4 non-overlapping obovate petals bearing somewhat straight-edged tips, in terminal racemes. Fruit a silique. Although the specific epithet *pinnatifidus* refers to pinnately divided leaves, certain plants may not possess them. IA, IL, IN, MN, MO, OH, WI. **Widespread in the middle to lower Midwest, mostly absent in the Northern Lakes.** *MAH*

Campanula rapunculoides
CAMPANULACEAE

creeping bellflower

Open woodlands, savannas, thickets, homesites, roadsides, railroads. Summer, fall.

A colony-forming, upright perennial to 3 ft. tall with rough-hairy stems. Sap milky. Leaves basal and alternate, coarsely toothed, with deeply impressed veins, cordate with long petioles at base, to 5 in. long × 2 in. wide, becoming smaller, sessile, and lanceolate above, containing milky sap. Flowers regular, nodding, bell-shaped, to 1 in. long and about as wide, with 5 spreading to recurved corolla lobes and protruding style, in many-flowered often 1-sided raceme at top of plant. Commonly used as an ornamental. IA, IL, IN, MI, MN, MO, OH, WI. **Introduced from Europe. Abundant in the Northern Lakes, scattered elsewhere.** *PD*

Campanula rotundifolia
CAMPANULACEAE

harebell

Well-drained gravel slopes, sand savannas,
cliffs, dunes, shores. Summer.

An erect perennial, mostly unbranched and
glabrous, 8–24 in. tall. Sap milky. Leaves
basal, broadly ovate to cordate, usually not
present during flowering, stem leaves alternate, commonly linear, entire, 1–3 in. long
× ⅛–¼ in. wide. Flowers regular, about 1 in.
long, bell-shaped with 5 pointed lobes shorter
than the tube, inflorescence a lax raceme of
nodding flowers. In addition to its main range,
harebell also occurs in a single county in the
Ozark Highlands where it is thought to be a
glacial relict, growing on relatively cool, moist,
north-facing dolomite cliffs. **IA, IL, IN, MI,
MN, MO, OH, WI. Occurs almost exclusively
in the Northern Lakes.** *SN*

Campanulastrum americanum
CAMPANULACEAE

American bellflower

Moist forests, ravines, forest edges, trails,
old roadbeds, stream terraces. Summer, fall.

An erect winter annual (when growing in considerable sunlight) or biennial (when in less
sunlight), mostly unbranched, with milky sap,
2–6 ft. tall. Basal leaves in a rosette, petiolate,
cordate, with toothed margins. Stem leaves
alternate, with somewhat winged petioles,
sessile in upper leaves, ovate-lanceolate,
sparsely hairy and finely toothed, 2–6 in. long
× 1–2½ in. wide. Flowers regular, about 1 in.
wide, flat and saucer-shaped with 5 pointed
lobes. Inflorescence a tall spike-like raceme.
Syn. *Campanula americana.* **IA, IL, IN, MI,
MN, MO, OH, WI. Common and widespread
throughout the southern two-thirds of the
Midwest, absent northward.** *MAH*

Lobelia inflata

CAMPANULACEAE

Indian tobacco

Upland forests, stream terraces, trails, disturbed sites. Summer, fall.

An erect annual, branching above, glabrous to hairy, 8–24 in. tall. Leaves alternate, sessile or short-petiolate, oblong-elliptic, toothed, 1–4 in. long × ½–1½ in. wide, containing milky sap. Flowers irregular, blue to white (mostly white in the lower Midwest), about ½ in. wide, tubular, 5-lobed with 3 positioned lowermost and 2 above, mostly glabrous with tufts of white hairs at base of lower lobes, numerous in a spike-like raceme. Fruit an inflated capsule. It is the only annual *Lobelia* species in the Midwest. **IA, IL, IN, MI, MN, MO, OH, WI. Common throughout but mostly absent in the northwest.** *SN*

Lobelia kalmii

CAMPANULACEAE

Kalm's lobelia, brook lobelia, bog lobelia

Fens, marshes, shores, moist meadows, pannes, conifer swamps, calcareous rock crevices. Summer, fall.

A weak-stemmed perennial to 1½ ft. tall. Leaves alternate, simple, entire to sparsely toothed, to 2 in. long × ⅓ in. wide (basal leaves slightly shorter and wider), containing white sap. Flowers irregular, tubular with 2 small upper lobes and 3 larger spreading lower lobes, pale blue with white on lower central lobe, sometimes white, to ½ in. long, few on short stalks in terminal raceme. An indicator of calcareous substrate. **IA, IL, IN, MI, MN, OH, WI. Primarily in the Northern Lakes and the adjacent Tallgrass Prairie, scattered in the Eastern Forests.** *SN*

Lobelia siphilitica
CAMPANULACEAE

great blue lobelia

Streambanks, fens, marshes, pond and lake borders. Summer, fall.

An erect perennial, mostly unbranched, glabrous to sparsely hairy, 1–3 ft. tall. Leaves alternate, sessile to short-petiolate, elliptic to lanceolate, with shallow teeth, 1–6 in. long × ¼–2½ in. wide, containing yellow-green sap. Flowers irregular, about 1 in. wide, tubular, 5-lobed, the 3 lowermost often white near the throat, the upper 2 ascending, numerous in a tall spike-like raceme. The more-southern downy lobelia (*L. puberula*) has densely hairy stems and flowers less intensely blue than those of *L. siphilitica*. IA, IL, IN, MI, MN, MO, OH, WI. **Common throughout.** *MAH*

Lobelia spicata
CAMPANULACEAE

pale-spiked lobelia

Prairies, glades, savannas, fens, open woodlands. Spring, summer.

An erect perennial, unbranched, glabrous to sparsely hairy, 1–3 ft. tall. Leaves alternate, sessile, oblong-obovate to oblanceolate, ½–3 in. long × ½–1 in. wide, containing milky sap. Flowers irregular, blue to white, about ⅓ in. wide, tubular, 5-lobed, with 3 positioned lowermost, upper 2 slightly recurved, mostly glabrous with white hairs at base of lower lobes, numerous in a tall spike-like raceme atop the plant. Fruit a capsule. Several varieties have been named but not all are universally accepted. IA, IL, IN, MI, MN, MO, OH, WI. **Widespread across the Midwest but often very local.** *SN*

Triodanis perfoliata
CAMPANULACEAE

clasping Venus' looking glass

Dry open woodlands, barrens, glades, savannas, fields. Spring, summer.

An erect annual, mostly unbranched, hairy, 4–24 in. tall. Leaves alternate, kidney-shaped, sessile and clasping, toothed, up to 1 in. long and wide. Flowers regular, about ½ in. wide, short-tubular and somewhat saucer-shaped with 5 pointed lobes and white throat, sessile, 1–3 per leaf axil. Fruit a capsule. Seeds disperse through an open pore on the side of the capsule near the middle. Syn. *Specularia perfoliata*. Small Venus' looking glass (*T. biflora*; syn. *T. perfoliata* var. *biflora*) has narrower, non-clasping leaves with pore on upper side of capsule. IA, IL, IN, MI, MN, MO, OH, WI. Widespread and common except for upper Northern Lakes. *MAH*

Commelina communis
COMMELINACEAE

Asiatic dayflower, common dayflower

Streambanks, woods, thickets, old homesites, vacant lots. Summer, fall.

A sprawling to erect, colony-forming annual to 2½ ft. tall (often much shorter), rooting at nodes. Leaves alternate, sheathing stem, entire, glabrous, to 5 in. long × 2 in. wide, with parallel veins. Flowers irregular, approximately 1 in. wide, composed of 2 stalked, round, blue petals and 1 much smaller white petal, with conspicuous exserted stamens and staminodes, in clusters in upper part of plant, each cluster subtended by a folded, green, leafy spathe that is open to its base. IA, IL, IN, MI, MN, MO, OH, WI. Introduced from Asia. Throughout, with fewer occurrences northward. *MAH*

Commelina diffusa
COMMELINACEAE

climbing dayflower, spreading dayflower

Moist sandy streambanks, forest openings, lawns. Summer, fall.

A sprawling to erect annual to 2½ ft., rooting at nodes. Leaves alternate, sheathing stem, entire, glabrous, to 5 in. long × 2 in. wide, with parallel veins. Flowers irregular, 1 inch wide, with 2 large petals, 1 smaller petal, and conspicuous exserted stamens and staminodes, in clusters in upper part of plant, each cluster in a folded, green, leafy spathe that is open to its base. Spathes in our other native dayflowers fused at base. IL, IN, MO, OH. Scattered in the Ozark Highlands and southern portions of the Tallgrass Prairie and Eastern Forests. Generally more tropical and considered by some to be introduced in the Midwest. *SN*

Commelina erecta
COMMELINACEAE

slender dayflower

Sand dunes, sand prairies, barrens, glades, gravel bars. Spring, summer, fall.

A weakly erect to ascending perennial with some branching, sparsely hairy, 1–2 ft. tall. Leaves alternate, entire, sessile, sheathing with small lobes at sheath apex, linear-lanceolate, 2–6 in. long × ¼–1 in. wide. Flowers irregular, about 1 in. wide, petals 3, upper 2 round and blue, the lower much smaller and white, emerging from green, folded spathes that are fused for the lower third of their length, occurring singly or clustered near stem tips. Each flower opens in the morning, withering by afternoon. IA, IL, IN, MI, MN, MO, WI. Most occurrences are in Ozark Highlands and sand deposits elsewhere. *MAH*

Commelina virginica
COMMELINACEAE

Virginia dayflower

Wet floodplain forests, swamp borders, riverbanks. Summer.

An erect or ascending perennial, unbranched or branched, glabrous to short-hairy, 1–3 ft. tall. Leaves alternate, entire, sessile, sheathing with short reddish or brownish hairs on sheaths, lanceolate-elliptic, 3–8 in. long × ½–2 in. wide. Flowers irregular, about 1 in. wide, petals 3, round and blue, the lower scarcely smaller than upper 2, emerging from folded green spathes that are fused for the lower third of their length, occurring singly or clustered near stem tips. Like our other *Commelina* species, each flower opens in the morning and withers by afternoon. **IL, IN, MO, OH. Occurs in the southernmost Midwest.** *SN*

Tradescantia ohiensis
COMMELINACEAE

common spiderwort, Ohio spiderwort

Prairies, glades, wet meadows, savannas, open forests, floodplains, roadsides, railroads. Spring, summer.

A gray-green perennial to 3½ ft. tall. Leaves alternate, sheathing, entire, arching, to 15 in. long × ¾ in. wide, parallel-veined, with clear sap. Flowers regular, to 1 in. wide, with 3 spreading blue petals (sometimes rose, pink, or white) and ascending feathery yellow-tipped stamens, stalked in clusters from leaf-like bracts at top of stem. Long-bracted spiderwort (*T. bracteata*) and western spiderwort (*T. occidentalis*) have gland-tipped hairs on sepals. **IA, IL, IN, MI, MN, MO, OH, WI. Widespread in the Ozark Highlands, southern portion of the Tallgrass Prairie, and eastern portion of the Northern Lakes, scattered elsewhere.** *SN*

Tradescantia subaspera
COMMELINACEAE

zigzag spiderwort

Rich mesic forests, ravines, stream terraces. Summer.

An erect perennial, unbranched, glabrous to sparsely hairy, stem somewhat zigzagged, 1½–3 ft. tall. Leaves alternate, sessile and sheathing or some short-petiolate, entire, narrowly lanceolate to lanceolate-elliptic, 4–8 in. long × 1½–2 in. wide, mid-stem leaves much wider than width of their sheaths. Flowers regular, ½–1 in. wide, with 3 ovate petals and 3 glabrous or hairy sepals, clustered atop stem and in axils of upper leafy bracts. Zigzag spiderwort often forms tall, multi-stemmed clumps. IL, IN, MO, OH. Common, mostly in the lower Midwest. *MAH*

Tradescantia virginiana
COMMELINACEAE

Virginia spiderwort

Dry upland woods, prairies, savannas, blufftops. Spring.

An erect perennial, unbranched, mostly glabrous, 6–16 in. tall. Leaves alternate, sessile, sheathing at the base, linear-lanceolate, 5–15 in. long × ¼–¾ in. wide, mid-stem leaves narrower than or equal to width of their sheaths. Flowers regular, about 1 in. wide, with 3 ovate petals, 3 hairy sepals, and hairy pedicels that lack glands, clustered at apex above 2 leafy bracts. Wild crocus (*T. longipes*) of the Ozark Highlands is similar but shorter, with sepals and pedicels bearing some gland-tipped hairs. IA, IL, IN, MI, MO, OH. Common in the lower Midwest, mostly rare or absent northward and in the far western counties. *SN*

Ipomoea hederacea
CONVOLVULACEAE

ivy-leaved morning-glory

Agricultural fields, fallow fields, fencerows.
Summer, fall.

An annual twining vine, hairy, branching,
growing to 6 ft. long or more. Leaves alternate, petiolate, unlobed to usually deeply
3- or 5-lobed, each lobe ovate and coming to a
sharp-pointed tip, with a cordate base, 1½–5
in. long and about as wide. Flowers regular,
about 2 in. wide, funnel-shaped, shallowly
5-lobed, with 5 linear-lanceolate sepals up to 1
in. long with long-tapering and sharp-pointed
linear tips, occurring in few-flowered clusters from leaf axils. IA, IL, IN, MI, MN, MO,
OH, WI. Introduced from tropical America.
Widespread but absent from much of the
Northern Lakes. *MAH*

Ipomoea purpurea
CONVOLVULACEAE

common morning-glory

Agricultural fields, fallow fields, fencerows.
Summer, fall.

An annual twining vine, branched, hairy, 6
ft. long or more. Leaves alternate, petiolate,
unlobed and cordate, or occasionally 3- or
5-lobed, each lobe ovate in shape coming to
sharp-pointed tip, 1½–5 in. long and about
as wide. Flowers regular, about 2 in. wide,
funnel-shaped, shallowly 5-lobed, ranging from purple to multi-colored, with 5
oblong-ovate and hairy sepals up to ⅔ in. long
with sharp-pointed triangular tips, occurring
in few-flowered clusters from leaf axils. IA, IL,
IN, MI, MN, MO, OH, WI. Introduced from
tropical America. Widespread but absent in
much of the upper Midwest. *PR*

Dipsacus fullonum
DIPSACACEAE

common teasel

Roadsides, old fields, disturbed sites.
Summer.

An erect biennial, branched above, with stiff
prickles, 2–6 ft. tall. Leaves opposite, sessile,
pairs fused at their bases and cup-like, entire
or with few teeth, prickly on midvein beneath,
oblanceolate to lanceolate, 5–15 in. long × 3–4
in. wide. Flowers irregular, about ⅓ in. long,
narrow, tubular, 4-lobed, densely arranged
in a thistle-like head 4–5 in. tall below which
are long, narrow, upward-curving bracts. Syn.
D. sylvestris. Common teasel and the similar
cut-leaved teasel (*D. laciniatus*) are very invasive.
Cut-leaved teasel has white flowers and deeply
pinnately lobed leaves. IA, IL, IN, MI, MN, MO,
OH, WI. Introduced from Eurasia. Mostly in
eastern and southern Midwest counties. *MAH*

Amorpha canescens
FABACEAE

leadplant

Prairies, glades, bluffs, open woodlands.
Summer.

A hairy shrub to 3½ ft. tall. Leaves pinnately
compound, alternate, to 12 in. long, with up to
50 leaflets to ¾ in. long × ¼ in. wide. Flowers
irregular, numerous, to ⅓ in. long, tubular, in
spreading arrays of spikes to 6 in. long at ends
of branches. Although usually functionally
herbaceous, becomes large and woody in
absence of fire. Entire plant has grayish cast
as result of dense grayish-white (canescent)
pubescence. IA, IL, IN, MI, MN, MO, WI.
Common in the Ozark Highlands and Tallgrass Prairie, uncommon in the southern
portion of the Northern Lakes, rare in the
western portion of the Eastern Forests. *SN*

Amphicarpaea bracteata

FABACEAE

American hog peanut

Upland forests, ravines, thickets. Summer.

An annual vine up to 8 ft. long, variously hairy with appressed or spreading hairs. Leaves alternate, petiolate, divided into 3 ovate-oval leaflets 2–4 in. long × 1½–3 in. wide. Flowers irregular with typical pea family shape, ½–¾ in. long, petals 5, lavender to white, in clusters of up to 15 on long axillary stalks. Fruit a legume. There are also self-fertile, petal-less flowers and fruit at the soil surface. Variety *comosa* is typically found in wet areas and has lavender flowers and spreading stem hairs; var. *bracteata* is often in drier areas and has white flowers and weakly hairy stems. IA, IL, IN, MI, MN, MO, OH, WI. Common throughout. *MAH*

Astragalus crassicarpus

FABACEAE

groundplum

Prairies, glades, dry open woods. Spring.

A hairy, sprawling perennial to 2 ft. Leaves alternate, short-petiolate, pinnately compound with up to 29 leaflets; leaflets ¾ in. long × ¼ in. wide, entire. Flowers irregular with typical pea family shape, pinkish-purple to blue-violet (cream in var. *trichocalyx*, with densely hairy calyx; see inset), ¾ in. wide, in dense, stalked racemes of up to 25 in leaf axils. Fruit a plum-like legume, 1 in. wide. Field milkvetch (*A. agrestis*) and Ozark milkvetch (*A. distortus*) produce longer, narrow legumes; the latter has glabrous upper leaflet surfaces. IA, IL, MN, MO, WI. Widespread in the Ozark Highlands, scattered in western portion of the Tallgrass Prairie. *SN*

Baptisia australis
FABACEAE

blue wild indigo, blue false indigo

Rocky riverbanks, prairies, glades. Spring, summer.

A perennial with erect or spreading stems, branched, glabrous, 2–4 ft. tall. Leaves alternate, divided into 3 short-stalked, oblance-olate leaflets 1–2½ in. long × ⅓–1¼ in. wide. Flowers irregular with typical pea family shape, about 1 in. long, densely arranged on upright terminal racemes. Fruit a legume. In the Midwest, var. *australis* generally occurs naturally near and along the Ohio River; var. *minor*, typically smaller and more spreading, is confined to the Ozark Highlands and southwestern Tallgrass Prairie. IA, IN, MO, OH. A few introduced occurrences exist elsewhere. *MAH*

Clitoria mariana
FABACEAE

butterfly pea

Dry upland woodlands, barrens. Summer.

A sprawling, climbing perennial vine, glabrous to sparsely hairy, up to 3 ft. long. Leaves alternate, compound, petiolate, leaflets 3, entire, ovate-lanceolate, the leaflets 1–3 in. long × ½–1½ in. wide. Flowers irregular with typical pea family shape, about 2 in. long, the large, obovate banner petal streaked within. Inflorescence of 1–3 flowers in leaf axils. Fruit a legume. Butterfly pea is unusual for the pea family as its flower is naturally positioned with the banner petal lowermost. The common name alludes to the butterfly-like appearance of the flowers. IL, IN, MO, OH. Occurs primarily in the far southern Midwest. *DT*

Dalea purpurea
FABACEAE

purple prairie clover

Dry to moist prairies, savannas, glades, woodland openings, often in sand or gravel. Summer.

A perennial to 3 ft. tall. Leaves pinnately compound, alternate, glabrous, dark green, to 5 in. long, with up to 7 linear leaflets. Flowers irregular with typical pea family shape, numerous, to ¼ in. wide, in cylindrical spikes to 2 in. long × ¾ in. wide at top of plant, each spike blooming from bottom up. The leaves are darker green with more linear leaflets than those of white prairie clover (*D. candida*), with which it sometimes grows. IA, IL, IN, MI, MN, MO, OH, WI. **Widespread in the Ozark Highlands and Tallgrass Prairie, uncommon in western portion of the Northern Lakes and Eastern Forests.** *SN*

Desmodium canadense
FABACEAE

showy tick-trefoil

Prairies, glades, floodplain meadows, railroads. Summer.

A hairy perennial to 6 ft. tall. Leaves compound with 3 leaflets each to 3½ in. long, alternate, short-petiolate. Flowers irregular with typical pea family shape, pinkish to purplish, numerous, to ½ in. wide, in several racemes to 1½ ft. long in upper part of plant; flowers large compared to most other species in the genus. Fruit a sticky-hairy segmented legume. Illinois tick-trefoil (*D. illinoense*) is overall more slender and unbranched with grayish cast and leaf undersides covered in hooked hairs that produce a Velcro-like sticky texture. IA, IL, IN, MI, MN, MO, OH, WI. **Common throughout, with fewer occurrences in the Eastern Forests, Ozark Highlands, and northern portion of the Northern Lakes.** *SN*

Desmodium cuspidatum
FABACEAE

large bracted tick-trefoil

Mesic to dry upland forests, clearings, edges, thickets. Summer, fall.

An erect perennial, branched above, glabrous or sparsely hairy, 2–5 ft. tall. Leaves alternate, long-petiolate with persistent lanceolate stipules ½–¾ in. long, compound with 3 long-pointed and entire lance-ovate leaflets, the larger terminal one 2–6 in. long × 1½–3 in. wide. Flowers irregular with typical pea family shape, about ½ in. long, in sparsely flowered terminal racemes. Fruit a legume of 1–7 triangular segments. Fruit of this and other *Desmodium* species have hooked hairs that allow them to stick to fur and clothing. IA, IL, IN, MI, MN, MO, OH, WI. **Widespread but generally most prevalent in our southwest.** *MJH*

Desmodium marilandicum
FABACEAE

smooth small-leaved tick-trefoil

Barrens, savannas, dry woodlands, old fields. Summer, fall.

An erect to ascending perennial, unbranched except above, mostly glabrous, 2–3 ft. tall. Leaves alternate, long-petiolate with deciduous linear stipules, compound with 3 oval and blunt-tipped entire leaflets ½–1½ in. long and almost as wide, terminal one about same size as lateral ones. Flowers irregular with typical pea family shape, about ¼ in. long, in sparsely flowered terminal racemes. Fruit a legume with 1–4 rounded segments bearing hooked hairs. Hairy small-leaved tick-trefoil (*D. ciliare*) is similar both in appearance and range but is obviously hairy throughout. IL, IN, MI, MO, OH. **Mostly in Ozark Highlands.** *SN*

Desmodium paniculatum
FABACEAE

panicled tick-trefoil

Forest edges, thickets, savannas, fields. Summer, fall.

An erect perennial, glabrous to sparsely hairy (hairs appressed and straight, but some also hooked), 2–4 ft. tall. Leaves alternate, long-petiolate with non-persistent linear stipules, compound with 3 entire, linear-lanceolate to lance-ovate leaflets 2–4 in. long × ⅓–¾ in. wide. Flowers irregular with typical pea family shape, about ¼ in. long, in widely spreading terminal racemes. Fruit a legume with 3–6 rounded to bluntly triangular segments bearing hooked hairs. Similar tick-trefoil species generally have wider, more densely hairy leaflets. IA, IL, IN, MI, MO, OH. Widespread and common in southern two-thirds of Midwest. SN

Desmodium rotundifolium
FABACEAE

round-leaved tick-trefoil

Dry upland forests and savannas, barrens, thickets. Summer, fall.

A prostrate, trailing perennial vine, branched, hairy, 3–5 ft. long. Leaves alternate, petiolate with persistent ovate stipules, compound with 3 round and entire leaflets 1–2 in. long and about as wide. Flowers irregular with typical pea family shape, about ⅓–½ in. long, upright in open terminal and axillary racemes. Fruit a legume with 3–6 somewhat triangular segments bearing hooked hairs. With its trailing habit and clearly round leaflets, this is one of the easiest tick-trefoil species in our region to identify. IL, IN, MI, MO, OH. Found mostly in the southern and eastern parts of the Midwest. MAH

Desmodium sessilifolium
FABACEAE

sessile-leaved tick-trefoil

Dry prairies, glades, sand savannas.
Summer, fall.

An erect perennial, branched above, hairy
with straight and hooked hairs, 2–5 ft. tall.
Leaves alternate, sessile or short-petiolate, with
deciduous lanceolate stipules, compound with
3 linear-oblong to elliptic-lanceolate entire leaflets 1½–3 in. long × ¼–½ in. wide. Flowers irregular with typical pea family shape, about ¼ in.
long, pale pink or white, in terminal racemes
of a few ascending branches. Fruit a legume
of 1–4 segments beset with hooked hairs and
rounded margins. This species has some of the
narrowest leaflets of our tick-trefoils, at least
4–6 times as long as wide. IA, IL, IN, MI, MO,
OH. Occurs mostly in MO, IL, northwestern
IN, and southern MI. *MJH*

Galactia regularis
FABACEAE

eastern milkpea

Dry open woodlands, glades, barrens.
Summer.

A trailing or somewhat climbing perennial
vine, with spreading or downwardly bent
hairs, 2–5 ft. long. Leaves compound, petiolate, with 3 sparsely hairy, oblong-elliptic
leaflets ½–1½ in. long × ⅓–¾ in. wide. Flowers
irregular with typical pea family shape, about
½ in. long, 1–5 in short racemes from leaf axils.
Fruit a legume. Both the genus and common
name allude to milky sap found in the stems,
but neither this nor any of the other 20 U.S.
species have it. IL, IN, MO, OH. Principally in
the Ozark Highlands and counties near the
Ohio River in the Eastern Forests. *AG*

Hylodesmum glutinosum
FABACEAE

pointed-leaved tick-trefoil, clustered tick-trefoil

Forests, forest borders, thickets, streambanks. Summer.

A perennial to 3½ ft. tall. Leaves compound, 5 or 6 closely alternate and appearing whorled, on petioles to 5 in. long, each with 3 leaflets to 5 in. long × 3 in. wide, terminal leaflet largest and abruptly sharp-pointed. Flowers irregular with typical pea family shape, to ⅓ in. wide, widely spaced in upright raceme or panicle to 2 ft. long above leaves. Fruit a legume with 1–4 triangular segments covered in tiny hooked hairs that allow them to attach to fur and clothing. Syn. *Desmodium glutinosum.* IA, IL, IN, MI, MN, MO, OH, WI. Common throughout. *SN*

Hylodesmum nudiflorum
FABACEAE

naked tick-trefoil

Forests. Summer.

A perennial to 3 ft. tall with separate leafy and flowering stems. Leaves compound, 5 or 6 closely alternate and appearing whorled atop 1-ft.-tall leafy stem, on petioles to 3 in. long, each with 3 leaflets to 3½ in. long × 2½ in. wide. Flowers irregular with typical pea family shape, on taller naked stem, pale pink to lavender, to ⅓ in. wide, in loose raceme or panicle. Fruit a legume with 1–4 triangular segments covered in tiny hooked hairs that allow them to attach to fur and clothing. Syn. *Desmodium nudiflorum.* IA, IL, IN, MI, MN, MO, OH, WI. Common; absent from northernmost and northwesternmost portions of the Midwest. *MAH*

Kummerowia stipulacea
FABACEAE

Korean clover

Fields, disturbed sites, roadsides. Summer.

An erect to reclining annual, branched with hairs pointing toward stem tip, 4–18 in. tall. Leaves alternate, petiolate, with 3 obovate leaflets ½ in. long × ⅓ in. wide, bearing noticeable parallel veins and a pair of papery stipules at base of petiole. Flowers irregular with typical pea family shape, about ⅛ in. long, 1 to few in leafy, axillary racemes. Fruit a flat, single-seeded legume. The similar-looking Japanese clover (*K. striata*) has stem hairs pointing downward. IA, IL, IN, MI, MO, OH, WI. **Introduced from eastern Asia. Occurs principally in the lower Midwest.** *ST*

Lathyrus palustris
FABACEAE

marsh vetchling, marsh pea

Marshes, sedge meadows, wet prairies, interdunal swales, shores, swamps, ditches. Spring, summer.

A clambering perennial herbaceous vine to 4 ft. long. Stems strongly winged (in var. *palustris*) or not (in var. *myrtifolius*). Leaves alternate, petiolate, with 2 leafy stipules at base, pinnately compound with 3 or 4 pairs of entire leaflets to 2 in. long × ¼ in. wide, terminating in branched tendril. Flowers irregular with typical pea family shape, purplish, pinkish, bluish, or two-toned with white, to ¾ in. wide, in racemes of up to 8 from leaf axils. Fruit a flattened legume to 2½ in. long. IA, IL, IN, MI, MN, MO, OH, WI. **Widespread and common northward, becoming scattered to rare southward.** *PR*

Lathyrus venosus
FABACEAE

veiny pea

Dry sandy or rocky forests, prairies.
Spring, summer.

A sprawling vine, stems 4-angled, mostly
glabrous, 1–3 ft. tall, often colonial. Leaves
alternate, stipulate, pinnately compound with
8–12 oval leaflets, finely hairy and terminated
by tendrils, about 6–8 in. long × 3–4 in. wide,
stipules narrow, less than ¼ in. wide. Flowers
irregular with typical pea family shape, about
½–¾ in. long, the banner petal with dark
veins, occurring in dense racemes from upper
leaf axils. Fruit a legume. The similar beach
pea (*L. japonicus*), found primarily along the
Great Lakes, has larger stipules, about ⅓–¾
in. wide. IA, IL, IN, MI, MN, MO, OH, WI.
Mostly in the upper Midwest. *CR*

Lespedeza frutescens
FABACEAE

violet bush-clover

Dry open woodlands, savannas, prairies.
Summer, fall.

A perennial to 2 ft. tall. Leaves compound
with 3 short-stalked leaflets (terminal slightly
larger) to 1½ in. long × ¾ in. wide, alternate,
on petioles to 1½ in. long. Flowers irregular
with typical pea family shape, numerous, to ⅖
in. wide, in several spreading racemes or panicles longer than leaves in upper part of plant,
also short-stalked in mid-stem leaf axils.
Slender bush-clover (*L. virginica*) has leaflets
narrower in relation to length. Wand-like
bush-clover (*L. violacea*) has short-stalked
inflorescences (shorter than leaves) resulting
in plant with wand-like appearance. IL, IN,
**MI, MO, OH. Widespread to scattered south-
ward and in southern MI.** *MAH*

Lespedeza procumbens

FABACEAE

trailing lespedeza

Dry upland woodlands, barrens. Summer.

A trailing, branched perennial, occasionally mat-forming, with spreading downy hairs, 1–4 ft. long. Leaves alternate, short-petiolate, with 3 oval leaflets ½–1 in. long × ¼–⅔ in. wide. Flowers irregular with typical pea family shape, ¼–⅓ in. long, the banner petal purple and streaked at base, 2–8 or more terminating axillary stalks that extend well above the leaves. Fruit a legume. Creeping lespedeza (*L. repens*) has a similar range and appearance, but its stems and leaves are nearly smooth or with closely appressed hairs. **IL, IN, MI, MO, OH, WI. Occurs mostly in the Ozark Highlands and Eastern Forests.** *DT*

Lespedeza violacea

FABACEAE

wand-like bush-clover

Dry open woodlands, savannas, prairies. Summer, fall.

A perennial to 3 ft. tall. Leaves compound with 3 short-stalked leaflets (terminal largest) to 1½ in. long × ¾ in. wide, alternate, on petioles to 1½ in. long. Flowers irregular with typical pea family shape, numerous, to ¼ in. wide, in short racemes shorter than leaves in upper part of plant. Syn. *L. intermedia*. Violet bush-clover (*L. frutescens*) has leaflets slightly broader in relation to length and flowers on stalks longer than leaves. Tall bush-clover (*L. stuevei*) has conspicuously spreading-hairy stems and leaves and is much less common. **IA, IL, IN, MI, MN, MO, OH, WI. Widespread southward and into southern half of MI.** *SN*

Lespedeza virginica
FABACEAE

slender bush-clover

Woodlands, savannas, barrens. Summer.

An erect perennial, mostly unbranched, hairy, 1–3 ft. tall. Leaves alternate, petiolate with 3 leaflets, the leaflets linear to narrowly elliptical or oblong, ½–1½ in. long × ¼–½ in. wide. Flowers irregular with typical pea family shape, about ¼ in. long, the banner petal with dark pink streaking at base, occurring 4–10 or more nested in axils of upper leaves. Fruit a legume. The species is variable, but typical specimens have linear leaflets 4–8 times as long as wide. IA, IL, IN, MI, MN, MO, OH, WI. Mostly in the Midwest's lower half. *EH*

Lupinus perennis
FABACEAE

wild lupine

Sand prairies, barrens, savannas, open woodlands, dunes. Spring, summer.

A perennial to 2 ft. tall. Leaves alternate, on petioles to 4 in. long, compound with 7–11 stalkless, gray-green, entire, palmately arranged leaflets to 2½ in. long × ½ in. wide. Flowers irregular with typical pea family shape, blue-violet (rarely white or pinkish), to ¾ in. long, numerous in erect terminal raceme to 10 in. long. Fruit a hairy legume to 2 in. long. The introduced bigleaf lupine (*L. poly-phyllus*) is larger in all aspects and has leaves with 9–17 leaflets. IA, IL, IN, MI, MN, OH, WI. Widespread in southern portion of the Northern Lakes and the adjacent Tallgrass Prairie. *SN*

Medicago sativa
FABACEAE

alfalfa, lucerne

Fields, pastures, roadsides, railroads. Spring, summer, fall.

A gray-green perennial to 2½ ft. tall. Leaves alternate, petiolate, compound with 3 minutely toothed leaflets to 1 in. long × ⅓ in. wide (many of which are blunt at tip), terminal leaflet short-stalked, lateral leaflets stalkless. Flowers irregular with typical pea family shape, to ⅓ in. long, lavender to deep violet (yellow in subsp. *falcata*), numerous in tight racemes to 2 in. long at ends of branches in upper part of plant. Fruit a tightly coiled legume to ⅓ in. long, in clusters. IA, IL, IN, MI, MN, MO, OH, WI. Introduced from Europe and cultivated as a crop. Widespread. *CR*

Orbexilum onobrychis
FABACEAE

French grass, lanceleaf scurfpea

Prairies, open woods, riverbanks, pond margins, ditches. Spring, summer.

A colony-forming perennial to 3½ ft. tall. Leaves alternate, on petioles to 6 in. long, compound with 3 pointed, ovate, entire leaflets to 4 in. long × 2 in. wide, the central on stalk to 1 in. long, the lateral on very short stalks. Flowers irregular with typical pea family shape, blue-violet (rarely white), to ¼ in. long, numerous and very short-stalked in dense, erect racemes to 6 in. long on stalks to 8 in. long from leaf axils. Fruit a short legume. Syn. *Psoralea onobrychis*. IA, IL, IN, MO, OH. Primarily in eastern portion of the Tallgrass Prairie and the Eastern Forests, rare elsewhere. *PR*

Orbexilum pedunculatum
FABACEAE

Sampson's snakeroot

Dry open woodlands, prairies, acidic glades, barrens. Summer.

An erect, glabrous to hairy perennial, 1–2½ ft. tall, somewhat colonial. Leaves alternate, petiolate, compound with 3 narrowly oblong to lanceolate, gland-dotted leaflets 1½–2¾ in. long × ¼–½ in. wide. Flowers irregular with typical pea family shape, about ¼ in. long, banner petal darker purple at base, clustered in cylindrical, congested racemes on long stalks from leaf axils. Fruit a small, rounded legume. Syn. *Psoralea psoralioides*. **IL, IN, MI, MO, OH. Found mostly in the Ozark Highlands and the southern Eastern Forests. Considered a waif in MI.** *CB*

Oxytropis lambertii
FABACEAE

Lambert's locoweed, Lambert's crazyweed

Prairies. Spring.

A silky-hairy, silvery-green perennial to 16 in. tall. Leaves in basal rosettes, erect to ascending, to 6 in. long, pinnately compound with 9–19 entire, sessile leaflets to 1 in. long × ¼ in. wide. Flowers irregular with typical pea family shape, to ¾ in. long, pinkish-purple (becoming blue with age), with beaked keel, numerous in erect terminal spike to 12 in. long on leafless stem from middle of or near basal rosette. Fruit an erect, hairy legume (becoming glabrous with age) to 1 in. long. Common names characterize the behavior of animals after ingesting the plant. **IA, MN, MO. Primarily in northwestern portion of the Tallgrass Prairie, with few occurrences in the Northern Lakes.** *JP*

Pediomelum argophyllum

FABACEAE

silverleaf scurfpea

Prairies. Summer.

A shiny, branching, silvery-hairy perennial to 3 ft. tall. Leaves alternate, on petioles to 1½ in. long, palmately compound with 3 (rarely to 5) very short-stalked, pointed, oval, entire leaflets of different sizes but to 1¼ in. long × ½ in. wide. Flowers irregular with typical pea family shape, to ⅓ in. long, dark blue to blue-violet, sessile, in 2–8 separated tight-whorled clusters on stalks from upper leaf axils. Fruit a small, silky legume. Syn. *Psoralea argophylla*. **IA, IL, MN, MO, WI. Primarily in western portion of the Tallgrass Prairie, rare in the Northern Lakes.** *SN*

Pediomelum esculentum

FABACEAE

prairie turnip, timpsila

Dry prairies, calcareous glades. Spring, summer.

An erect perennial from a thickened taproot, branched, hairy, 4–16 in. tall. Leaves alternate, petiolate, palmately compound with 5 narrowly obovate to lanceolate leaflets, those 1–2½ in. long × ¼–½ in. wide. Flowers irregular with typical pea family shape, ½–¾ in. long, several in a crowded leafy spike atop stem. Fruit a 1-seeded legume. The tuberous root of prairie turnip is a favored food of the Lakota and other Plains tribes, known to them as timpsila. Syn. *Psoralea esculenta*. **IA, MN, MO, WI. Occurs mostly in our western counties.** *PD*

Pediomelum tenuiflorum

FABACEAE

slenderleaf scurfpea, gray scurfpea, wild alfalfa

Prairies, glades, savannas, open woods. Spring, summer, fall.

A branching perennial to 3 ft. tall. Leaves alternate, petiolate, palmately compound with 3–5 (rarely 7) sessile or short-stalked, entire leaflets to 1¾ in. long × ⅓ in. wide, pitted or gland-dotted on top surface, densely hairy below. Flowers irregular with typical pea family shape (see inset), to ¼ in. long, numerous in loose, erect racemes to 3 in. long on stalks to 4 in. long from upper stems and leaf axils. Syns. *Psoralea tenuiflora*, *Psoralidium tenuiflorum*. IA, IL, IN, MN, MO, WI. Primarily in northern portion of the Ozark Highlands and southern portion of the Tallgrass Prairie, rare elsewhere. *PD/BS*

Pueraria montana

FABACEAE

kudzu

Forest edges, roadsides, fencerows, disturbed sites. Summer.

A climbing and twining perennial vine, branched, hairy, 30–100+ ft. long. Leaves alternate, long-petiolate, compound with 3 ovate, commonly 2- or 3-lobed leaflets 4–6 in. long and almost as wide. Flowers irregular with typical pea family shape, ½–1 in. long, with a scent of grapes, on axillary racemes. Fruit a legume. This species is famous for its ability to rapidly overtake almost anything in its path, including houses. Widely promoted for erosion control, it is a prime example of good intentions gone awry. IL, IN, MI, MO, OH. Introduced from Asia. Most common in the far southern counties. *DT*

Vicia americana

FABACEAE

American vetch

Prairies, forests, swamps, thickets, roadsides. Spring, summer.

A climbing perennial herbaceous vine to 3 ft. Leaves alternate, petiolate, pinnately compound with up to 8 pairs of subopposite, entire leaflets, terminating in tendril, leaflets to 1½ in. long × ¼ in. wide, shorter toward tendril; 2 leafy, 3-toothed, pointed-tipped stipules to ⅓ in. at base of leaf. Flowers irregular with typical pea family shape, pinkish-purple or bluish, to ¾ in. long, in racemes of up to 9 in leaf axils. Fruit a flattened legume, to 1 in. long. IA, IL, IN, MI, MN, MO, OH, WI. Widespread in western portion of the Northern Lakes and northern portion of the Tallgrass Prairie, becoming scattered eastward and rare to the south. *SN*

Vicia sativa

FABACEAE

common vetch

Fields, roadsides, railroads, waste areas. Spring, summer.

A climbing, annual herbaceous vine to 3 ft. Leaves alternate, short-petiolate, pinnately compound with up to 8 pairs of subopposite, entire leaflets with spine-like tip from shallowly indented apex, terminating in branched tendril, leaflets to ¾ in. long × ¼ in. wide, shorter toward tendril; 2 leafy, pointed-tipped, ¼-in.-long stipules with 2 or more teeth on margins at base of leaf. Flowers irregular with typical pea family shape, to 1 in. long, short-stalked, solitary or in pairs in leaf axils. Fruit a flattened legume to 3 in. long. IA, IL, IN, MI, MN, MO, OH, WI. Introduced from Europe. Scattered throughout, most frequent in MI and southern MO. *MJH*

Vicia villosa
FABACEAE

hairy vetch

Prairies, fields, forests, streambanks, roadsides, waste areas. Spring, summer, fall.

A sprawling, spreading-hairy, annual herbaceous vine to 3 ft. Leaves alternate, petiolate, pinnately compound with up to 12 pairs of subopposite, entire leaflets to 1 in. long × ¼ in. wide, terminating in tendril, leaflets becoming shorter toward tendril; 2 leafy, entire, pointed-tipped stipules to ⅓ in. at base. Flowers irregular with typical pea family shape, pinkish-purple to blue-violet, to ¾ in. long with bulbous base, short-stalked, up to 20 in 1-sided racemes in leaf axils. Fruit a flattened legume to 2 in. long. Cow vetch (*V. cracca*), perennial, has appressed hairs and flowers lack bulbous base. **IA, IL, IN, MI, MN, MO, OH, WI. Introduced from Eurasia. Throughout.** *CR*

Gentiana andrewsii
GENTIANACEAE

bottle gentian, closed gentian

Prairies, fens, thickets, swamps, floodplains. Summer, fall.

A perennial to 2 ft. tall. Leaves opposite, entire, glabrous, glossy, to 4 in. long × 2 in. wide, larger above than below, sessile. Flowers regular, blue to purple, sometimes pink or very pale, to 1½ in. long, tubular, pleated, closed at top, with 5 fused corolla lobes interspersed with fringed appendages (plaits) longer than lobes, clustered at top of plant and in upper leaf axils. Soapwort gentian (*G. saponaria*) and Great Lakes gentian (*G. rubricaulis*) have more blunt-tipped flowers and plaits not longer than corolla lobes; plaits fringed in former and mostly triangular in latter. **IA, IL, IN, MI, MN, MO, OH, WI. Widespread but not abundant.** *SN*

Gentiana puberulenta
GENTIANACEAE

prairie gentian, downy gentian

Prairies, barrens, glades. Summer, fall.

A perennial to 1½ ft. tall. Leaves opposite
(in whorl of 3–7 just below inflorescence),
entire, glabrous, glossy, to 3 in. long × 1¼ in.
wide, sessile. Flowers regular, brilliant blue
to violet, to 2 in. long and wide, bell-shaped,
upright, corolla fused below with 5
spreading-ascending, pointed lobes connected
by short fringed appendages (plaits), clustered
at top of plant and from upper leaf axils. Prairie gentian is often an indicator of a remnant
natural area; rarely planted. IA, IL, IN, MI,
MN, MO, OH, WI. Uncommon, primarily in
the Ozark Highlands, Tallgrass Prairie, and
adjacent portions of the Northern Lakes,
with scattered occurrences elsewhere. *CB*

Gentianella quinquefolia
GENTIANACEAE

stiff gentian, agueweed

Prairies, seeps, savannas, thin woods, bluffs,
cliffs, streambanks, often in calcareous
substrates. Summer, fall.

An annual to 2 ft. tall with square, somewhat
winged stems. Leaves opposite, entire, glabrous, to 2 in. long × 1 in. wide, sessile with
impressed parallel veins. Flowers regular, light
blue to violet, sometimes whitish, to 1 in. long
× ½ in. wide, upright, tubular with 5 lobes,
pointed at tip when closed or barely open
with lobes erect, appendages (plaits) between
corolla lobes lacking, clustered at top of plant
and from upper leaf axils. Maintains blooms
late into the fall. IA, IL, IN, MI, MN, MO, OH,
WI. Scattered throughout, but mostly absent
in northernmost and westernmost portions
of Midwest. *ST*

Gentianopsis crinita
GENTIANACEAE

greater fringed gentian

Wet prairies, sedge meadows, fens, seeps, interdunal wetlands, open swamps. Summer, fall.

An annual or biennial to 2½ ft. tall. Leaves opposite, entire, glabrous, glossy, to 2½ in. long × 1 in. wide, sessile. Flowers regular, brilliant blue to violet, to 2 in. long × 2 in. wide, trumpet-shaped, upright, fused below with 4 spreading, conspicuously fringed lobes (fringing irregular and uneven), solitary on numerous stalks in upper part of plant. Lesser fringed gentian (*G. virgata*) is smaller overall, with narrower leaves (to ⅖ in.) and more even fringing on petals. IA, IL, IN, MI, MN, OH, WI. Primarily in the Northern Lakes and adjacent Tallgrass Prairie, with few occurrences in the Eastern Forests. Not abundant. *SN*

Geranium maculatum
GERANIACEAE

wild geranium

Rich mesic forests, stream terraces, ravines. Spring.

An erect perennial, unbranched or branched, hairy, 1–2 ft. tall. Stem and basal leaves similar, the former opposite, both petiolate, rounded, palmately divided into 3–5 wedge-shaped lobes toothed at tips, 3–5 in. long and wide. Flowers regular, 1–1½ in. wide, with 5 obovate petals, on long stalks in an open inflorescence of 1–4 flowers. Fruit narrow with a long crane-like beak. The greenhouse potted geranium, while of the same family as wild geranium, belongs to *Pelargonium*, most species of which are native to South Africa. IA, IL, IN, MI, MN, MO, OH, WI. Widespread and mostly common. *MAH*

Geranium robertianum
GERANIACEAE

herb-Robert, Robert's geranium, mountain cranesbill

Moist forests, forest openings and edges, gravelly shores. Spring, summer, fall.

A weak, hairy annual to 1½ ft. tall. Stems reddish. Leaves opposite to whorled below on long stalks, alternate and sessile above, compound into 3 segments with terminal segment stalked, leaflets pinnately lobed, and lobes again lobed, overall leaf shape generally pentagonal (below) to triangular (above). Flowers regular, to ¾ in. wide, with 5 lavender or pink to white petals, in stalked, branched clusters at top of plant. Fruit long-beaked. IA, IL, IN, MI, MN, MO, OH, WI. Primarily in eastern portion of the Northern Lakes, with more scattered representation in the Eastern Forests and Tallgrass Prairie. *BS*

Camassia scilloides
HYACINTHACEAE

wild hyacinth, eastern camas

Mesic upland forests, glades, prairies, floodplain terraces. Spring.

An erect perennial, unbranched, glabrous, 1–2 ft. tall, often in colonies. Leaves a basal rosette, linear, strap- or grass-like, emanating from a bulb, 8–18 in. long × ¼–¾ in. wide. Flowers regular, about 1 in. wide, with 6 radiating linear to elliptic tepals, up to 50 in a terminal raceme. A close relative, prairie hyacinth (*C. angusta*), has deeper purple tepals and a tall, narrow inflorescence; it is quite rare in the Midwest. IA, IL, IN, MI, MO, OH, WI. Common in the south, rare to absent in the Northern Lakes. *MAH*

Muscari botryoides
HYACINTHACEAE

grape hyacinth

Homesites, woods, lawns, fields, pastures. Spring.

A glabrous perennial to 12 in. tall from a bulb. Leaves basal, linear, to 8 in. long × ⅕ in. wide, U-shaped in cross section. Flowers regular, to ¼ in. long, the fertile nodding, the sterile few and ascending to erect at top of flowering cluster, spherical to urn-shaped, inflated, with 6 tiny whitish lobes at tip, numerous in dense spike-like raceme at top of solitary stem, reminiscent of a cluster of grapes. Fruit a 3-winged capsule on spreading stalk. This and other *Muscari* species are available in the nursery trade and sometimes escape. IA, IL, IN, MI, MO, OH, WI. Introduced from Europe. Scattered. *SN*

Othocallis siberica
HYACINTHACEAE

Siberian squill

Homesites, woods, fields, ditches, gardens, roadsides. Spring.

A glabrous, colony-forming perennial to 6 in. tall, from a bulb. Leaves basal, lanceolate, entire, shiny, to 5 in. long × ¾ in. wide, often U-shaped in cross section. Flowers regular, to 1 in. wide, with 6 spreading-ascending tepals, solitary or up to 3 in short raceme, nodding at top of leafless stems, sometimes several stems per plant. Fruit a capsule. Syn. *Scilla siberica*. IA, IL, IN, MI, MN, MO, OH, WI. Introduced from Eurasia. Absent from the Ozark Highlands but scattered elsewhere. *MJH*

Iris cristata
IRIDACEAE

dwarf crested iris

Mesic and dry-mesic forests, streambanks. Spring.

A low-growing perennial from horizontal rhizomes, glabrous, colonial, flowering stalk 3–5 in. tall. Leaves basal, sword-shaped, arching, at flowering 4–8 in. long × 1 in. wide, lengthening somewhat afterward. Flowers regular, 3–4 in. wide, with 3 unmarked spreading petals and 3 larger petal-like sepals, each marked with a yellow and white patch bordered by dark purple and containing 3 parallel ridges (crests). Flowers single or in pairs atop stem. Dwarf lake iris (*I. lacustris*), a near-endemic of the Midwest, is smaller and occurs along the northern shores of Lakes Michigan and Huron. **IL, IN, MO, OH. Occurs locally southeast of a line from southcentral MO to northeastern OH.** *SN*

Iris virginica
IRIDACEAE

southern blueflag

Open to forested wetlands. Spring, summer.

A glabrous, gray-green perennial to 3½ ft. tall. Leaves mostly basal and fan-like, some alternate and smaller along flowering stem, entire, lanceolate, erect, to 3½ ft. long × 1 in. wide. Flowers regular, to 3½ in. wide, with 3 ascending petals slightly shorter than 3 spreading to reflexed petal-like sepals with bright yellow splotch at base, in several-flowered terminal clusters. Fruit an erect, 3-angled capsule to 3 in. long. Northern blueflag (*I. versicolor*) has shorter leaves and pale greenish-yellow splotch on sepals. Blue marsh iris (*I. brevicaulis*) has 6-angled fruit. German iris (*I. germanica*), introduced, has bearded sepals and broader leaves. **IA, IL, IN, MI, MN, MO, OH, WI. Widespread, less frequent at extremes.** *SN*

Sisyrinchium angustifolium
IRIDACEAE

stout blue-eyed grass

Prairies, fields, savannas, woodland edges, floodplains, thickets. Spring, summer.

A gray-green perennial to 1 ft. tall with broadly winged stems, branched above. Leaves basal in fan-like arrangement, entire, sword-shaped, erect, to 10 in. long × ⅙ in. wide. Flowers regular, to ½ in. wide, with 6 spreading tepals with abrupt tips in shallow notches, stalked, several in umbel from 2 short bracts partially contained within 2-in.-long leaf-like bract. Eastern blue-eyed grass (*S. atlanticum*) and strict blue-eyed grass (*S. strictum*) have narrowly winged stems; the latter has minutely toothed stems. IA, IL, IN, MI, MN, MO, OH, WI. Frequent in southern half of the Midwest and in MI, scattered elsewhere. *ST*

Sisyrinchium campestre
IRIDACEAE

prairie blue-eyed grass

Prairies, glades, fields, savannas, open forests. Spring, summer.

A gray-green perennial to 1 ft. tall. Leaves basal in fan-like arrangement, entire, sword-shaped, erect, to 10 in. long × ⅛ in. wide. Flowers regular, to ½ in. wide, with 6 spreading white to blue tepals with yellow throat and abrupt tips from notches, stalked, in umbel from 2 bracts to 1 in. long at top of naked stem, outer bract barely fused at base. Mountain blue-eyed grass (*S. montanum*) and slender blue-eyed grass (*S. mucronatum*) have outer bract fused at base; the latter has narrower stems. IA, IL, MI, MN, MO, WI. Widespread westward, rare to absent eastward. *SN*

Agastache scrophulariifolia
LAMIACEAE

purple giant hyssop

Thickets, moist to dry woodlands. Summer.

A mostly glabrous perennial, branched above, stems 4-angled, 3–5 ft. tall. Leaves opposite, petiolate, ovate to lanceolate, coarsely toothed, 3–6 in. long × 2–3 in. wide. Flowers irregular, ⅓–½ in. long, tubular, 5-lobed with 2 upper lobes and 3 lower, the central one downcurved, calyx lobes 5, glabrous and sharply pointed, mostly more than ⅛ in. long. Inflorescence of compact terminal spikes 3–8 in. long. Blue giant hyssop (*A. foeniculum*) of our northwestern counties is distinguished from other giant hyssops by its hairy calyx. **IA, IL, IN, MI, MN, MO, OH, WI. Occurs throughout except in far northern and far southern counties.** *BS*

Blephilia ciliata
LAMIACEAE

downy wood mint, downy pagoda-plant

Glades, barrens, rocky woodlands, prairies, especially calcareous sites. Summer.

An erect perennial, unbranched, stems 4-angled, short-hairy, 1–2 ft. tall, somewhat stoloniferous and colonial. Basal leaves petiolate, stem leaves sessile or short-petiolate, opposite, ovate-lanceolate, to 3 in. long × 1–1½ in. wide. Leaf undersurface commonly dark purple. Flowers irregular, ¼–½ in. long, tubular, upper lip 2-lobed, lower lip 3-lobed, in 2–4 whorled head-like clusters atop the stem (arranged like a pagoda). Corolla and calyx tube long-hairy, floral bracts ovate. Besides typical habitats, some populations occur on moist stream terraces. **IA, IL, IN, MI, MO, OH, WI. Occurs mostly in the southern Northern Lakes and southward.** *CB*

Clinopodium arkansanum
LAMIACEAE

low calamint, limestone calamint

Glades, fens, cliffs, gravel prairies, savannas, pannes. Spring, summer, fall.

A sprawling, colony-forming perennial to 9 in. tall. Leaves opposite, simple, linear, entire, sessile, to 1 in. long × ⅛ in. wide, sometimes reddish-purple, pleasantly fragrant (minty) when bruised. Flowers irregular, pale lavender to white, to ½ in. long, tubular and flaring with cleft upper lobe and 3 lower lobes, solitary on stalks to ⅓ in. in leaf axils. Sometimes lumped with *C. glabellum*; syns. *Calamintha arkansana, Satureja arkansana*. Often found by stepping on it (without first seeing it) and smelling the pleasant mint odor. IL, IN, MI, MO, OH, WI. **Throughout but most abundant in the Ozark Highlands.** *SN*

Clinopodium vulgare
LAMIACEAE

wild basil

Dry upland woodlands, savannas, old fields. Summer.

An erect perennial, unbranched, hairy and square-stemmed, 8–24 in. tall. Leaves opposite, short-petiolate to sessile, ovate to ovate-lanceolate, mostly entire, 1–2 in. long × ½–1 in. wide. Flowers irregular, about ½ in. long, lavender to pink, tubular, 2-lipped with upper lip slightly notched in middle, the lower one 3-lobed, sepal tube quite hairy, with 5 needle-like lobes. Inflorescence of dense heads at stem tips and upper leaf axils. Wild basil is not the herb commonly used in cooking; that is sweet basil (*Ocimum basilicum*), native to Africa and Asia. IA, IL, IN, MI, MN, OH, WI. **Occurs primarily in OH, MI, and northern WI.** *SN*

Cunila origanoides
LAMIACEAE

dittany

Dry, often rocky and sparsely vegetated upland forests, barrens. Summer.

An erect perennial, freely branched, stems 4-angled, glabrous to sparsely hairy, 8–18 in. tall. Leaves opposite, mostly sessile, ovate-lanceolate, punctate, very fragrant when bruised, 1–2 in. long × ⅓–1½ in. wide. Flowers irregular, about ¼–½ in. long, tubular, hairy, 2-lipped, the upper lip erect and lower 3-lobed, in rounded to flat-topped clusters from leaf axils and branch tips. The smell of its leaves is reminiscent of the Eurasian oregano (*Origanum vulgare*) used in cooking, hence an alternative common name, wild oregano. IA, IL, IN, MO, OH. **Common in the Ozark Highlands and unglaciated Eastern Forests.** *MAH*

Glechoma hederacea
LAMIACEAE

ground ivy, creeping Charlie

Rich mesic forests, floodplains, lawns. Spring, summer.

A trailing perennial, mat-forming with long, creeping, square-stemmed stolons that root at the nodes, flowering stems to 6 in. tall when vigorous, glabrous to hairy, 1–2 ft. long. Leaves opposite, petiolate, rounded to kidney-shaped with scalloped margins, ½–2 in. long and about as wide. Evergreen to semi-evergreen and strongly scented when bruised. Flowers irregular, tubular, about ½ in. long, 2-lipped, the upper shallowly 2-lobed, the lower deeply 3-lobed, bearded and marked with dark purple spots, situated 2–4 in leaf axils. IA, IL, IN, MI, MN, MO, OH, WI. **Introduced from Eurasia. Abundant in most of the Midwest.** *MAH*

Hedeoma hispida

LAMIACEAE

rough pennyroyal, false pennyroyal

Prairies, glades, fields, clearings, roadsides, railroads, in sand and gravel. Spring, summer.

A square-stemmed hairy annual to 10 in. tall. Leaves opposite, simple, linear, entire, sessile, to ¾ in. long × ⅛ in. wide, slightly mint-scented when bruised. Flowers irregular, pale blue to purple, to ¼ in. long, tubular and flaring with cleft upper lip and 3-lobed lower lip (see inset), very short-stalked in whorls of 2–12 in leaf axils, the lowest often cleistogamous. American pennyroyal (*H. pulegioides*) has strongly scented foliage and broader leaves. IA, IL, IN, MI, MN, MO, OH, WI. Common in the Northern Lakes, Tallgrass Prairie, and Ozark Highlands; nearly absent from the Eastern Forests. *DT/CR*

Hedeoma pulegioides

LAMIACEAE

American pennyroyal

Dry to dry-mesic upland forests and savannas, rock outcrops, barrens. Summer, fall.

An erect annual, commonly branched, glandular-hairy, 5–15 in. tall. Leaves opposite, short-petiolate, bluntly toothed or entire, oblong-ovate, ½–1 in. long × ⅛–¼ in. wide. Flowers irregular, ⅛–¼ in. long, tubular, 2-lipped, the upper shallowly 2-lobed, the lower 3 lobes spreading and marked with purple spots, bearded within, occurring in axillary whorls. This plant often indicates its presence more by smell than appearance; its bruised foliage has a strong mint scent. It has been likened to European pennyroyal (*Mentha pulegium*). IA, IL, IN, MI, MN, MO, OH, WI. Generally common but rare in the Midwest's northern third. *MAH*

Lamium amplexicaule
LAMIACEAE

henbit

Agricultural fields, roadsides, gardens, lawns. Spring.

A weakly erect to sprawling annual, branched, stems 4-angled, sparsely hairy, 5–15 in. tall. Leaves opposite, the lower petiolate, becoming sessile and clasping above, rounded with scalloped margins, ¾–1 in. long and wide. Flowers irregular, ½–¾ in. long, tubular, hairy externally, 2-lipped, the upper rounded to form a hood, the lower 3-lobed and spotted, the central lobe drooping and cleft, in axillary clusters in upper leaves. Large populations of this species give agricultural fields a purple hue in early spring. IA, IL, IN, MI, MN, MO, OH, WI. Introduced from Eurasia and Africa. Common, mostly in the lower Midwest. *MJH*

Lamium purpureum
LAMIACEAE

purple deadnettle

Agricultural fields, roadsides, gardens. Spring.

A weakly erect to sprawling annual, branched, stems 4-angled, sparsely hairy, 6–12 in. tall. Leaves opposite, long- to short-petiolate, rounded to broadly triangular, drooping, with rounded teeth, 1–2 in. long and wide. Flowers irregular, ½–¾ in. long, tubular and hairy externally, 2-lipped, the upper rounded to form a hood, the lower 3-lobed and spotted, the central lobe drooping and cleft, in axillary clusters in upper leaves. Purple deadnettle commonly grows with henbit (*L. amplexicaule*). IA, IL, IN, MI, MO, OH, WI. Introduced from Europe. Widespread and common, infrequent to absent northward. *MAH*

Perilla frutescens
LAMIACEAE

beefsteak plant

Moist upland forests, creek banks, gravel bars, roadsides, disturbed sites. Summer.

An erect annual, branched, stems 4-angled and glabrous or with downward-pointing hairs, 1–3 ft. tall. Leaves opposite, petiolate, broadly ovate, often wrinkled with coarsely toothed margins, distinctively aromatic, ranging in color from green to deep purple, 3–5 in. long × 1½–3 in. wide. Flowers irregular, about ¼ in. long, tubular, 2-lipped and notched, the corolla barely extending from the 5-lobed calyx. Beefsteak plant is highly invasive and increasing in range. IA, IL, IN, MI, MO, OH, WI. Introduced from India. Occurs mostly in the lower Midwest. *MAH*

Prunella vulgaris
LAMIACEAE

self-heal

Open woodlands, trails, disturbed sites. Summer.

An erect or prostrate perennial, mostly unbranched, glabrous or hairy, stems 4-angled, 4–18 in. tall. Leaves opposite, petiolate, lanceolate to elliptic, entire or with few shallow teeth, 1–3 in. long × ½–1½ in. wide. Flowers irregular, about ½ in. long, tubular, 2-lipped, the upper one purplish and hood-like, lower lip 3-lobed, the lowermost lobe fringed and often white, several in a dense cylindrical spike atop the stem. Some consider the upright var. *lanceolata* as native here and the smaller, creeping var. *vulgaris* an introduction from Europe; the latter has ovate leaves. IA, IL, IN, MI, MN, MO, OH, WI. Common, probably in every county. *KC*

Salvia azurea

LAMIACEAE

blue sage

Prairies, glades, barrens, blufftops.
Summer.

An erect perennial, mostly unbranched,
hairy, stems square in cross section, 1–4
ft. tall. Leaves opposite, short-petiolate,
the axils bearing clusters of small leaves,
oblong-lanceolate to linear with serrated
margins, 2–5 in. long × ½–1½ in. wide, with
aromatic sage scent when bruised. Flowers
irregular, ¾–1 in. long, tubular, 2-lipped,
the upper forming a hood, lower obscurely
3-lobed, middle lobe much enlarged, in whorls
of 10 or more per node in terminal and axillary
spikes. Our plants are var. *grandiflora*. IA, IL,
IN, MI, MO, OH, WI. Most native populations occur in MO; elsewhere, many are likely
introductions. *MJH*

Salvia lyrata

LAMIACEAE

lyre-leaved sage

Open woodlands, meadows, glades. Spring.

An erect perennial, hairy, stems square, 1–2 ft.
tall. Leaves mostly in a basal rosette, petiolate,
blades obovate in overall outline, lyre-shaped
with deep irregular and wavy lobes often with
deep purple veins, 3–5 in. long × 1–2½ in. wide,
lacks aromatic sage scent when bruised. Flowers irregular, about 1 in. long, long-tubular,
2-lipped, upper lip small, the larger lower lip
tipped with pair of rounded lobes, in whorls
of 6 or more per node positioned on terminal
and axillary spikes. Some individuals also
form cleistogamous flowers. IL, IN, MO, OH.
**Occurs mostly in unglaciated territory of the
lower Midwest.** *MAH*

Salvia reflexa
LAMIACEAE

Rocky Mountain sage, lanceleaf sage

Prairies, pastures, roadsides, railroads, waste areas. Summer, fall.

A branching, square-stemmed annual to 2 ft. tall. Leaves opposite, simple, sparsely toothed with upcurved margins, to 2 in. long × ½ in. wide, on petioles to ¾ in. long, becoming scale-like in inflorescence. Flowers irregular, to ⅖ in. long, tubular with short, unlobed arching upper lip and lower lip with 2 lateral lobes and basal lobe shallowly cleft, with deeply veined, conspicuous, 2-lobed green calyx covering tube, opposite and short-stalked in raceme at top of plant. IA, IL, IN, MI, MN, MO, OH, WI. Scattered, more frequent in the Tallgrass Prairie and the Ozark Highlands. *PD*

Scutellaria elliptica
LAMIACEAE

hairy skullcap

Dry rocky woods. Spring, summer.

A hairy, square-stemmed perennial to 2½ ft. tall. Leaves opposite, bluntly toothed, to 3 in. long × 1¾ in. wide, narrowing above, rounded to flat or wedge-shaped at base, petiolate. Flowers irregular, to ¾ in. long, tubular with unlobed hood-like upper lip, 2 small whitish lateral lobes, and shallowly lobed spreading lower lip, subtended by 2-lipped calyx with protuberance on top lip, opposite in leafy spike-like raceme to 4 in. long, also with shorter racemes from upper leaf axils. Fruit with distinctive bicycle-seat shape. IL, IN, MI, MO, OH. Widespread in the Ozark Highlands and Eastern Forests, rare in the Tallgrass Prairie, disjunct in southern MI and northern IN. *BS*

Scutellaria galericulata
LAMIACEAE

marsh skullcap

Marshes, sedge meadows, conifer swamps, bogs, shores, ditches. Summer.

A weak, square-stemmed perennial to 2½ ft. tall. Leaves opposite, simple, bluntly toothed, to 2¼ in. long × ¾ in. wide, deeply veined, progressively smaller into inflorescence, sessile or nearly so. Flowers irregular, blue-violet (rarely white or pink) with blue-violet-spotted white throat, to 1 in. long, tubular, wider and inflated toward opening, with upper 3 lobes forming hood and lower lip broad and spreading, subtended by 2-lipped calyx with protuberance on top lip, opposite and short-stalked in leaf axils. Fruit with distinctive bicycle-seat shape. IA, IL, IN, MI, MN, MO, OH, WI. **Widespread northward, scattered to scarce southward.** *PS*

Scutellaria incana
LAMIACEAE

downy skullcap

Mesic to mostly dry forests, savannas, barrens. Summer.

An erect perennial, branched above, short-hairy, stems 4-angled, 2–3 ft. tall. Leaves opposite, petiolate, blades lance-ovate to ovate with rounded teeth, 2–4 in. long × 1–2 in. wide. Flowers irregular, ½–1 in. long, corolla tube S-shaped, 2-lipped, the upper forming a beaklike hood over the larger, wider lower lip, the latter bearing a white center, numerous in racemes atop the stem. Fruit resembles bicycle or old-fashioned tractor seat. IL, IN, MI, MO, OH. **Common, mostly in the lower Midwest.** *SN*

Scutellaria lateriflora
LAMIACEAE

mad dog skullcap

Floodplain forests, swamps, marshes.
Summer.

An erect perennial, becoming ascending to
sprawling, branched, mostly glabrous, stems
4-angled, 1–2 ft. tall. Leaves opposite, petiolate, blades ovate to lance-ovate, toothed,
1–3 in. long × ⅔–2 in. wide. Flowers irregular,
⅛–⅓ in. long, the corolla tube nearly straight
and covered with minute hairs, 2-lipped, the
upper hood-like, arching over the longer lower
lip. Flowers often in pairs along one side of
1- to 4-in.-long axillary racemes. The calyx of
flower and fruit has a raised crest or ridge.
IA, IL, IN, MI, MN, MO, OH, WI. Common,
possibly in every county. *MAH*

Scutellaria nervosa
LAMIACEAE

veiny skullcap

Moist forests, flatwoods, floodplain forests.
Spring, summer.

A weak, square-stemmed perennial to 20 in.
tall. Leaves opposite, simple, coarsely toothed,
deeply veined, to 2 in. long × 1¼ in. wide,
rounded to cordate at base, sessile or nearly
so. Flowers irregular, pale blue with darker
blue spots in throat, to ½ in. long, tubular
with hood-like upper lip and 4-lobed spreading lower lip, subtended by 2-lipped calyx
with protuberance on top lip, opposite and
short-stalked in leaf axils. Fruit with distinctive bicycle-seat shape. IA, IL, IN, MI, MO,
OH. Scattered in the Eastern Forests and the
Tallgrass Prairie, rare in the Northern Lakes
and the Ozark Highlands. *EM*

Scutellaria ovata
LAMIACEAE

heart-leaved skullcap

Mesic to dry woodlands, often with a rocky substrate, bluffs. Spring, summer.

An erect perennial, mostly unbranched, hairy, stems 4-angled, 1–2 ft. tall. Leaves opposite, petiolate, blades ovate-cordate with deeply wrinkled surface and blunt-toothed margins, 1–4 in. long and almost as wide. Flowers irregular, ½–1 in. long, corolla tube S-shaped, 2-lipped, the upper forming a hood over the larger, wider lower lip, the latter bearing a white center variously speckled with purple, the calyx of flower and fruit with a raised crest or ridge. Flowers numerous in racemes atop the stem. IA, IL, IN, MI, MN, MO, OH, WI. Occurs mostly in the lower Midwest. *CB*

Scutellaria parvula
LAMIACEAE

small skullcap

Dry, rocky or sandy prairies, glades, blufftops. Spring, summer.

An erect perennial, simple to branched, hairy, stems 4-angled, 3–8 in. tall. Leaves opposite, mostly sessile, ovate to somewhat triangular, entire or with few small teeth, ¼–¾ in. long × ⅛–⅓ in. wide. Flowers irregular, ¼–⅓ in. long, corolla tube S-shaped, 2-lipped, the upper forming a hood over the larger, wider and lobed lower lip that bears a white center variously speckled with purple, calyx with a raised crest, 1 per leaf axil. The similar Bush's skullcap (*S. bushii*), found in the Ozark Highlands, has larger flowers and gradually narrowed leaf bases. IA, IL, IN, MI, MN, MO, OH, WI. Common and widespread but sparse in the Northern Lakes. *SN*

Teucrium canadense

LAMIACEAE

American germander

Prairies, forested floodplains, swamps, wet meadows. Summer.

An erect perennial, mostly unbranched, stems 4-angled, hairy, rhizomatous and colonial, 1–3 ft. tall. Leaves opposite, petiolate on lower stem, sessile above, elliptic, lanceolate to ovate, toothed, glabrous above with impressed veins, 2–5 in. long × ¾–2 in. wide. Flowers irregular, ⅓–½ in. long, tubular, 2-lipped, the upper lip reduced to 2 pointed lobes (see inset), the lower lip comparatively broad, 3-lobed and commonly purple-spotted basally, in terminal racemes. The common name is a nod to the related wall germander (*T. chamaedrys*) of Europe. IA, IL, IN, MI, MN, MO, OH, WI. Common, possibly in every county. *SN/PR*

Trichostema brachiatum

LAMIACEAE

false pennyroyal

Limestone glades, gravel bars, rocky prairies, alvars, calcareous waste areas. Summer.

An erect annual, branched, short-hairy, glandular, 6–12+ in. tall. Leaves opposite, short-petiolate, elliptic to lanceolate, entire, 1–1½ in. long × ⅓–½ in. wide. Flowers nearly regular, about ¼ in. wide, tubular, 2-lipped, 5-lobed with the lowermost lobe slightly longer, stamens arching and curling slightly downward, in spreading terminal and axillary panicles. Unlike its near-relative bluecurls (*T. dichotomum*), this species mostly prefers calcareous substrates. Syn. *Isanthus brachiatus*. IA, IL, IN, MI, MN, MO, OH, WI. Throughout but patchy, with only a few occurrences in the Northern Lakes. *BS*

Trichostema dichotomum
LAMIACEAE

bluecurls

Dry and acidic rocky slopes, glades, sand barrens, railroads. Summer.

An erect annual, branched, minutely but densely glandular-hairy, 6–18 in. tall. Leaves opposite, petiolate, oblong to elliptic, entire, 1–2½ in. long × ¼–1 in. wide. Flowers irregular, ¼–½ in. long, 2-lipped, the upper with 4 triangular lobes, the lower 1-lobed, whitish at base, purple-spotted and drooping, stamens curling forward and down at tip (see inset), in widely spreading terminal and axillary clusters. This species typically grows in areas with sparse competition. **IA, IL, IN, MI, MO, OH. Widespread, mostly in unglaciated terrain of the lower Midwest.** *SN*

Pinguicula vulgaris
LENTIBULARIACEAE

common butterwort

Calcareous rocky shores, alkaline interdunal flats, fens. Summer.

A carnivorous perennial to 6 in. tall. Leaves few, basal, yellow-green, succulent, upper surface covered in tiny sticky hairs that trap and digest prey, to 2 in. long × 1¼ in. wide, spatulate, sessile, with margins curled upward. Flowers irregular, tubular with spur at end, to ¾ in. long, with 2 upper lobes and 3 lower lobes, solitary and oriented horizontally at end of leafless, reddish-purple stem. Up to 9 stems from a rosette. Fruit an upright ovoid capsule to ⅓ in. long. **MI, MN, WI. Restricted to northern portion of the Northern Lakes, generally in proximity to the Great Lakes.** *SN*

Utricularia purpurea
LENTIBULARIACEAE

purple bladderwort

Submergent marshes, lakes, riverbanks.
Summer.

A carnivorous, glabrous, colonial, submerged
aquatic perennial with stalks emerged to 6 in.
Stem leaves in numerous whorls of up to 7, to
2½ in. long and wide, each leaf with several
whorls of filiform segments, bearing bladders
at their tips that trap and digest tiny animals;
flowering stem naked. Flowers irregular, to
½ in. long, 2-lipped, with lower lip 3-lobed
and bearing white-rimmed orange-yellow
spot toward base, upper lip shorter, spur
short, yellowish, and inconspicuous; up to 4
short-stalked in terminal raceme. Small purple
bladderwort (*U. resupinata*) has solitary flower
to ⅓ in. long and unsegmented, filiform leaves.
IN, MI, MN, WI. Scattered to rare in the
Northern Lakes. *SN*

Linum usitatissimum
LINACEAE

common flax

Fields, gardens, roadsides, railroads,
waste areas. Spring, summer, fall.

A glabrous annual to 3 ft. tall. Leaves numerous, alternate, sessile, entire, 3-nerved, to 1½
in. long × ⅛ in. wide, smaller above. Flowers
regular, with 5 spreading petals, to 1 in. wide,
with stamens with blue anthers, stalked at
ends of branches in upper part of plant. Fruit
a spherical, hardened capsule to ⅓ in. wide
that splits into segments like an orange. This
species is the source of linen fabric and linseed
oil. The leaves of garden flax (*L. perenne*) and
the native blue flax (*L. lewisii*) have a single
vein; the former has some flowers with long
styles and some with short styles. IA, IL, IN,
MI, MN, MO, OH, WI. Introduced from the
Old World. Scattered. *PD*

Cuphea viscosissima
LYTHRACEAE

clammy cuphea, blue waxweed

Open woodlands, barrens, glades, roadsides. Summer.

An erect annual, branching, with dense glandular (and sticky) hairs, 8–18 in. tall. Leaves opposite, petiolate, entire, ovate-lanceolate, up to 2 in. long × ¾ in. wide. Flowers irregular, about 1 in. long, tubular, with 4 spreading lower petals and 2 similar but larger upper ones, calyx densely glandular, 1 or 2 per upper leaf axils. Clammy cuphea is being investigated for agronomic use, as its seeds contain an oil with characteristics of coconut and palm oils. IA, IL, IN, MO, OH. Occurs mostly in the lower Midwest. *MAH*

Decodon verticillatus
LYTHRACEAE

swamp loosestrife, whorled loosestrife

Marshes, bogs, fens, swamps, lake and pond margins. Summer, fall.

An arching herb-like shrub to 6 ft. tall with a spongy base. Rooting at tips. Leaves usually in whorls of 3 (sometimes 4 or opposite), short-petiolate, simple, entire, lanceolate to elliptic, shiny, to 8 in. long × 2 in. wide. Flowers regular, to ¾ in. wide, with 4–7 ascending petals, appearing messy, in dense clusters in leaf axils. The arching form and dense axillary flower clusters differentiate it from other pinkish- or purplish-flowered loosestrifes. IA, IL, IN, MI, MN, MO, OH, WI. Common in eastern portion of the Northern Lakes with fewer occurrences elsewhere. *SN*

Lythrum alatum
LYTHRACEAE

winged loosestrife

Moist prairies, marshes, ditches. Summer.

An erect perennial, branched, stems 4-angled and slightly winged, glabrous, 1–2½ ft. tall. Leaves mostly alternate (lowermost opposite), sessile, lanceolate, entire, ¾–1½ in. long × ¼–½ in. wide. Flowers regular, ⅓–½ in. wide, bearing 6 obovate, often wrinkled-looking petals, each with a dark midvein, occurring singly in upper leaf axils. The introduced insects used for biocontrol of the invasive purple loosestrife (*L. salicaria*) also feed upon winged loosestrife; the long-term impact is unknown. IA, IL, IN, MI, MN, MO, OH, WI. Common and widespread except in the Midwest's northern third. *SN*

Lythrum salicaria
LYTHRACEAE

purple loosestrife

Various treeless wetlands, marshes, bogs, fens, streambanks, ditches. Summer.

An erect perennial, branched, glabrous or hairy, stems 4-angled, rather woody at base, 2–7 ft. tall. Leaves opposite or rarely in whorls of 3, sessile, entire, lanceolate, 1–4 in. long × ½–1 in. wide. Flowers regular, ½–1 in. wide, bearing 6 narrowly obovate, often wrinkled-looking petals, each with a central dark midvein, occurring in whorls on long, wand-like branches atop the plant. Insects native to the homeland of this highly invasive plant and specializing in feeding on it are being released as a control measure. IA, IL, IN, MI, MN, MO, OH, WI. Introduced from Eurasia. Common in the Northern Lakes and continuing to increase southward. *SN*

Chamaenerion angustifolium
ONAGRACEAE

fireweed

Forests, clearings, meadows, rocky ground, gravel pits, roadsides. Summer.

A perennial to 5 ft. tall. Leaves alternate, to 8 in. long × 1 in. wide, with entire to irregularly toothed margins and strong white midvein. Flowers regular, to 1 in. wide, with 4 petals and conspicuous 4-parted exserted stigma, in erect terminal raceme. Fruit a skinny, erect, pink capsule opening from tip to expose small seeds topped with long silky hairs. Several plants with a positive response to burning and other disturbances share the common name. Syns. *Chamerion angustifolium*, *Epilobium angustifolium*. **IA, IL, IN, MI, MN, OH, WI. Nearly exclusively in the Northern Lakes, with few occurrences in the nearby Tallgrass Prairie and Eastern Forests.** *SN*

Cypripedium arietinum
ORCHIDACEAE

ram's head lady's slipper

Dry to moist calcareous forests, swamps. Spring.

A short-hairy perennial to 1 ft. tall. Leaves alternate and sheathing, usually 3, to 4 in. long × 1¼ in. wide, with conspicuous parallel veins. Flowers irregular, to 1 in. long and nearly as wide, with 3 purplish-brown sepals, the upper broader and hood-like, the lower 2 narrow and spreading downward, 2 lateral petals similar to sepals, lip petal white at opening and white with conspicuous purple pattern on veins below (rarely all white), slipper-like but with a downward-pointed protuberance at end, flowers 1 or rarely 2 at top of plant immediately subtended by erect bract. **MI, MN, WI. Becoming exceedingly rare, restricted to the Northern Lakes.** *AG*

Galearis spectabilis
ORCHIDACEAE

showy orchis

Rich mesic forests, ravines, stream terraces.
Spring.

An erect perennial, glabrous, 5–10 in. tall.
Leaves basal, 1 or 2, oblong-obovate to elliptic, entire, somewhat fleshy, 3–8 in. long
× 1–4 in. wide. Flowers irregular, about 1
in. long, lip petal lowermost, white, ovate,
wavy-margined, constricted at base, covered
by a hood of 3 pinkish-purple sepals and 2
similar petals, 4–10 in a spike-like terminal
raceme. Syn. *Orchis spectabilis*. In the far north
the round-leaved orchid (*G. rotundifolia*; syn.
Amerorchis rotundifolia) occurs in wet conifer
swamps; it typically has a single leaf and
purple-spotted white flowers (see inset). IA, IL,
IN, MI, MN, MO, OH, WI. **Locally common
except for far northern counties.** *AG/SN*

Liparis liliifolia
ORCHIDACEAE

lily-leaved twayblade, purple twayblade

Upland woodlands, thickets, early
successional regrowth. Spring.

An erect perennial, unbranched, glabrous, 4–7
in. tall. Leaves 2, basal, elliptic to ovate, shiny,
2–6 in. long × 1–3 in. wide. Flowers irregular,
about 1 in. long, petals 3, the lip petal obovate,
purplish and translucent, with 2 thread-like
petals and 3 linear-lanceolate sepals, 15–25 in
a terminal raceme. Fruit a capsule equal to or
shorter than its stalk. The species appears to
be increasing in abundance and range in the
Midwest. IA, IL, IN, MI, MN, MO, OH, WI.
**Widespread but rare or absent in far north-
ern and western counties.** *CB*

Platanthera peramoena
ORCHIDACEAE

purple fringeless orchid

Floodplain forests, ephemeral pools,
wet meadows, ditches. Summer.

An erect perennial, unbranched, glabrous,
1½–2½ ft. tall. Leaves alternate, lanceolate
to elliptic, sheathing, 3–7 in. long × 1–2 in.
wide. Flowers irregular, about 1 in. long, with 3
petals and 3 ovate sepals, lip petal lowermost,
broadly ovate with 3 fan-shaped, shallowly
toothed lobes and a basal spur about 1 in. long,
upper petals broadly wedge-shaped, 15–40
flowers in loose raceme. From a distance this
species resembles a wild phlox of similar color
and size. IL, IN, MO, OH. **Occurs mostly in
southern portions of the Eastern Forests.** *BS*

Platanthera psycodes
ORCHIDACEAE

purple fringed orchid

Sedge meadows, moist woodlands,
stream borders, fens. Summer.

An erect perennial, unbranched, glabrous, 1–3
ft. tall. Leaves alternate, lanceolate to elliptic,
sheathing, 4–7 in. long × 1–3 in. wide. Flowers
irregular, about ½ in. long, with 3 petals and
3 ovate sepals, lip petal lowermost, broadly
ovate with 3 fan-shaped, deeply fringed lobes
and a basal spur about ½ in. long, other petals
obovate with finely toothed upper margins.
Inflorescence 20–50 flowers in a terminal
raceme. Its lip is clearly more deeply fringed
than purple fringeless orchid (*P. peramoena*),
and their ranges rarely overlap. IA, IL, IN,
MI, MN, OH, WI. **Occurs principally in the
Northern Lakes.** *PG*

Agalinis auriculata
OROBANCHACEAE

eared false foxglove

Wet to mesic prairies, savannas. Summer.

An annual to 2½ ft. tall. Leaves opposite,
lanceolate, sessile, sometimes (especially
toward top of plant) with 1 or 2 small lobes
at base, to 2 in. long × ¾ in. wide, sometimes
reddish-purple (especially with age), covered
in small white hairs. Flowers irregular, tubular with 5 spreading-ascending lobes, to 1 in.
long × ¾ in. wide, sessile in axils of upper
leaves. Grows in remnant natural areas but
requires disturbance to persist. Hemiparasitic.
Syn. *Tomanthera auriculata*. **IA, IL, IN, MI,
MN, MO, OH, WI. Rare, most abundant in
the Tallgrass Prairie and Ozark Highlands,
scattered in the Northern Lakes and Eastern
Forests.** *SN*

Agalinis gattingeri
OROBANCHACEAE

Gattinger's false foxglove

Glades, barrens, old fields, rocky or sandy
open woodlands. Summer.

A slender, highly branched, minutely hairy
annual 1–2 ft. tall with a round (vs. angled)
main stem. Leaves opposite (some upper
leaves alternate), sessile, entire, linear, 1–2 in.
long × ¹⁄₁₆ in. wide, yellowish-green and commonly few in number. Flowers about 1 in. long,
irregular and somewhat bell-shaped, with 5
lobes, the upper ones flaring upward, inner
surfaces hairy, pedicels ½–1 in. long, in axils
of upper bracteal leaves. This species is a more
narrow-leaved, delicate plant than slender
false foxglove (*A. tenuifolia*). **IA, IL, IN, MI,
MN, MO, OH, WI. Widely scattered, generally not common in the Midwest.** *EH*

Agalinis purpurea
OROBANCHACEAE

purple false foxglove

Wet sandy or organic soils in prairies, meadows, fens. Summer.

A minutely hairy, branched annual, 2–3 ft. tall with 4-angled stems. Leaves opposite (some upper leaves alternate), sessile, linear, entire, 1–2 in. long × 1/16–1/8 in. wide, often somewhat curved and deep green or purplish-tinged. Flowers irregular, about 1 in. long, bell-shaped, 5-lobed, the upper lobes flaring upward, inner surfaces hairy, on pedicels 1/8–1/4 in. long from leaf axils. Pauper false foxglove (*A. paupercula*) is sometimes considered a small-flowered variety of purple false foxglove. IA, IL, IN, MI, MN, MO, OH, WI. Widely scattered, most occurrences in the Northern Lakes. *MAH*

Agalinis tenuifolia
OROBANCHACEAE

slender false foxglove

Prairies, old fields, barrens, open woodlands. Summer.

A slender, minutely hairy annual with several branching stems, 1–2 ft. tall. Leaves opposite (some upper leaves alternate), sessile, entire, linear, 1–2 in. long × 1/16–1/8 in. wide, often somewhat curved and not uncommonly deep green or purplish-tinged. Flowers irregular, 1–1½ in. long, somewhat bell-shaped with 5 lobes, the upper ones arching forward and down over the flower's mouth, inner surfaces of lobes glabrous, on slender pedicels ¾–1 in. long in axils of upper bracteal leaves. IA, IL, IN, MI, MN, MO, OH, WI. Widespread and common. *MAH*

Buchnera americana
OROBANCHACEAE

American blue hearts

Barrens, glades, dry prairies. Summer, fall.

An erect perennial, unbranched, rough-hairy, 1–3 ft. tall. Leaves opposite, sessile, lanceolate, coarsely toothed, 1–3 in. long × ¼–¾ in. wide. Flowers nearly regular, about ½ in. wide, tubular with 5 oblong, flared lobes bearing small notches on rounded tips. Inflorescence of mostly paired flowers on terminal spikes. The plants are hemiparasites. This species is much more common in the southern United States, perhaps not surprising given that most members of the genus are tropical. **IL, IN, MI, MO, OH. Occurs mostly in Ozark Highlands.** *MAH*

Oxalis violacea
OXALIDACEAE

violet wood sorrel

Mesic to (more commonly) dry-mesic forests, savannas, prairies. Spring, summer.

A perennial with basal leaves from bulbs, flowering stalk glabrous to hairy, 3–6 in. tall. Leaves basal, petiolate, compound, bearing 3 obovate to cordate leaflets with notched tips, these ¼–⅔ in. long and slightly wider, often folded and marked with purple splotches or chevrons. Flowers regular, ⅓–½ in. wide, with 5 oblong petals, in clusters of 3–10 extending above the leaves. This is our only oxalis with solidly purple flowers. Oddly, occasional leafless plants appear and bloom in late summer. **IA, IL, IN, MI, MN, MO, OH, WI. Common except for northern MN and WI, and almost totally absent from MI.** *EH*

Passiflora incarnata
PASSIFLORACEAE

passionflower, maypop

Thickets, roadsides, railroads, open woodlands. Summer.

A perennial vine, sprawling to climbing by tendrils, branched, glabrous to minutely hairy, 10–20 ft. long. Leaves alternate, petiolate, palmately and deeply 3-lobed, minutely toothed, 2–5 in. long and about as wide. Flowers regular, about 3 in. wide, with a 3-styled pistil and 5 stamens encircled by 5 oblong sepals, 5 similar-looking petals, and a series of wavy, thread-like filaments. Flowers solitary in leaf axils. Fruit a round berry 2 in. in diameter with seeds whose mucilaginous coatings are quite fragrant and tasty. IL, IN, MO, OH. Occurs in the southern quarter of the region. *MAH*

Mimulus ringens
PHRYMACEAE

monkey-flower

Floodplain forests, shallow swamps, streambanks, ditches. Summer.

An erect perennial, glabrous, stems 4-angled, 1½–3 ft. tall. Leaves opposite, sessile and clasping, lanceolate, margins with small teeth, 2–4 in. long × ½–1 in. wide. Flowers irregular, about 1 in. long and wide, tubular, with a yellowish center, 2-lipped with wavy margins, the upper 2-lobed and lower 3-lobed, singular on pedicels 1–2 in. long from leaf axils. Mostly in the region's southern half is the similar winged monkey-flower (*M. alatus*); pedicels of its flowers are ½ in. long or less, and leaves have distinct petioles. IA, IL, IN, MI, MN, MO, OH, WI. Widespread and common throughout. *MJH*

Chaenorhinum minus
PLANTAGINACEAE

dwarf snapdragon, small toadflax

Roadsides, railroads, gravel pits, pavement cracks, dumps. Summer.

A glandular-pubescent annual to 16 in. tall (often shorter). Leaves alternate (sometimes opposite below), simple, to 1¼ in. long × ¼ in. wide, often purple beneath. Flowers irregular, tubular, with 2 upper and 3 lower spreading lobes and short spur on back, pale bluish, purplish, or white, to ¼ in. long and wide, on stalks to 1 in. long in upper part of plant. IA, IL, IN, MI, MN, MO, OH, WI. Introduced from Europe. Widespread in the Eastern Forests, Northern Lakes, and eastern portion of the Tallgrass Prairie; less common in western portion of the Tallgrass Prairie and the Ozark Highlands. *MJH*

Collinsia verna
PLANTAGINACEAE

blue-eyed Mary

Floodplain forests, ravines, stream terraces. Spring.

A winter annual, weakly erect to reclining, unbranched or branched, glabrous to hairy, 4–15 in. tall, often occurring in large masses. Leaves opposite (or whorled above), mostly sessile and partly clasping, lanceolate and shallowly toothed, 1–2 in. long × ½–1 in. wide. Flowers irregular, about ½ in. wide, tubular, with 2 double-lobed, shallowly notched lips, the upper one white, the lower usually blue, on 1-in. stalks in whorls of 4–6 atop stems. The lower lip of Ozark blue-eyed Mary (*C. violacea*; see inset) is mostly violet-colored with more deeply notched lobes. IA, IL, IN, MI, MO, OH, WI. Primarily in the southern Midwest. *SN*

Leucospora multifida

PLANTAGINACEAE

narrowleaf paleseed, Obi-Wan-Conobea

Mudflats, streambanks, roadsides, wet prairies. Summer, fall.

An erect annual, branched, short-hairy, 3–8 in. tall. Leaves opposite, ¾–1¼ in. long × ¼–⅔ in. wide, short-petiolate, triangular-ovate, deeply pinnately lobed, the segments linear or narrowly oblong. Flowers irregular, about ¼ in. long, tubular, 2-lipped, upper one 2-lobed, lower 3-lobed, solitary in upper leaf axils. This easily overlooked plant shows up in some unexpected places, such as cracks in highway and sidewalk pavement. Syn. *Conobea multifida*. IA, IL, IN, MI, MO, OH. **Common in the lower Midwest, rare to absent northward.** *BS*

Nuttallanthus canadensis

PLANTAGINACEAE

blue toadflax, old-field toadflax

Sand prairies, fields, barrens, savannas, rocky glades, roadsides. Spring, summer.

A glabrous, spindly annual to 1½ ft. tall. Leaves mostly alternate, simple, entire, to 1½ in. long × ⅛ in. wide. Flowers irregular, to ⅓ in. wide, tubular with 2 small rounded upper and 3 larger rounded lower lobes, with slender spur to ¼ in. on back, pale blue to blue-violet with white hump on middle lower lobe, short-stalked in sparse terminal raceme to 8 in. long. Syn. *Linaria canadensis*. Sometimes confused with Kalm's lobelia (*Lobelia kalmii*). Texas toadflax (*N. texanus*) has larger flowers with longer spurs. IA, IL, IN, MI, MN, MO, OH, WI. **Scattered, most frequent in the Northern Lakes.** *SN*

Penstemon calycosus
PLANTAGINACEAE

smooth beardtongue

Open woodlands, streambanks, fields, prairies. Spring, summer.

An erect perennial, unbranched, mostly glabrous, 1½–2½ ft. tall. Leaves basal and petiolate, stem leaves opposite, sessile and clasping, linear-lanceolate, with a few scattered teeth, 3–6 in. long × 1–1½ in. wide. Flowers irregular, about 1 in. long, violet to whitish, tubular, widest toward the opening, 2-lipped with 5 spreading lobes, 2 upper and 3 lower, sepals linear-lanceolate, about ⅓ in. long, anthers glabrous. The similar foxglove beardtongue (*P. digitalis*) has predominantly white flowers, and its anthers possess hairs on back near attachment to filament. IL, IN, MI, OH. **Mostly common in these states except rare in MI.** *MAH*

Penstemon grandiflorus
PLANTAGINACEAE

large beardtongue

Prairies, roadsides. Spring, summer.

A glabrous, glaucous perennial to 3½ ft. tall. Leaves in basal rosette first year, to 5 in. long × 3 in. wide, thick-textured, entire, petiolate. Stem leaves opposite, sessile to clasping, similar to basal, to 4 in. long × 2½ in. wide. Flowers irregular, tubular, pink to lavender, to 2 in. long, upper lip with 2 lobes, lower lip with 3 lobes, in groups of 2–6 subtended by bracts notably smaller than leaves, forming terminal raceme to 1½ ft. long. Fruit a teardrop-shaped capsule to 1 in. that matures dark brown. IA, IL, IN, MI, MN, MO, OH, WI. **Scattered throughout, more prevalent northwest, likely introduced eastward.** *CR*

Penstemon hirsutus

PLANTAGINACEAE

hairy beardtongue

Prairies, fields, savannas, streambanks. Spring, summer.

A hairy perennial to 2½ ft. tall. Leaves opposite, sessile to clasping, sharply toothed, to 4¾ in. long × 1¼ in. wide. Flowers irregular, tubular, flattened, white to lavender with yellow throat, to 1⅓ in. long, upper lip 2-lobed, lower lip nearly closing tube and 3-lobed, in terminal raceme to 8 in. long. Pale beardtongue (*P. pallidus*) has smaller, open-throated, white flowers on stalks over ⅖ in. long; slender beardtongue (*P. gracilis*) is similar to pale beardtongue but with shorter-stalked, pale violet flowers. IA, IL, IN, MI, OH, WI. Widespread in southeastern portion of the Northern Lakes and eastern portion of the Eastern Forests, scattered in eastern portion of the Tallgrass Prairie. *SN*

Veronica americana

PLANTAGINACEAE

American brooklime, American speedwell

Marshes, swamps, seeps, streams, rivers, ponds. Spring, summer, fall.

A glabrous, upright to creeping perennial to 20 in. tall, often rooting at nodes and growing in standing water. Leaves opposite, short-petiolate, to 3 in. long × 1½ in. wide, shallowly toothed to entire, pointed at tips. Flowers slightly irregular, pale blue to lavender, often with darker streaks, to ⅓ in. wide, with 4 spreading, rounded lobes surrounded by 4 sepals about same length, in racemes of 10–15 from leaf axils and at top of plant. Syn. *V. beccabunga* var. *americana*. IA, IL, IN, MI, MN, MO, OH, WI. Scattered in the Northern Lakes and eastern portion of the Eastern Forests, rare elsewhere. *PR*

Veronica anagallis-aquatica
PLANTAGINACEAE

water speedwell, brook pimpernel

Marshes, seeps, streams, rivers, ponds.
Spring, summer.

A glabrous perennial to 2½ ft. tall, often growing in standing water. Leaves opposite, sessile,
clasping, to 4 in. long × 1½ in. wide, shallowly
toothed to entire, pointed at tips. Flowers
slightly irregular, pale blue to lavender, often
with darker streaks, to ¼ in. wide, with 4
spreading, rounded lobes surrounded by 4
sepals, in racemes of 20–65 from leaf axils and
at top of plant. IA, IL, IN, MI, MN, MO, OH,
WI. Native and introduced elements from
Eurasia. Scattered to somewhat frequent
throughout. *CR*

Veronica arvensis
PLANTAGINACEAE

field speedwell, corn speedwell

Prairies, fields, glades, rock ledges, gardens,
lawns, forest trails, roadsides, vacant lots.
Spring, summer.

A spreading-hairy annual to 8 in. tall. Leaves
opposite, petiolate below, sessile above, to ½
in. long and nearly as wide, bluntly toothed,
becoming narrower, alternate bracts lacking marginal teeth at top of plant. Flowers
slightly irregular, blue-violet, often with
darker streaks, to ⅛ in. wide, with 4 spreading, rounded lobes surrounded by 4 sepals,
sessile or nearly so in axils of bracts, forming
leafy spike or raceme at top of plant. Fruit a
heart-shaped capsule. IA, IL, IN, MI, MN,
MO, OH, WI. Introduced from Eurasia.
Widespread throughout except in IA, MN,
and WI, where scattered. *SN*

Veronica officinalis
PLANTAGINACEAE

common speedwell

Fields, forests, swamps, pine plantations, rock outcrops. Spring, summer.

A gray-hairy, glandular, mat-forming, creeping perennial to 1 ft. tall, rooting at nodes. Leaves opposite, petiolate, to 2 in. long × 1¼ in. wide, toothed. Flowers slightly irregular, pale blue to whitish with darker streaks, to ⅓ in. wide, with 4 spreading, rounded lobes surrounded by 4 sepals, short-stalked from axils of short bracts in upright racemes to 6 in. long from axils of upper leaves. Fruit a heart-shaped capsule. IA, IL, IN, MI, MN, OH, WI. Introduced from Eurasia. Widespread in eastern portion of the Eastern Forests, becoming scattered westward, scattered in the Northern Lakes, rare in the Tallgrass Prairie. *SN*

Veronica polita
PLANTAGINACEAE

wayside speedwell

Lawns, mowed grassy fields, disturbed sites. Spring.

A low-growing to ascending annual, branched and forming patches, hairy, 3–12 in. long. Leaves opposite (upper commonly alternate), nearly sessile, broadly ovate with deep veins and coarsely toothed margins, ⅓ in. long and about as wide. Flowers irregular, ⅛–¼ in. wide, tubular, with 4 rounded lobes, one noticeably smaller, usually all with dark blue striping and white bases presenting a white "eye," pedicels ¼–⅓ in. long, on a spike-like terminal raceme. Fruit a heart-shaped capsule. The similar birds-eye speedwell (*V. persica*) has slightly larger flowers and noticeably longer pedicels. IA, IL, IN, MI, MO, OH, WI. Introduced from Europe. Common, mostly in the Eastern Forests and Ozark Highlands. *MAH*

Veronica scutellata
PLANTAGINACEAE

marsh speedwell

Wet meadows, marshes, swamps, seeps, swales, thickets, ditches, streams, rivers, ponds. Spring, summer, fall.

A weak perennial to 2 ft. tall. Leaves opposite, sessile to clasping, to 3 in. long × ⅜ in. wide with parallel margins, shallowly and sparsely toothed or entire, pointed at tips. Flowers slightly irregular, pale blue to lavender or white, often with darker streaks, to ¼ in. wide, with 4 spreading, rounded lobes surrounded by 4 sepals, on filiform stalks to ¾ in. in weak, slightly zigzag racemes of 8–20 from upper leaf axils. IA, IL, IN, MI, MN, OH, WI. Scattered in the Northern Lakes, becoming less frequent in adjacent portions of the Eastern Forests and Tallgrass Prairie. *CB*

Phlox bifida
POLEMONIACEAE

sand phlox, cleft phlox

Sand barrens, glades, dry rocky slopes, cliff ledges. Spring.

A reclining to ascending perennial, branched and sometimes mat-forming, hairy, hairs gland-tipped or not, 6–12 in. high. Leaves opposite, linear to linear-lanceolate, ½–2 in. long × ⅛–¼ in. wide. Flowers regular, ⅔–¾ in. wide, light blue to nearly white, tubular with 5 spreading lobes each with a ⅛- to ¼-in.-deep notch, in few-flowered terminal branches. Sand phlox is the western counterpart of the commonly cultivated moss-pink phlox (*P. subulata*), which may be native in our region only in eastern OH (naturalized elsewhere); it has smaller leaves and less deeply cleft lobes. IA, IL, IN, MI, MO, WI. Occurs mostly in IL, IN, and MO. *MAH*

Phlox divaricata

POLEMONIACEAE

wild blue phlox

Rich mesic forests, stream terraces, ravines. Spring.

An erect to reclining perennial, hairy, 1–1½ ft. tall. Leaves opposite, sessile, lanceolate to oblong, entire, 1–2 in. long × ½–1 in. wide. Non-flowering stems with persistent winter leaves arise during and after flowering. Flowers regular, about 1 in. wide, tubular, glabrous, with 5 abruptly spreading lobes, notched (subsp. *divaricata*) or un-notched (subsp. *laphamii*), on widely spreading terminal branches. IA, IL, IN, MI, MN, MO, OH, WI. Common throughout except for the far northern counties. *SN*

Polemonium reptans

POLEMONIACEAE

Jacob's ladder, Greek valerian

Rich mesic forests, stream terraces, ravines, moist prairies and savannas. Spring.

An erect to spreading perennial, branched, mostly glabrous (var. *reptans*), 8–18 in. tall. Leaves alternate, petiolate, pinnately compound with 5–15 elliptic-ovate leaflets, 4–7 in. long × 2–3 in. wide. Flowers regular, about ½ in. long, bell-shaped with 5 rounded lobes, occurring few to several in loosely flowered panicles. Variety *villosum*, with its densely glandular-villous stems and inflorescence of smaller flowers, occurs in southern OH. IA, IL, IN, MI, MN, MO, OH, WI. Occurs commonly in the southern Midwest, northward mostly in WI, eastern IA, and southeastern MN. *SN*

Heteranthera limosa
PONTEDERIACEAE

blue mudplantain

Ponds, ditches, mudflats. Summer, fall.

A sprawling emergent or floating-leaved aquatic annual with elongate stems. Leaves simple, basal, oblong to ovate, rounded to wedge-shaped at base, sessile (when submerged) or with sheathing petioles, entire, to 2½ in. long × 1¼ in. wide, thick, glabrous. Flowers slightly irregular, to 1 in. wide, tubular with 6 spreading tepals (upper 1–3 with yellow spot at base), solitary at end of tubular stalk from sheathing spathe. Leaves of roundleaf mudplantain (*H. rotundifolia*) are rounded to cordate at base. Inflorescences of bouquet mudplantain (*H. multiflora*) have more than 1 strongly irregular flower. IA, IL, IN, MN, MO. Scattered, **primarily in western and southern portions of Midwest, with most occurrences in MO. *CB***

Pontederia cordata
PONTEDERIACEAE

pickerelweed

Marshes, swamps, bogs, lakes, ponds, rivers. Summer, fall.

A colony-forming perennial to over 3 ft. tall. Leaves mostly basal, often with 1 along stem, simple, entire, cordate, to 10 in. long × 5 in. wide, long-petiolate, thick-textured, glabrous, with distinctive parallel veins generally following shape of leaf. Flowers slightly irregular, to ½ in. wide, tubular with 6 spreading blue-violet lobes, uppermost lobe with 1 or 2 yellow spots, numerous, spreading in dense glandular-pubescent terminal spike to 6 in. long; spike subtended by spathe-like sheath. IA, IL, IN, MI, MN, MO, OH, WI. Widespread in much of the Northern Lakes, scattered elsewhere. *SN*

Primula mistassinica
PRIMULACEAE

bird's-eye primrose, Mistassini primrose

Calcareous rocky shores, alkaline interdunal flats, fens. Spring.

A perennial to 6 in. tall. Leaves in basal rosette, spreading to ascending, to 1¼ in. long × ½ in. wide, sessile, with margins wavy and irregularly sharply toothed; each rosette producing a solitary, naked flowering stem from middle. Flowers regular, tubular with 5 spreading notched to sometimes ragged lobes, to ¾ in. wide, pinkish-lavender at tip, fading to whitish at base, with yellow ring around tube opening, up to 5 on short branches at top of stem. IA, IL, MI, MN, WI. **Uncommon and mostly restricted to northern portion of the Northern Lakes, rare southward into the Tallgrass Prairie.** *SN*

Anemone caroliniana
RANUNCULACEAE

Carolina anemone

Prairies, glades, barrens, in sandy or rocky soils. Spring.

A perennial with a single unbranched, hairy stem, 4–8 in. tall. Basal leaves 3- to 6-parted and deeply divided, petiolate. Mid-stem with 1 whorl of leafy involucral bracts, these 3-parted, linear-lobed, and sessile. Flowers regular, 1–1¾ in. wide, with mostly 10–15 linear petal-like sepals, white often with tinges of rose-purple. Carolina anemone is somewhat similar in appearance to Grecian windflower (*A. blanda*), a non-native known to occasionally escape from plantings. The latter differs by its petiolate floral bracts bearing oblanceolate-obovate lobes. IA, IL, IN, MN, MO, WI. **Occurs primarily in the Tallgrass Prairie and Ozark Highlands.** *CB*

Clematis occidentalis
RANUNCULACEAE

purple clematis, western blue virginsbower

Rocky woods, thickets, streambanks, openings. Spring.

A perennial woody vine to 12 ft. Leaves opposite, compound into 3 leaflets to 3½ in. long × 2 in. wide, entire or with a few small teeth on margins, sometimes lobed, with deeply impressed veins producing wrinkled appearance. Flowers regular, on stalks to 4 in. long in leaf axils, lacking true petals, with 4 showy, elliptic sepals to 2½ in. long that droop and spread slightly. Fruit a fluffy-headed cluster. Requires disturbance (fire or clearing, naturally through windthrow). **IA, IL, MI, MN, OH, WI. Uncommon in the Northern Lakes, with a few occurrences in the Tallgrass Prairie. SN**

Clematis pitcheri
RANUNCULACEAE

Pitcher's leather flower

Thickets, woodlands, streambanks. Spring, summer.

A climbing perennial vine, branching, sparsely hairy to glabrous, up to 12 ft. long. Leaves opposite, petiolate, pinnate with 3–5 pairs of simple or deeply lobed leaflets, these lanceolate to ovate, somewhat leathery, about 1–4 in. long × ½–2½ in. wide. Flowers regular, about 1 in. long, urn-shaped and nodding, petals lacking, sepals 4, thick, ovate-lanceolate, tips recurved, 1–5 on long stalks from leaf axils. Fruit an achene with a short-hairy to glabrous persistent style. Style of northern leather flower (*C. viorna*) has noticeably plume-like hairs; it occurs primarily in IN and OH. **IA, IL, IN, MO. Relatively common save for IN, where quite local. ST**

Delphinium carolinianum
RANUNCULACEAE

Carolina larkspur

Prairies, glades, savannas, openings in rocky woods. Spring, summer.

A short-hairy perennial to 3 ft. tall. Leaves basal and alternate, with 3–5 palmate lobes divided again into linear lobes, to 4 in. long and wide, smaller with shorter petioles along stem. Flowers irregular with 5 spreading lobes and long spur, to 1 in. wide and long, in raceme at top of plant. Two subspecies often recognized: subsp. *carolinianum* (pictured), usually with blue flowers and shorter-petiolate stem leaves, and subsp. *virescens*, with white flowers and longer-petiolate stem leaves. IA, IL, MN, MO, WI. Widespread in the Ozark Highlands and western Tallgrass Prairie, with few occurrences in western extremes of the Northern Lakes and Eastern Forests. *EH*

Delphinium tricorne
RANUNCULACEAE

dwarf larkspur

Rich mesic forests, ravines, stream terraces. Spring.

An erect perennial, unbranched, mostly glabrous, 10–18 in. tall. Leaves basal and on lower stem, petiolate, alternate, 1–3 in. long and almost twice as wide, palmately divided into narrow linear-lanceolate segments. Flowers irregular, 1–1½ in. long, sepals 5, petal-like with uppermost bearing a long spur, petals 4, smaller, lighter colored, surrounded by sepals, in a terminal raceme of 6–20 flowers. Dwarf larkspur flowers range in color from shades of purple to white. The common name alludes to the spur-like sepal, which superficially resembles the hind claw of a lark. IA, IL, IN, MO, OH. Common in the lower Midwest. *MAH*

Pulsatilla patens
RANUNCULACEAE

American pasqueflower

Dry prairies, open gravelly or sandy woodland slopes, barrens. Spring.

An erect perennial, unbranched, hairy, 5–15 in. tall. Leaves basal, petiolate, barely developed when flowering, kidney-shaped, palmately compound into deeply incised linear segments 1–3 in. long × 2–5 in. wide at maturity. The whorl of leaf-like bracts below flower are smaller and clasping. Flowers regular, 1–2 in. wide, with 5–7 elliptic petal-like sepals, pale blue to white, occurring singly atop stem. Fruit an achene with 1–1½ in. long persistent and plume-like styles (see inset). One of the earliest prairie species to bloom. Syn. *Anemone patens*. IA, IL, MN, WI. Occurs principally in the northern Tallgrass Prairie. *CR/SN*

Geum rivale
ROSACEAE

purple avens, water avens

Wet meadows, marshes, swamps, bogs, fens, often in alkaline soils. Spring, summer.

A hairy perennial to 2 ft. tall. Leaves variable, pinnately compound with coarsely toothed leaflets of variable size and large leafy stipules at connection to stem, basal leaves petiolate, stem leaves alternate on lower part of stem, short-petiolate to sessile, terminal leaflet largest. Flowers regular with 5 pale yellow to purplish petals shorter than 5 reddish-purple pointed sepals, to 1 in. wide, nodding, few on stalks at top of plant. Fruiting heads upright, round, feathery-hairy. IL, IN, MI, MN, OH, WI. Primarily in the Northern Lakes, with few occurrences in adjacent portions of the Tallgrass Prairie and Eastern Forests. *SN*

Galium lanceolatum
RUBIACEAE

lance-leaf wild licorice

Mesic to dry-mesic upland forests.
Spring, summer.

An erect to ascending perennial, simple to
branched at base and/or above, stems mostly
glabrous, 8–18 in. tall. Leaves 4, whorled, sessile, elliptic to lanceolate to a long-tapering
tip, glabrous or margins and veins minutely
hairy, 1–3 in. long × ⅓–1¼ in. wide. Flowers
regular, ⅛ in. wide, yellowish-green to purple,
short-tubular with 4 pointed, spreading lobes,
on widely spreading branches. Fruit 2-lobed,
rounded and bristly. In our region this is the
largest species in the genus with respect to leaf
size. IL, IN, MI, OH, WI. **Most occurrences in
eastern OH, MI, and WI.** *RS*

Houstonia caerulea
RUBIACEAE

Quaker ladies, azure bluet

Upland forests, woodland openings,
sand prairies. Spring.

An erect perennial, wiry, branched, glabrous,
rhizomatous, 2–5 in. tall. Leaves basal in a
rosette, petiolate, oblanceolate, about ½ in.
long, stem leaves opposite, sessile, bract-like.
Flowers regular, about ½ in. wide, tubular
with 4 flared ovate lobes and a yellow center,
solitary from upper leaf axils. This bluet is our
only one whose flowers have a yellow center.
Although individually small, large populations can color the forest floor with light blue.
IA, IL, IN, MI, MO, OH, WI. **Most occurrences concentrated in southern IN, eastern
OH, and southeastern MO.** *SN*

Houstonia pusilla
RUBIACEAE

tiny bluet, small bluet

Sandstone glades, pastures, fields, lawns.
Spring.

An erect annual, somewhat branched,
mostly glabrous, 1–4 in. tall. Leaves opposite,
short-petiolate to sessile, oval to ovate, entire
with marginal hairs, ¼–⅓ in. long × ⅛–¼
in. wide. Flowers regular, about ¼ in. wide,
tubular with 4 flared ovate-lanceolate lobes,
the basal portions of which are reddish, presenting the look of a red "eye." Flowers occur
singly from middle and upper leaf axils. True
to its common names, this bluet is one of the
shortest *Houstonia* species in the Midwest. IA,
IL, IN, MO. Occurs mostly in IL and MO, also
eastern IA and southern IN. *CB*

Solanum dulcamara
SOLANACEAE

bittersweet nightshade, climbing nightshade

Marshes, swamps, floodplains, thickets,
ponds, disturbed sites. Spring, summer, fall.

A sprawling to climbing woody vine to 8 ft.
long, lacking tendrils. Leaves simple, alternate,
cordate with large, pointed terminal lobe and
2 much smaller, spreading basal lobes (lobes
absent on young leaves), to 4 in. long × 2½ in.
wide, petiolate. Flowers regular, violet (rarely
white), to ½ in. long, with short tube and 5
recurved lobes beneath protruding stamens,
dangling, stalked, in branched clusters. Fruit
a roundish ¼-in. red berry. Matrimony-vine
(*Lycium barbarum*) has unlobed leaves and
spreading corolla lobes with cream-colored
anthers. IA, IL, IN, MI, MN, MO, OH, WI.
Introduced from Eurasia. Widespread north
and east, less frequent west and south. *SN*

Glandularia canadensis
VERBENACEAE

rose vervain

Glades, rocky slopes, barrens, prairies.
Spring, summer.

A creeping to ascending perennial, rooting
at nodes, branched, hairy, 1–2 ft. long. Leaves
opposite, petiolate, ovate to lanceolate,
deeply lobed, 1–3 in. long and almost as wide.
Flowers regular, about ½ in. wide, tubular,
with 5 notched lobes, in a congested terminal
spike. Rose vervain is quite showy and used in
landscaping. Syn. *Verbena canadensis*. IA, IL,
IN, MI, MO, OH, WI. Widespread but many
occurrences are escapes from cultivation. In
the Midwest perhaps native only in MO and
southern IL. *EH*

Verbena bracteata
VERBENACEAE

creeping vervain, prostrate vervain, bigbract verbena

Prairies, fields, roadsides, railroads,
parking lots. Spring, summer, fall.

An inconspicuous, branching, sprawling, abundantly hairy annual to 8 in. tall.
Leaves opposite, simple, to 3 in. long × 1 in.
wide, deeply lobed and coarsely toothed
below, becoming smaller above, sessile to
short-petiolate. Flowers slightly irregular, to ⅛
in. wide, tubular with 5 spreading-ascending
lobes, in whorls in dense terminal spike to 6
in. long × ⅔ in. wide, each flower subtended
by entire, unlobed bract to ¾ in. long; only a
few terminal-end flowers open at a time. IA,
IL, IN, MI, MN, MO, OH, WI. Widespread
to scattered throughout, less frequent to the
east. *SN*

Verbena hastata
VERBENACEAE

blue vervain, swamp verbena, simpler's joy

Wet prairies, sedge meadows, moist fields, marshes, shores, floodplains, thickets, riverbanks, ditches. Summer, fall.

A hairy, upright perennial to 5 ft. tall. Stem often purplish-red, angled. Leaves opposite, simple, to 6 in. long × 1 in. wide, deeply reticulate-veined, sharply toothed, petiolate, sometimes with a pair of narrow, ascending basal lobes. Flowers slightly irregular, blue-lavender (rarely white), to ¼ in. wide, tubular with 5 spreading-ascending lobes, in whorls along several to numerous dense terminal spikes to 5 in. long at ends of ascending, leafless branches in upper part of plant. IA, IL, IN, MI, MN, MO, OH, WI. Widespread throughout, with fewer occurrences in the Ozark Highlands. *MJH*

Verbena simplex
VERBENACEAE

narrowleaf vervain

Prairies, fields, glades, limestone outcrops, roadsides, railroads. Spring, summer, fall.

An upright, multi-stemmed perennial to 2½ ft. tall with square stems. Leaves opposite, simple, to 4 in. long × ½ in. wide, deeply reticulate-veined, coarsely toothed (at least in outer half), tapering to base or short-petiolate. Flowers slightly irregular, lavender to whitish, to ¼ in. wide, tubular with 5 spreading-ascending lobes, in whorls along 1 to few loose to fairly dense terminal spikes to 12 in. long in upper part of plant. Fruit a cluster of 4 nutlets. IA, IL, IN, MI, MN, MO, OH, WI. Widespread in the Ozark Highlands, scattered elsewhere, least frequent northward. *EH*

Verbena stricta
VERBENACEAE

hoary vervain

Sand barrens, savannas, glades, prairies, railroads. Summer.

An erect perennial, branched above, densely gray-hairy, 2–4 ft. tall. Leaves opposite, sessile or short-petiolate, ovate to elliptic, sharply toothed, 1–4 in. long × ¾–2 in. wide. Flowers irregular, about ¼ in. wide, tubular, with 5 spreading and rounded lobes, several arranged on stiff, compact spikes at stem tips. Fruit composed of 4 nutlets. IA, IL, IN, MI, MN, MO, OH, WI. **Common in most of the Midwest.** *SN*

Viola bicolor
VIOLACEAE

American field pansy, wild pansy

Sandy fields, roadsides. Spring, summer.

An annual to 10 in. tall. Basal leaves round, long-petiolate, to 1¼ in. long and wide, bluntly toothed; stem leaves alternate, narrower, to 2 in. long, short-petiolate to sessile above, with pair of leaf-like, pinnately lobed stipules to 1 in. long from bases of petioles. Flowers irregular, to ⅗ in. wide, with 5 spreading petals, pale blue-violet with white bases and dark veins, with bright yellow spot on petal with short spur on back, subtended by sepals shorter than petals; solitary on erect stalks from upper leaf axils. Johnny-jump-up (*V. tricolor*), introduced, has flowers over ⅗ in. wide, tricolored, with tips of upper petals purple. IA, IL, IN, MO, OH. **Widespread south, scattered to rare north.** *MAH*

Viola labradorica
VIOLACEAE

dog violet, Labrador violet

Forests, swamps, streambanks. Spring.

A glabrous perennial to 8 in. tall. Leaves basal and alternate, cordate, to 2½ in. long × 1½ in. wide, petiolate, bluntly toothed, with pair of narrow, toothed stipules to ½ in. where petioles attach to stem. Flowers irregular, to ¾ in. wide, pale blue-violet with dark veins in whitish throat, with 5 spreading to recurved petals, 1 with stout, straight to curved spur to ¼ in. long, lateral petals bearded at base, solitary on stalks from leaf axils. Syn. *V. conspersa*. Sand violet (*V. adunca*) has deeper blue flowers and hairy leaves. IL, IN, MI, MN, OH, WI. Widespread in the Northern Lakes, uncommon in adjacent areas. *SN*

Viola missouriensis
VIOLACEAE

Missouri violet

Floodplain woods, moist thickets. Spring.

A mostly glabrous perennial to 6 in. tall. Leaves basal, cordate, to 3½ in. long × 2½ in. wide, long-petiolate, bluntly toothed. Flowers irregular, to ¾ in. wide, pale blue-violet with dark veins in whitish throat, with 5 spreading petals, 1 short-spurred, lateral petals bearded at base, spurred petal glabrous at base; sepals ciliate; solitary on naked stalks from base. LeConte's violet (*V. affinis*) lacks hairs on sepals and often has spurred petal bearded. Marsh violet (*V. cucullata*) has leaves as wide as long and lateral petals bearded with very short knob-tipped hairs. IA, IL, IN, MN, MO, OH, WI. Scattered in the Ozark Highlands, the Tallgrass Prairie, and adjacent areas. *DT*

Viola palmata
VIOLACEAE

wood violet, cleft violet, three-lobed violet

Prairies, forests, thickets. Spring.

A perennial to 6 in. tall. Leaves basal, variable, cordate and unlobed early in season, becoming sagittate and deeply 5- to 7-lobed later in season, to 3 in. long and wide, long-petiolate, bluntly toothed. Flowers irregular, to 1 in. wide, blue-violet with dark veins in whitish throat, with 5 spreading petals, 1 short-spurred, lateral petals bearded at base, spurred petal glabrous at base; solitary on long, naked stalks from base, taller than leaves. Syn. *V. triloba*. IL, IN, MI, MO, OH, WI. Widespread to scattered in the Ozark Highlands, Eastern Forests, and Tallgrass Prairie, scattered in southeastern portion of the Northern Lakes. *CR*

Viola pedata
VIOLACEAE

birdfoot violet

Prairies, fields, glades, savannas, barrens, dunes, open woods. Spring.

A perennial to 6 in. tall. Leaves basal, long-petiolate, to 2 in. long and wide, deeply palmately lobed, the 3–5 nearly linear lobes lobed. Flowers irregular, to 1½ in. wide, pale blue-violet, or with upper 2 petals dark purple, rarely entirely white, sometimes with dark veins in white throat, with 5 spreading to recurved petals, 1 short-spurred, petals glabrous at base; solitary on long, naked stalks from base, often slightly taller than leaves. IA, IL, IN, MI, MN, MO, OH, WI. Widespread in the Ozark Highlands, eastern portion of the Tallgrass Prairie, and southern portion of the Northern Lakes; scattered to rare elsewhere. *SN*

Viola pedatifida
VIOLACEAE

prairie violet

Prairies, savannas, open woods. Spring.

A perennial to 6 in. tall. Leaves basal,
long-petiolate, to 3 in. long and wide, deeply
palmately lobed, the 5–9 linear lobes lobed.
Flowers irregular, to ¾ in. wide, pale to deep
blue-violet with dark veins in whitish throat,
with 5 spreading petals, 1 short-spurred, lateral
petals bearded at base, spurred petal usually
bearded at base; solitary on long, naked stalks
from base, often slightly taller than leaves.
Lobed blue violet (*V. subsinuata*) has leaves
divided into 5–16 lobes of varying shapes and
widths. IA, IL, IN, MI, MN, MO, OH, WI.
Widespread in the Tallgrass Prairie, scattered
to rare in the Ozark Highlands, Eastern Forests, and Northern Lakes. *SN*

Viola rostrata
VIOLACEAE

long-spurred violet

Forests. Spring.

A glabrous perennial to 8 in. tall. Leaves basal
and alternate, cordate, to 1¾ in. long × 1¼ in.
wide, petiolate, bluntly to sharply toothed,
with pair of narrow, shallowly toothed stipules
to ½ in. where petioles attach to stem. Flowers
irregular, to 1 in. wide, solitary on stalks from
leaf axils, pale blue-violet with dark spots
just above whitish throat and dark veins into
throat, with 5 spreading to recurved petals, 1
with straight to curved spur up to ⅗ in. long,
lateral petals glabrous at base. IN, MI, OH,
WI. Widespread in eastern portion of the
Northern Lakes and eastern portion of the
Eastern Forests, scattered westward. *MJH*

Viola sagittata
VIOLACEAE

arrowleaf violet

Sand prairies, fields, glades, savannas, barrens, open woods. Spring, summer.

A perennial to 6 in. tall. Leaves basal, sagittate with coarsely toothed small basal lobes, to 4 in. long × 2½ in. wide, long-petiolate, sparsely shallowly toothed. Flowers irregular, to ¾ in. wide, blue-violet with dark veins in white throat, with 5 spreading petals, 1 short-spurred, lateral petals bearded at base, spurred petal bearded at base; solitary on long, naked stalks from base. Fruit a capsule. Plants with short-petiolate, densely hairy leaves sometimes treated as var. *ovata*. **IA, IL, IN, MI, MN, MO, OH, WI. Scattered to widespread throughout.** *KB*

Viola sororia
VIOLACEAE

common blue violet

Prairies, fields, glades, savannas, forests, swamps, thickets, lawns. Spring.

A perennial to 6 in. tall. Leaves basal, cordate, to 3 in. long and wide, long-petiolate, bluntly toothed. Flowers irregular, ¾ in. wide, blue-violet with dark-veined cream throat, with 5 spreading petals, 1 short-spurred, lateral petals bearded at base, spurred petal glabrous at base; solitary on naked stalks from base. Northern bog violet (*V. nephrophylla*) has bearded spurred petal and leaves purplish-iridescent beneath. Great-spurred violet (*V. selkirkii*) has glabrous petals and large spur. Sweet violet (*V. odorata*), introduced and sweetly fragrant, has hooked style. New England blue violet (*V. novae-angliae*) has leaves longer than wide. **IA, IL, IN, MI, MN, MO, OH, WI. Widespread.** *SN*

BROWN TO MAROON FLOWERS

Matelea obliqua
APOCYNACEAE

climbing milkvine

Limestone glades, rocky woods.
Spring, summer.

A perennial twining vine, hairy, branched, up to 10 ft. long with milky sap. Leaves opposite, entire, cordate, up to 6 in. long × 3½ in. wide. Flowers regular, about 1 in. wide, with 5 minutely hairy and fleshy linear lobes about ½ in. long × ⅛ in. wide, occurring in loose axillary umbels (see inset). Fruit a warty follicle. Found in the Ozark Highlands, oldfield milkvine (*M. decipiens*) has flower lobes up to ¼ in. wide. Angular-fruit milkvine (*Gonolobus suberosus*) of the southwestern Eastern Forests has smooth follicles. IL, IN, OH. **Restricted to the southern counties of each state.** *CB/SN*

Vincetoxicum nigrum
APOCYNACEAE

black swallowwort

Fencerows, gardens, forests, thickets.
Summer.

A twining perennial herbaceous vine, sparsely hairy, to 8 ft. long. Leaves opposite and entire, petiolate, lance-ovate to subcordate, 1–4 in. long × ½–3 in. wide. Flowers regular, about ¼ in. wide, dark purple-black, thick, with 5 triangular lobes, minutely hairy above, in umbels of 10+ flowers from leaf axils. Black swallowwort has the potential to become a serious widespread weed. Likewise for the similar European swallowwort (*V. rossicum*). The latter tends to have somewhat paler flowers that are glabrous above. IL, IN, MI, MN, MO, OH, WI. Introduced from Europe. **Currently known only from a few counties in each state.** *MAH*

Symplocarpus foetidus
ARACEAE

skunk cabbage

Swamps, floodplains, seeps, fen and
bog margins. Winter, spring.

A glabrous, colony-forming perennial to 2 ft.
tall, entire plant malodorous. Leaves in basal
rosette, developing at end of flowering period,
ascending, to 2 ft. long × 1 ft. wide, cordate to
ovate, entire, petiolate. Flowers regular, individually inconspicuous, lacking petals, numerous
on thick, oblong, yellowish or purple spadix to
2 in. long within conspicuous hood-like, conical, purplish-brown spathe streaked or mottled
with yellowish-green, to 6 in. long × 3 in. wide.
Fruit in stalked, brown, segmented cluster to 4
in. long × 3 in. wide. IA, IL, IN, MI, MN, OH,
WI. Most frequent in the Northern Lakes,
scattered southward into the Tallgrass Prairie
and the Eastern Forests. *SN*

Asarum canadense
ARISTOLOCHIACEAE

wild ginger

Rich mesic forests, ravines,
stream terraces. Spring.

A creeping rhizomatous perennial,
unbranched, 2–6 in. tall. Leaves alternate,
crowded, kidney-shaped, sparsely hairy,
entire, petiolate, 1½–3 in. long × 3–5 in. wide.
Flowers regular, single, prostrate between
paired leaves and often hidden under leaf
litter, petals absent, sepals maroon to tannish,
short- to long-tapered, recurved or spreading
from a short tube that is white internally, the
outer surface hairy, the lobes ¼–1 in. long.
Variety *reflexum* (pictured) and var. *acumina-
tum* (inset), among others, have been desig-
nated based on floral characteristics. IA, IL,
IN, MI, MN, MO, OH, WI. Common in most
of the region. *BS/SN*

Endodeca serpentaria
ARISTOLOCHIACEAE

Virginia snakeroot

Mesic upland forests, stream terraces.
Spring, summer.

An erect or ascending perennial, stem zigzag,
mostly glabrous, 6–12 in. tall. Leaves few,
alternate, petiolate, ovate-lanceolate with
cordate base, glabrous or hairy, 2–6 in. long
× ½–2½ in. wide. Flowers irregular, ½–1 in.
long, tubular and S-shaped, like a pipe with
a flaring mouth, 3-lobed, usually 1 per plant,
typically hidden under leaf litter a few inches
from the main stem (see inset). Its leaves are a
food source for caterpillars of Pipevine Swal-
lowtail, a common woodland butterfly. Syn.
Aristolochia serpentaria . IA, IL, IN, MI, MO,
OH. Common, mostly in the Eastern Forests
and Ozark Highlands. *PR*

Lechea mucronata
CISTACEAE

hairy pinweed

Barrens, glades, savannas, in sandy or
rocky soils. Summer.

An erect perennial, branched above, stems
with dense spreading hairs, 8–18 in. tall.
Leaves alternate, opposite, or whorled, sessile
or short-petiolate, entire, elliptic, hairy below,
glabrous above, ½–1 in. long × ⅛–⅓ in. wide.
Flowers inconspicuous, petals 3 but minuscule, sepals 5 with the outermost 2 narrow
and innermost 3 broad and keeled, best seen
in fruit. Inflorescence on terminal and lateral
branches. Fruit a capsule. This is our largest
pinweed and the only one with spreading
hairs. It and most pinweeds produce overwin-
tering prostrate branches. IA, IL, IN, MI, MO,
OH, WI. Occurs mostly in the southern Great
Lakes and Ozark Highlands. *NP*

Lechea tenuifolia
CISTACEAE

narrowleaf pinweed

Dry upland forests, glades, barrens, savannas. Summer, fall.

An erect perennial, branching above, glabrous or with few appressed hairs, 4–12 in. tall. Leaves alternate to subopposite, sessile, linear and entire, ¼–¾ in. long × ¹⁄₁₆ in. wide. Flowers inconspicuous, petals 3 but minuscule, sepals 5 with the outer 2 narrow and longer than the 3 broader inner ones. Inflorescences on terminal and lateral branches. The outer sepals of the mostly northern-ranging prairie pinweed (*L. stricta*) and intermediate pinweed (*L. intermedia*) are shorter than the inner ones. IA, IL, IN, MN, MO, OH, WI. Occurs mostly in the Ozark Highlands and eastern Tallgrass Prairie. *PD*

Veratrum woodii
MELANTHIACEAE

Wood's false hellebore

Rich mesic to dry-mesic forests, often on steep rocky slopes. Summer.

An erect perennial, unbranched, stems short-hairy, 3–5 ft. tall. Leaves principally basal in a rosette, elliptic to oblanceolate, parallel-veined and pleated, a few alternate leaves occur on flowering stalk (see inset), glabrous, 8–18 in. long × 1–4 in. wide. Flowers regular, about ¾ in. wide, with 6 oblanceolate, spreading tepals, numerous in a rather narrowly branched terminal inflorescence 1–2 ft. long. The flower, among the darkest in pigment of our Midwestern wildflowers, may appear almost black. Syn. *Melanthium woodii*. IA, IL, IN, MO, OH. Occurs locally in the lower Midwest. *BS/DT*

Aplectrum hyemale
ORCHIDACEAE

puttyroot, Adam and Eve

Mesic deciduous forests. Spring.

A glabrous, unbranched perennial 10–15 in.
tall. Leaf solitary, basal, approximately 6 in.
long × 3–4 in. wide, oval and pleated, light
green with white veins (see inset). Flowers
irregular, about ⅔–1 in. long, with 3 petals and
3 petal-like sepals, the lowest lip petal distinct
with white and purple markings. Puttyroot's
leaf typically emerges in late summer, over-
winters above ground, and then deteriorates
around the time the plant blooms. Although
uncommon, forms with yellow or green flowers exist. IA, IL, IN, MI, MN, MO, OH, WI.
**Throughout, but local to absent in the west
and north.** *PG/SN*

Hexalectris spicata
ORCHIDACEAE

crested coral-root

Dry limestone woodlands and glade
margins. Summer.

An erect perennial, unbranched, glabrous,
fleshy, tan to purplish, 8–18 in. tall. Leaves
absent except for ¼-in.-long, non-green
sheathing bracts. Flowers irregular, about 1
in. wide and somewhat fleshy, the lip petal
ovate to obovate, 3-lobed with purple ridges,
surrounded by 3 yellowish sepals with purple
stripes and 2 similar petals, 8–15 in a spike-like
raceme. The rocky sites where this orchid
grows are some of the hottest and driest environments in the Midwest. IL, IN, MO, OH.
**Occurs locally in Ozark Highlands and deep
southern counties of IL, IN, and OH.** *CB*

Isotria verticillata
ORCHIDACEAE

large whorled pogonia

Dry-mesic upland forests, sphagnum bogs.
Spring.

An erect perennial, glabrous, 5–10 in. tall,
colonial. Leaves elliptic-obovate, 1–3½ in. long
× ½–1½ in. wide, in a whorl atop a purplish
hollow stem. Flowers irregular, single, lip petal
oblong and about ¾ in. long, 3-lobed with
middle lobe bearing a greenish ridge, upper
2 petals yellow-green, with 3 greenish-purple
and widely spreading sepals 1½–2½ in. long.
The foliage of this and Indian cucumber-root
(*Medeola virginiana*) are similar, but the latter
has a solid wiry stem. Flowers of the extremely
rare small whorled pogonia (*I. medeoloides*)
are green overall and smaller. IL, IN, MI, MO,
OH. Occurs mostly in southern IN, southern
MI, and eastern OH. *PG*

Neottia cordata
ORCHIDACEAE

heartleaf twayblade

Moist forests, swamps, bogs, in moss.
Spring, summer.

An erect perennial to 13 in. tall. Leaves 2, opposite, bluntly triangular, to 1½ in. long and wide,
sessile, at midstem or slightly below, glabrous,
entire. Flowers irregular, ¼ in. long, with 3
spreading yellow-green to brownish-purple
sepals and 3 spreading yellow-green to
brownish-purple petals, lip petal spreading
downward, narrow and deeply cleft into pointed
lobes (see inset); up to 25 in 4-in. terminal
raceme. Broad-leaved twayblade (*N. convallarioi-
des*) has yellow-green to whitish flowers with lip
petal narrow at base, broad and barely notched,
with round lobes at tip. Syn. *Listera cordata*.
MI, MN, OH, WI. Restricted to the Northern
Lakes, more frequent northward. *PD*

Tipularia discolor
ORCHIDACEAE

cranefly orchid

Moist forests, wooded dunes. Summer.

A glabrous perennial to 2 ft. tall. Single ovate basal leaf to 3½ in. long × 2 in. wide develops in fall, entire, green with purplish blistered spots on top surface (see inset), purple below, petiolate, present through winter, withering in spring; stem leaves absent. Flowers irregular, reminiscent of a crane fly, inconspicuous, ¾ in. wide, with 3 greenish-purple sepals, 2 greenish-purple petals, and 1 lighter-colored lip petal, subtended by minute scale-like bracts, spreading to drooping in terminal raceme to 1 ft. long. IL, IN, MI, MO, OH. **Common in southern portion of the Eastern Forests, disjunct and rare in the Northern Lakes near the ends of Lakes Michigan and Erie.** *BS/CR*

Epifagus virginiana
OROBANCHACEAE

beech drops

Mesic forests. Summer, fall.

An erect annual, branched, glabrous, tan, 5–15 in. tall. Leaves scale-like and brown, alternate, lanceolate, about ⅛ in. long. Flowers irregular, ⅓–⅜ in. long, tubular with 4 triangular lobes, tan with purplish-brown streaks, of two types, 1 budlike and closed (cleistogamous) and self-fertilized, the other open (chasmogamous) but sterile (see inset), all occurring alternately along the stem. This unusual plant is parasitic on roots of American beech (*Fagus grandifolia*) and found only where the tree is found. It is not harmful to the tree. *Epifagus* translates to "upon beech." IL, IN, MI, MO, OH, WI. **Occurs primarily in the Eastern Forests and eastern Northern Lakes.** *EH/SN*

Sarracenia purpurea
SARRACENIACEAE

purple pitcherplant

Bogs, fens, conifer swamps. Spring, summer.

A carnivorous perennial to 2 ft. tall. Leaves basal, modified into swollen, water-filled, insect-trapping "pitchers" to 8 in. long with rimmed opening at top below erect hood, interior covered with downward-pointing hairs, with keel vertically along front, green with or without reddish-purple veins, or reddish-purple. Flowers regular, 2½ in. wide, nodding, with 5 arching, soon-deciduous petals (yellowish in forma *heterophylla*) enclosing large umbrella-like stigmatic disk that conceals numerous stamens, subtended by 5 large, persistent sepals, solitary at end of naked stem. IL, IN, MI, MN, OH, WI. Uncommon in the Northern Lakes and nearby portions of the Tallgrass Prairie and Eastern Forests. *SN*

Scrophularia lanceolata
SCROPHULARIACEAE

early figwort, lanceleaf figwort

Forest openings and edges, fields, thickets, roadsides. Summer.

A square-stemmed perennial to 6 ft. tall. Leaves opposite, to 8 in. long × 3 in. wide, irregularly toothed, round to flat at base, on petioles to 1¼ in. long, often with smaller axillary leaves. Flowers irregular, reddish-brown to reddish-green, to ⅓ in. long × ¼ in. wide, tubular with longer erect 2-lobed upper lip, 2 erect lateral lobes, and single spreading lower lobe, with yellowish, flattish, sterile stamen inside at top of flower (see inset), numerous in opposite stalked clusters atop stem. IA, IL, IN, MI, MN, MO, OH, WI. Widespread and common in northern half of the Midwest, scattered to rare southward. *KB/PR*

Scrophularia marilandica
SCROPHULARIACEAE

late figwort

Mesic woodlands, forest edges and openings. Summer.

An erect perennial, unbranched to branched above, glabrous to glandular-hairy, stems square with grooved sides, 4–7 ft. tall. Leaves opposite, petiolate, lance-ovate to ovate with a somewhat cordate base and evenly spaced marginal teeth, 4–6 in. long × 2–3 in. wide. Flowers irregular, ¼–⅓ in. long, 2-lipped, dull and urn-shaped, with upper lip lobed, the lower reflexed, interior with a purple, flattish, sterile stamen (see inset); inflorescence a widely branched panicle. IA, IL, IN, MI, MN, MO, OH, WI. Mostly in southern two-thirds of Midwest, somewhat common. *PR/PD*

Trillium recurvatum
TRILLIACEAE

purple trillium, prairie trillium

Rich mesic to dry-mesic upland and floodplain forests, thickets. Spring.

An erect perennial, unbranched, glabrous, 6–15 in. tall. Leaf-like bracts 3 atop stalk, whorled, short-petiolate, elliptic to lance-ovate, entire, commonly mottled, 2–4 in. long × 1–3 in. wide. Flowers regular, about 1 in. long, with 3 lanceolate to ovate and erect petals narrowed at base and 3 recurved lanceolate sepals, solitary and sessile atop the leaf-like bracts. Sepals are distinctive in their position but may not be fully recurved in early development. IA, IL, IN, MI, MO, OH, WI. Most common in IL, IN, and eastern MO. *SN*

Trillium sessile
TRILLIACEAE

sessile trillium, toadshade

Rich mesic forests, floodplains, flatwoods, ravines. Spring.

An erect perennial, unbranched, glabrous, 5–10 in. tall. Leaf-like bracts 3 atop stalk, whorled, sessile, elliptic to ovate, entire, mottled or not, 2–3 in. long × 1½–2½ in. wide. Flowers regular, about 1 in. long, with 3 oblanceolate to elliptic and erect petals and 3 lanceolate sepals with a sessile base, somewhat ascending or on the same plane as the leaf-like bracts. Compare with purple trillium (*T. recurvatum*). IL, IN, MI, MO, OH. **Most common in IN, MO, and OH.** *DT*

Typha latifolia
TYPHACEAE

broad-leaved cattail

Marshes, swamps, seeps, bogs, fens, ponds, ditches. Summer.

A glabrous, colonial perennial, to 9 ft. tall. Leaves appearing basal, simple, to 1 in. wide, to as long as stem, flat, blue-green. Flowers regular, inconspicuous, sessile in separate male and female spikes to 10 in. long, male spike terminal, tan, to ¾ in. thick, female spike dark brown, to 1½ in. thick. Narrow-leaved cattail (*T. angustifolia*), introduced, has dark green leaves to ⅓ in. wide, one side convex in cross section, with gap between male and female spikes, female spikes to ⅓ in. thick. Hybrid cattail (*T. ×glauca*), invasive, is intermediate or resembles parents; it is widespread, particularly in Great Lakes region. IA, IL, IN, MI, MN, MO, OH, WI. **Widespread, becoming less frequent due to hybridization.** *SN*

BIBLIOGRAPHY

Cohen, J. G., et al. 2015. *A Field Guide to the Natural Communities of Michigan.* East Lansing: Michigan State University Press.

Curtis, J. T. 1959. *The Vegetation of Wisconsin.* Madison: University of Wisconsin Press.

Deam, C. C. 1940. *Flora of Indiana.* Indianapolis: Indiana Department of Conservation.

Fernald, M. L. 1950. *Gray's Manual of Botany,* 8th ed. New York: American Book Company.

Flora of North America Editorial Committee, eds. 1993–. *Flora of North America,* 21+ vols. New York: Oxford University Press.

Gleason, H. A., and A. Cronquist. 1991. *Manual of Vascular Plants of Northeastern United States and Adjacent Canada,* 2d ed. Bronx: New York Botanical Garden.

Homoya, M. A. 1993. *Orchids of Indiana.* Indianapolis: Indiana Academy of Science.

———. 2012. *Wildflowers and Ferns of Indiana Forests.* Bloomington: Indiana University Press.

Homoya, M. A., et al. 1985. "The Natural Regions of Indiana." *Proceedings of the Indiana Academy of Science* 94:245–268.

Jackson, M. T., ed. 1997. *The Natural Heritage of Indiana.* Bloomington and Indianapolis: Indiana University Press.

Kurz, D. 1999. *Ozark Wildflowers.* Helena: Falcon Press.

Ladd, D. 1995. *Tallgrass Prairie Wildflowers.* Helena: Falcon Press.

———. 2001. *North Woods Wildflowers.* Helena: Falcon Press.

Mohlenbrock, R. H. 2002. *Vascular Flora of Illinois.* Carbondale: Southern Illinois University Press.

Nelson, P. 2010. *The Terrestrial Natural Communities of Missouri.* Jefferson City: Missouri Department of Natural Resources.

Schwegman, J. 1973. *The Natural Divisions of Illinois.* Springfield: Illinois Nature Preserves Commission.

Smith, W. 2012. *Native Orchids of Minnesota.* Minneapolis: University of Minnesota Press.

Steyermark, J. A. 1963. *Flora of Missouri.* Ames: Iowa State University Press.

Tester, J. R. 1995. *Minnesota's Natural Heritage.* Minneapolis: University of Minnesota Press.

Voss, E. G., and A. A. Reznicek. 2012. *Field Manual of Michigan Flora.* Ann Arbor: University of Michigan Press.

Wilhelm, G., and L. Rericha. 2017. *Flora of the Chicago Region.* Indianapolis: Indiana Academy of Science.

Yatskievych, G. 1999, 2006, and 2013. *Steyermark's Flora of Missouri.* Vols. 1–3. Jefferson City and St. Louis: Missouri Department of Conservation and Missouri Botanical Garden Press.

Yatskievych, K. 2000. *Field Guide to Indiana Wildflowers.* Bloomington and Indianapolis: Indiana University Press.

WEB RESOURCES

Biota of North America Program: bonap.org

Consortium of Midwest Herbaria: midwestherbaria.org

Flora of North America: eFloras.org

Go Botany: gobotany.nativeplanttrust.org

Illinois wildflowers: illinoiswildflowers.info

Jepson eFlora: ucjeps.berkeley.edu/eflora
Michigan flora: michiganflora.net
Minnesota wildflowers:
    minnesotawildflowers.info
Missouri plants: missouriplants.com
Native plant societies: michbotclub.org/
    national-botanical-organizations
Natural heritage programs: natureserve.org/
    natureserve-network/united-states
North American Native Plant Society:
    nanps.org
Wisconsin flora: Wisflora.herbarium.wisc.edu

PHOTOGRAPHERS

AB Adam Balzer
CB Christopher David Benda,
    illinoisbotanizer.com
KB Keith Board
JB Jim Brighton
LC Lee Casebere
KC Katy Chayka,
    minnesotawildflowers.info
PC Pete Curtis, Minneflora
JD Jacqueline Donnelly, Saratoga Woods
    and Waterways
PD Peter Dziuk,
    minnesotawildflowers.info
ME Michael Eason
CG Casey Galvin
AG Andrew Gibson
BG Bill Glass
OG Otto Gockman,
    minnesotawildflowers.info
PG Peter Grube
AH Arthur Haines
DH Darel Hess
SH Steven Hill
MAH Michael Homoya
MJH Michael Huft
EH Eric Hunt, Arkansas Native Plant Society
BJ Bart Jones
THK Thomas Kent, florafinder.org
TRK Thomas Klein, Minnesota Department
    of Natural Resources
GL Glen Lee, saskwildflower.ca
BL Brian Lowry
EM Elizabeth Miller,
    elizabethswildflowerblog.com

KM Keir Morse
JM James Mundy, Nature's Ark
    Photography
SN Scott Namestnik
PN Paul Nelson
CN Christopher Noll
JP John Pearson, Iowa Department
    of Natural Resources
NP Nathanael Pilla
CR Corey Raimond
PR Paul Rothrock
RS Russ Schipper
PS Perry Scott
BS Brad Slaughter
DT Dan Tenaglia,
    missouriplants.com
JT John Thayer,
    minnesotawildflowers.info
TT Tony Troche
ST Steve Turner
EU Eric Ungberg
KY Kay Yatskievych

INDEX

ABOUT THE AUTHORS

Michael Homoya was the Heritage Program Botanist and Plant Ecologist for the Indiana Department of Natural Resources Division of Nature Preserves for 37 years before retiring in 2019 and is currently an adjunct faculty member in biology at Indiana University–Purdue University Indianapolis. Michael is a Fellow, Distinguished Scholar, and former president of the Indiana Academy of Science and also served as president of the Indiana Native Plant Society. He has published several books on Indiana plants, among them *Orchids of Indiana* and *Wildflowers and Ferns of Indiana Forests*, and contributed to several others.

Scott Namestnik is the Heritage Program Botanist for the Indiana Department of Natural Resources Division of Nature Preserves and a founding member of the North Chapter of the Indiana Native Plant Society. He is also a Fellow of the Indiana Academy of Science, serving on the Biodiversity and Natural Areas Committee and as the chair of the Plant Systematics and Biodiversity Section. He is co-author of *Wildflowers of Indiana Dunes National Park* and enjoys botanizing, birding, and everything nature-related.

LEAF FORM

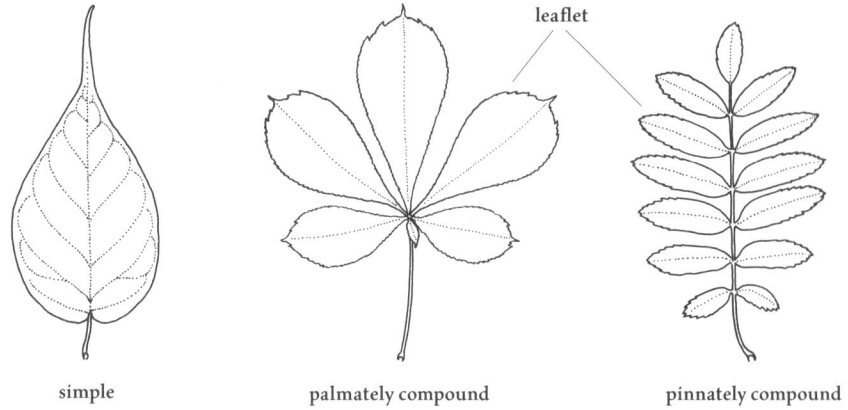

leaflet

simple   palmately compound   pinnately compound

LEAF SHAPE

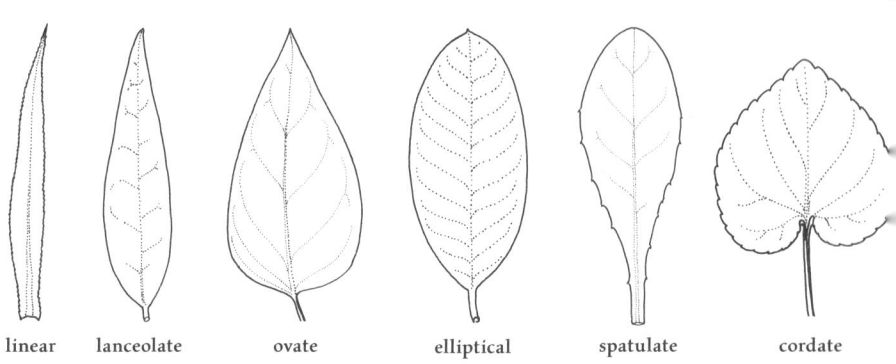

linear   lanceolate   ovate   elliptical   spatulate   cordate

LEAF MARGINS

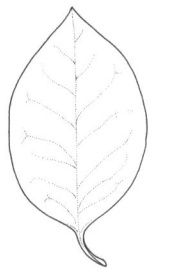 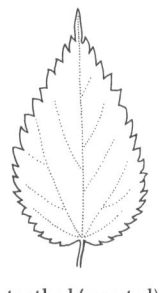 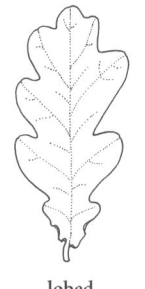

entire   toothed (serrated)   lobed

LEAF ARRANGEMENT

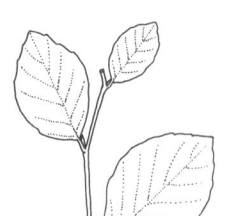

alternate